# Academic Entrepreneurship in Asia

NEW HORIZONS IN ENTREPRENEURSHIP

**Series Editor: Scott Shane**, *SBC Professor of Economics, Weatherhead School of Management, Case Western Reserve University, USA*

This important series is designed to make a significant contribution to the development of Entrepreneurship Studies. As this field has expanded dramatically in recent years, the series will provide an invaluable forum for the publication of high-quality works of scholarship and show the diversity of issues and practices around the world.

The main emphasis of the series is on the development and application of new and original ideas in Entrepreneurship. Global in its approach, it includes some of the best theoretical and empirical work with contributions to fundamental principles, rigorous evaluations of existing concepts and competing theories, historical surveys and future visions. Titles include original monographs, edited collections and texts.

Titles in the series include:

A General Theory of Entrepreneurship
The Individual–Opportunity Nexus
*Scott Shane*

Academic Entrepreneurship
University Spinoffs and Wealth Creation
*Scott Shane*

Economic Development Through Entrepreneurship
Government, University and Business Linkages
*Edited by Scott Shane*

Growth Oriented Women Entrepreneurs and their Businesses
A Global Research Perspective
*Edited by Candida G. Brush, Nancy M. Carter, Elizabeth J. Gatewood, Patricia G. Greene and Myra M. Hart*

Women Entrepreneurs Across Racial Lines
Issues of Human Capital, Financial Capital and Network Structures
*Andrea E. Smith-Hunter*

Effectuation
Elements of Entrepreneurial Expertise
*Saras D. Sarasvathy*

The Entrepreneur in Youth
An Untapped Resource for Economic Growth, Social Entrepreneurship, and Education
*Marilyn L. Kourilsky and William B. Walstad*

Academic Entrepreneurship in Asia
The Role and Impact of Universities in National Innovation Systems
*Edited by Poh Kam Wong*

# Academic Entrepreneurship in Asia

## The Role and Impact of Universities in National Innovation Systems

*Edited by*

Poh Kam Wong

*Professor, NUS Business School and Director, NUS Entrepreneurship Centre, National University of Singapore*

NEW HORIZONS IN ENTREPRENEURSHIP

**Edward Elgar**
Cheltenham, UK • Northampton, MA, USA

Published by
Edward Elgar Publishing Limited
The Lypiatts
15 Lansdown Road
Cheltenham
Glos GL50 2JA
UK

Edward Elgar Publishing, Inc.
William Pratt House
9 Dewey Court
Northampton
Massachusetts 01060
USA

A catalogue record for this book
is available from the British Library

Library of Congress Control Number: 2011927320



ISBN 978 1 84980 307 6

Typeset by Servis Filmsetting Ltd, Stockport, Cheshire
Printed and bound by MPG Books Group, UK

# Contents

# Contributors

**Erik Baark**, Division of Social Science, Hong Kong University of Science and Technology, Hong Kong SAR.

**Zong-Tae Bae**, Graduate School of Management, Korea Advanced Institute of Science and Technology, Korea.

**René Carraz**, Bureau d'Économie Théorique et Appliqué, Strasbourg University, France.

**Nimesh Chandra**, Centre for Studies in Science Policy, School of Social Sciences, Jawaharlal Nehru University, India.

**Dar-Zen Chen**, Department of Mechanical Engineering and Institute of Industrial Engineering, National Taiwan University, Taiwan.

**Ming-Yu Cheng**, Universiti Tunku Abdul Rahman, Malaysia.

**Jen-Diann Chiou**, Graduate Institute of Intellectual Property, National Chengchi University, Taiwan.

**Yixin Dai**, School of Public Policy and Management, Tsinghua University, China.

**Yuen-Ping Ho**, Entrepreneurship Centre, National University of Singapore, Singapore.

**Robert Kneller**, Research Center for Advanced Science and Technology, University of Tokyo, Japan.

**V.V. Krishna**, Professor in Science Policy, Jawaharlal Nehru University, India and Visiting Professor, IAS, United Nations University, Japan; Visiting Fellow, University of New South Wales, Australia.

**Paul C.B. Liu**, Graduate Institute of Intellectual Property, National Chengchi University, Taiwan.

**Sangmoon Park**, Department of Management, Kangwon National University, Korea.

**Hiromi Sasaki**, Intellectual Property Management Center, Kyushu University and New Energy and Industrial Technology Development Organization, Japan.

**Naubahar Sharif**, Division of Social Science, Hong Kong University of Science and Technology, Hong Kong SAR.

**Annette Singh**, Entrepreneurship Centre, National University of Singapore, Singapore.

**Makiko Takahashi**, Center for Research Strategy and Support, Tohoku University, Japan.

**Megumi Takata**, Faculty of Economics and Intellectual Property Management Center, Kyushu University, Japan.

**Toru Tanigawa**, Art, Science and Technology Center for Cooperative Research and Intellectual Property Management Center, Kyushu University, Japan.

**Thanaphol Virasa**, College of Management, Mahidol University, Thailand.

**Poh-Kam Wong**, Entrepreneurship Centre, National University of Singapore, Singapore.

**Lan Xue**, School of Public Policy and Management, Tsinghua University, China.

# 1. Contribution of universities to national innovation systems in Asia: technology commercialization and academic entrepreneurship

**Poh-Kam Wong, Yuen-Ping Ho and Annette Singh**

## 1.1 INTRODUCTION

### 1.1.1 Background and Context

The traditional view of universities as seats of learning and basic research has expanded to embrace the 'third mission' (Laredo 2007), with universities becoming increasingly involved in the industrial and commercialization spheres of the economy. As a source of new knowledge and technology, a role inherent to the third mission, universities influence the dynamics of the national innovation system (NIS) and contribute to national and regional economic growth (Etzkowitz et al. 2000).

Globally, there has been a rise in technology commercialization activities in universities, spurring interest in university technology transfer and commercialization as a source of economic growth (see, for example, AUTM 2001; Shane and Stuart 2002; Thursby and Kemp 2002; Paytas, et al. 2004). Notwithstanding the growing literature on university technology commercialization, it is still largely dominated by work on universities in the USA (see, for example, Etzkowitz 2002; Mowery et al. 2004) and to a smaller extent Europe (see, for example, Clark 1998; Wright et al. 2007), with scant published scholarly research on universities in Asia. An exception is the recent special issue of *World Development* on 'University-Industry Linkages in Metropolitan Areas in Asia' (Hershberg et al. 2007), which however primarily examined the issues from the regional policy perspective and did not provide in-depth case studies of specific universities. Moreover while a number of international comparative studies have been made on university technology commercialization, these tend to focus on universities in the large, advanced G7 economies (for example, Mowery et al. 1999; Shane 2002; Thursby and Kemp 2002; Heher 2006).

The lack of research on this region limits the ability of Asian policy makers to share with each other relevant policy lessons and institutional frameworks for university roles in their national innovation systems. While several Asian countries have implemented major new policy initiatives to encourage greater university-industry interactions (for example, the Japanese government has introduced major policy incentives to encourage university technology commercialization in recent years), most have not yet embarked on major policy reforms. In the absence of familiarity with the approaches adopted by other Asian economies, the experience of the USA (for example, the Bayh-Dole Act and the Small Business Innovation Research (SBIR) scheme) is often referenced by policy makers in Asian countries as possible models for adoption in reforming the university system in their respective countries in recent years. However the lack of research on the experiences of Asian universities hinders policy makers' ability to accurately evaluate which elements of the US university technology commercialization model are appropriate for adoption by Asian universities.

It is with this in mind that the NUS Entrepreneurship Centre (NEC) embarked on a multi-country comparative study of university technology transfer and commercialization, funded by the Sasakawa Peace Foundation of Japan. Researchers from major universities around Asia examined the emerging role of selected universities in their national and regional innovation systems. Spanning a two-year period, this study brings together researchers from Japan, China, India, Korea, Hong Kong, Taiwan, Malaysia, Thailand and Singapore to study 13 selected leading universities in these economies. These efforts have yielded the current volume, which we hope will contribute richer insights on how each individual university contributes to its national and regional innovation system, while simultaneously drawing comparisons between the different university approaches. This will further allow researchers to draw policy implications for senior university administrators, taking into account the diversity of Asian universities and their national contexts.

### 1.1.2 Research Question

Among Asian universities there is evidence of an emerging shift towards the 'third mission', although this phenomenon is more nascent and less widespread than in their Western counterparts. When discussing Asian universities, one has to keep in mind the heterogeneity of the region, with great diversity among the economies in terms of stages of economic development, capacity of the innovation ecosystem and the role of the state in higher education. This necessarily means that universities in different Asian economies would likely adopt different approaches to accomplishing the 'third mission'.

One key question addressed in this research project is whether there is a convergence in the approaches towards the 'third mission' among the focal universities. This begins from establishing a basic understanding of how the 'third mission' role is conceptualized in each university and ascertaining the key drivers behind the shift towards this role. The country chapters will describe how each focal university has institutionalized its third mission, examine the organization and implementation of these efforts, assess the impact of the efforts and identify the issues and challenges faced.

Through the focal universities in this project, we also aim to investigate differences in approaches compared to the established models from North American and European universities. A number of interesting questions are raised with reference to the relative importance attached to technology commercialization in the mission of the universities, institutional mechanisms for managing technology commercialization (including policy on faculty and student involvement in technology commercialization activities), and the relative emphasis on industry collaboration and technology licensing versus entrepreneurial spin-off formation and incubation. We also seek to address the fundamental issue of whether Asian universities are tending towards the adoption of mimetic approaches, or are attempting to develop unique approaches based on their respective unique contexts.

### 1.1.3 Methodology

A comparative research approach was adopted for this project. The activities under this project entailed forming a group of researchers from major universities in Asia. Participating researchers from the selected countries studied their selected focal universities' role in the NIS, highlighting the universities' performance in technology transfer and related activities that contribute to the innovation systems in these economies. To provide the relevant context, the national innovation systems are described, with particular attention paid to the historical and current roles played by the higher education sector. Each focal university is given in-depth treatment, including analysis of how it organizes and implements its third mission, the impact and results of such efforts and the key challenges faced. The key technology commercialization activities such as technology licensing, spin-offs formation and university-industry collaboration are examined. Other relevant factors such as entrepreneurship education, industrial consulting and training, and internationalization are also investigated.

## 1.2 OVERVIEW OF FOCAL UNIVERSITIES

### 1.2.1 Overview

Thirteen universities from nine different Asian economies are included in this volume. The economies covered in the study represent the spectrum of industrialization levels that is found in Asia: (i) Japan as Asia's most mature industrialized economy; (ii) the group of four Newly Industrialized Economies (NIEs) (Korea, Taiwan, Singapore and Hong Kong) that developed rapidly over the last 30 years to achieve relatively high levels of per capita income by the mid-2000s; (iii) the developing economies of the Association of South East Asian Nations (ASEAN) region, represented by Thailand and Malaysia; and (iv) the emerging giant economies of China and India.

Within the nine selected Asian economies, between one to three universities were covered in our comparative study. The selected universities are among the leading universities in their respective economies, with significant research output and graduating the most sought-after science and technology graduates by industry. The universities were also selected to provide diversity in terms of size, age, structure and disciplinary focus. As can be seen in Table 1.1, the focal universities range considerably in size from under 2000 students enrolled, as in the case of the two Indian Institutes of Technology (IITs) to very large student bodies over 30 000 in the case of Tsinghua University in China and the National Taiwan University (NTU). Correspondingly the faculty size ranges from fewer than 500 to close to 10 000 academic staff. All the universities offer both undergraduate and graduate degree programmes although several of the universities are more focused on postgraduate education (IITs, Korea Advanced Institute of Science and Technology (KAIST)). In terms of disciplinary coverage, the selected universities fall into two groups: seven are comprehensive, research-intensive universities (Kyushu, Tohoku, Tokyo, NTU, National University of Singapore (NUS), Mahidol and Tsinghua); the other six are more specialized universities with a strong focus on science and technology disciplines.[1] With the exception of Malaysia's Multimedia University (MMU), all the universities are public universities although several have been corporatized or are in process of being corporatized.

There is also variation among the focal universities in terms of research development (R&D) expenditure. As evident from Table 1.1, the universities from Japan, Singapore, China and Korea have relatively large R&D budgets. At the other end of the spectrum, the two ASEAN universities of MMU and Mahidol spend very little on R&D (US$2 million and US$20

*Table 1.1 Profile of focal universities, FY 2006*

| | Total academic staff FY 2006 | Students enrolled FY 2006 | % graduate students FY 2006 | No. of graduates FY 2006 (% national graduates in parentheses) | R&D expenditure FY 2006 in USD million[a] | No. of research publications FY 2006 (SSCI) |
|---|---|---|---|---|---|---|
| Kyushu University | 2274 | 18622 | 36.50 | 4183 | 68.6 | 3885 |
| Tohoku University | 2675 | 17849 | 39 | 4796 | 234.2 | 4312 |
| Tokyo University | 4444 | 28952 | 49 | 7662 (7.6%) | 468.6 | 6887 |
| HK University of S&T | 450[b] | 9000 | 36 | na | 39.5 | na |
| Korean Advanced Institute of S&T | 745 | 7336 | 58.8 | 1687 (0.1%) | 123.2 | 1407 |
| National Taiwan University | 1869 | 32233 | 45.3 | 7494 | 170.8[c] | 3546 |
| National Taiwan University of S&T | 883 | 9246[d] | 44.1[d] | 3153 | 6.0[e] | na |
| National University of Singapore | 1944 | 29305 | 22.6 | 8559 (24%) | 117.1[f,g] | 3367 |
| Indian Institute of Technology Bombay | 402[b] | 1451 | 64.7 | 1451 | 33.1 | 650[h] |
| Indian Institute of Technology Madras | 360[b] | 1566[g] | 64.8[g] | 1267 | 43.6 | 650[h] |
| Tsinghua University (China) | 6945 | 31786 | 56.6 | na | 186.5 | 2874[g,i] |
| Mahidol University (Thailand) | 9562 | 23815 | 34.8 | na | 19.7 | 698 |
| Multimedia University (Malaysia) | 715 | 20410 | 3.5 | 2983 (6.84%) | 2.2 | 168[j] |

*Notes:*
a. Converted to USD from various national currencies using exchange rates as at January 2009.
b. Tenure-track only.
c. FY 2007, research funding from academia-third party collaboration only (includes funding from government, industry and other sources).
d. 2008
e. Definition of R&D spending at the National Taiwan University of Science and Technology is not consistent with that used for other focal universities, hence figures are not directly comparable.
f. Data is for research funding.
g. Data is 2005.
h. Data is for SCI-extended publications.
i. Data is for SCI publications only.
j. Data is for ISI publications.

million, respectively). (Note: comparable data on the two Taiwanese universities are not available.) Even after adjusting for differences in purchasing power parity (PPP) and faculty size, such cross-country differences remain large, reflecting the different stages of technological development of the host economies. The universities also exhibit large variations in research output as measured by research publications in international refereed journals, with the Japanese universities leading the way and the two ASEAN universities having the lowest output.

### 1.2.2 Forms of Technology Commercialization

University technology commercialization takes multiple forms. While all the university case studies identify a common set of technology commercialization mechanisms, significant variations in the composition and relative importance of the different mechanisms are observed. At the most basic level, all universities transfer knowledge to industry through their core education mission, by imparting graduates and researchers with relevant skills to contribute to economic development. All the universities also engage in more direct interactions with industry through staff consultancy and student internship placements.

While acknowledging the importance of such basic mechanisms for knowledge interactions and transfer between universities and industry, the intended aim of the case research in this study is a more narrow one, focusing on activities that involve more explicit technology commercialization goals, where issues concerning the return or share of potential economic value of the potential intellectual capital being created by the university as a corporate body need to be managed more formally. In particular, each of the university case studies examines the nature and extent of university technology commercialization activities that involve issues of intellectual property (IP) creation and transfer, in one or more of the following direct forms: (i) joint/sponsored research with industry; (ii) transfer of usage right to university inventions and knowledge materials to external private firms; and (iii) the creation of new spin-off companies by university faculty, researchers or students to bring the university invention to market.

Because leading universities in the USA and Europe have already established formalized practice models and institutional mechanisms for managing each of these three forms of technology commercialization – joint/sponsored research, technology licensing and spin-off formation – the starting point for the Asian case studies is to examine the extent to which the individual focal universities have adopted similar practices, whether deliberately (mimetic approach) or otherwise. To the extent that the focal universities are found to have in fact implemented practices or mechanisms that are different from those in the Western universities, the respective researchers examine the possible reasons for such variations, and whether and how the variations have been caused by the need to adapt to the unique national socio-economic and policy context of the host economies of the respective universities. These contexts necessitate a broader array of university institutions and activities that provide direct or indirect support to facilitate technology commercialization, including institutions or activities that may not have similar counterparts in the Western university model. The case studies have revealed numerous such indirect forms of commercialization activities. For example, besides examining the different variants of the Technology Transfer Office (TTO), several of the university case studies have identified physical infrastructure support programmes such as incubator facilities and even large scale science park developments, while others have highlighted support programmes that involve 'software' and other intangibles, including entrepreneurship education programmes targeted at fostering students and faculty interest in commercializing their invention and equipping them with relevant knowledge on entrepreneurial start-up formation. Other commercialization supporting activities identified include seed-funding programmes, schemes for mentoring and informal networking, and university-owned enterprises to market the university's knowledge assets directly.

Table 1.2 presents several comparative benchmarks on the direct forms of technology commercialization adopted by the focal universities. It is observed that all the focal universities (for which data are available) involve industry in their research activities, through sponsored/contract research or collaborative research projects. Traditional technology transfer activities through licensing of patents are also practised, at vastly varying levels of intensity.

There are considerable variations among the universities in terms of emphasis on the different forms of direct technology commercialization. The Japanese universities engage heavily in collaborative research with industry, while universities from the emerging and developing economies appear to emphasize sponsored and contract research. Technology licensing features prominently in the two Taiwanese universities and is also an important form of commercialization for KAIST. These comparisons highlight the differences between the approaches adopted by Asian universities, which are further reflected in the different types of indirect supporting activities and institutional frameworks that have been developed, as discussed in the next section.

*Table 1.2 Indicators on direct technology commercialization activities in focal universities, FY 2006*

| | Joint/sponsored research with industry | | | Technology licensing to external parties | | | Spin-off formation | |
|---|---|---|---|---|---|---|---|---|
| | % of research sponsored by industry | Value of contract research from industry per $'000 R&D spending | Value of collaborative research from industry per $'000 R&D spending | Licenses per 100 patents | Licenses per 100 academic staff | Growth in number of licenses, FY 2004–06 (%) | Spin-offs per 100 patents | Spin-offs per 100 academic staff |
| Kyushu | 4.0 | 53.5 | 365.1 | 357.1 | 1.1 | 104.1 | 71.4 | 0.2 |
| Tohoku | 5.7 | 291 | 74.4 | 204.3 | 1.8 | 159.1 | 0 | 0 |
| Tokyo | 24.8 | 8.1 | 57.6 | 574.6 | 8.1 | 63.8 | 0 | 0 |
| HKUST | 17.1 | na | na | na | na | na | na | na |
| KAIST | 13.6 | na | na | 8.8 | 9.3 | 61.2 | na | na |
| NTU | na | na | na | 72.9 | 2.3 | na | na | na |
| NTUST | 9.1[a] | na | na | 133.3 | 3.6 | 3.2[e] | na | na |
| NUS | 7[b] | na | 41.8 | 13.3 | 0.3 | -52.9 | 4.4 | 0.1 |
| IIT Bombay | 47[b] | 247.7 | 66.5 | na | 10.7 | na | na | 1.2 |
| IIT Madras | 33 | 228.4 | 111.7 | na | na | na | 28.6 | 1.7 |
| Tsinghua | 39.5 | na | na | na | na | na | na | na |
| Mahidol | 22.1 | 220.9 | na | 37.5 | 0.1 | 73.2 | 12.5 | 0.03 |
| MMU | 21.2 | na | na | na[c] | 0 | na | na[c] | 0.6[d] |

*Notes:*

a. FY 2007, basis for computation is research funding from academia-third party collaboration only (includes funding from government, industry and other sources).

b. Basis for computation is research funding in FY 2005.

c. No patents granted to MMU in 2006 (cumulatively, MMU only has one issued patent as at end of 2006).

d. Does not involve any formal licensing of technology or intellectual property from the university.

e. Growth based on FY 2005–06.

### 1.2.3 Institutional Organization for Technology Commercialization

All 13 focal universities had established some forms of organizational unit to manage the core industry relations, IP management and technology transfer functions associated with industry R&D collaboration and technology licensing activities. Similar to many Western universities, which have established TTOs either as stand-alone organizations or as affiliates or divisions of Industrial Relations Offices (IROs), all the universities covered in our study (with the possible exception of MMU) have established the equivalent of the TTO, although the specific institutional forms taken by the different TTOs differ quite considerably. For example, in Japan the organizational configuration for technology transfer usually involves a Technology Licensing Office (TLO) and separately an IP Management Office. The TLO may be internal or external to the university. As a case in point, the TLO at the University of Tokyo is an independent for-profit corporation, although in effect it shares dual authority with the IP Management Office in the university's central administration. Outside Japan, the focal universities have established TTOs that have different positions in the universities' overall structures, from being a part of the administration cluster to stand-alone organizational units reporting directly to the university's senior management. In MMU, the TTO function is nominally subsumed under an umbrella organization handling IRO and entrepreneurship promotion functions, with little real activity.

Unsurprisingly, there are also differences in the roles played by the TTOs, although there are basic similarities across the board. Apart from the ubiquitous technology transfer functions, most of the TTOs also undertake to educate the universities' faculty and researchers on issues such as IP and the challenges of working with industry partners. The majority of TTOs also manage university-industry collaborations in the form of industry sponsored research. Less commonly, a number of the TTOs have a wider scope of responsibilities that extend to entrepreneurial activities such as spin-off formation and incubation. Typically the incubation function in the focal universities is organized separately from the TTO, although the two institutional units often work closely together.

### 1.2.4 University Policy on Faculty Involvement in Technology Commercialization

To incentivize participation by faculty members in the technology transfer process, most of the universities reported having instituted income-sharing policies for royalties arising from commercialized technologies. The share of net income that accrues to individual faculty inventors ranges from 20

*Table 1.3 University policy on sharing of net income from commercialized technologies*

| | N[a] | Minimum (%) | Maximum (%) | Mean (%) | Standard deviation |
|---|---|---|---|---|---|
| Individual Inventor(s) | 13 | 10 | 70 | 42.3 | 16.6 |
| Institution | 13 | 0 | 50 | 27.2 | 12.4 |
| Inventor's Department | 13 | 0 | 33.3 | 15.7 | 13.4 |
| Tech Transfer Office | 13 | 0 | 33.3 | 11.2 | 13.2 |

*Note:* a Where a university has more than one distribution policy, each policy has been included separately. Excludes Tsinghua, whose distribution varies on a case-by-case basis; 12 universities were included.

per cent to a very high 70 per cent (Table 1.3). The TTO takes a relatively small cut, averaging slightly more than 10 per cent of net income, with a few focal universities reporting that their TTO receives zero income.

The focal universities have quite liberal policies on faculty involvement in corporate activities. In most cases, faculty members are able to serve on the board of directors of existing companies, although approval must be sought from the universities. Similarly most universities also allow, with prior approval, individual faculty to serve on the board of start-up companies that were established to commercialize the faculty member's own inventions. Nonetheless there remains a degree of control and regulation somewhat higher than in the USA, as approval is still required in most cases, even for industry consultancy by individual faculty.

## 1.3 KEY THEMES FROM THE INDIVIDUAL UNIVERSITIES' CASE STUDIES

### 1.3.1 Mature Industrialized Economy – Japan

Three chapters of this report are devoted to universities from Japan, as Asia's most advanced developed economy. The three cases illustrate different aspects of technology transfer in Japan, which has a long history of university-industry linkages and has recently undergone a number of legislative and policy reforms to promote university technology commercialization.

The chapter on Kyushu University begins with a comprehensive account of the roles played by universities in the national innovation systems of Japan, and finds that the influence of universities through traditional technology transfer mechanisms is less pronounced compared with the USA. Instead of technology licensing and university start-ups, the dominant mode for knowledge transfer to Japanese industry appears to be university-industry R&D collaborations. A survey of companies that conduct joint research with Kyushu University reveals that companies choose the joint-research mode in order to consolidate their resources in applied research, relying on university expertise for basic and advanced research. To expand joint and sponsored research, Kyushu University adopts a model termed 'organization-level cooperation'. Introduced in 2004, the approach allows for large-scale and interdisciplinary joint research between the university and businesses. The implementation framework emphasizes cooperative decision making and strives to eliminate discrepancies in the recognition of research by the parties involved. The authors illustrate how the collaboration framework is implemented in practice with a successful case study of an organizational-level research collaboration between Kyushu University and Nissan Chemical. This chapter also dwells into other university-industry collaboration activities in Kyushu, including consultation and cluster-based projects aimed at transferring knowledge to the local region and community.

In his chapter on the University of Tokyo, Robert Kneller affirms joint/sponsored research to be the main mechanism for technology transfer. This is cast in the context of the past and present legal framework for technology transfer in Japan, with four laws implemented between 1998 and 2004 that have changed the governance of IP management and university-industry cooperation. Despite universities being allowed to assert ownership over commercially viable inventions by faculty, and the perceived competence of its TLO, the technology transfer system at Tokyo University appears to be biased in favour of large companies. This is particularly true in the non-life science fields such as engineering and software, where a large proportion of patent applications are for joint-research inventions sponsored by large companies. Such university discoveries are automatically exclusively controlled by the large companies. This has placed a drag on academic entrepreneurship in the form of spin-off firms, reflected in the declining number of start-ups being formed annually at the University of Tokyo. Macro-level data confirm this to be a phenomenon that occurs throughout Japan, which is unique among technologically advanced economies in the high proportion of university inventions that are co-assigned to private companies. This raises the issue of appropriation of public-funded research by large companies as joint/sponsored research accounts for a disproportionately small share of total R&D funding in universities. Kneller argues that the technology transfer system, together with environmental factors that contribute to the weakness of high-tech start-ups in Japan, has perpetuated a closed innovation system which may be to the long-term detriment of Japan's industry and economy.

The chapter on Tohoku University focuses on the issue of academic patenting, which has been on a dramatic upswing in Japan since the legal reforms of the last decade. The authors examine in particular the organizational structure of IP management institutions in Japanese universities and identify two common entities, the IP office and the Technology Liaison Office, as having major roles in promoting technology transfer activities, in particular IP-related activities such as patenting and patent licensing. The inherent complexities of this two-structure system present a challenge for certain universities too in coordinating and managing commercialization processes. The Technology Liaison Office structures receive subsidies from the government and have been promoted as a way to develop greater university-industry relations. However these organizations have not been able to cover operating costs. In the context of Tohoku University, the authors describe the framework for managing IP rights and show how patent applications are linked to contractual research, which is the main source of income from technology transfer. The university has emphasized patenting as a channel for technology transfer despite the high costs and relatively limited returns, as the use of intellectual property rights (IPR) formalizes the relationship between industry and university, rendering it easier to monitor. The concern then is whether the emphasis on patenting has been detrimental to the research output of faculty members. To examine this question in more depth, the authors report findings from their empirical analysis of academic patenting by Tohoku University staff. Patenting and publishing are found to be complementary, although the emphasis placed on publishing by individual department or units may be a mitigating factor.

### 1.3.2 Newly Industrialized Economies (NIEs)

This section of the report comprises chapters on five universities from the four East Asian NIEs – Singapore, Hong Kong, Korea and Taiwan.

The chapter on the Korean Advanced Institute of Science and Technology (KAIST) explains how Korea's most prestigious technological university leads technological entrepreneurship in Korea with more than 700 new ventures founded by students and alumni. The authors argue that Korea is currently changing her Science and Technology (S&T) policies from fast-catching-up to creativity-oriented innovation, and consequently the Korean government is being encouraged to invest more resources in building overall university research capabilities and exploring new fields. Since the Asian Financial Crisis of the late 1990s, a number of policy measures have been implemented to promote entrepreneurship in universities and public research institutes, including the introduction of a Korean Bayh-Dole Act. These have borne results in the form of increased technology commercialization overall. However Korean universities are still under-performing and in order to fulfil the national vision of creativity-oriented innovation, the importance of universities in the Korean economy is expected to be significantly enhanced. KAIST is well positioned to fulfil this role, having successfully set up institutions and policies for technology commercialization from early on. In recent years, these processes have been upgraded and streamlined; for example, maximizing licensing income from the pre-patent-filing stage, rather than focusing on licensing patents. The authors find that there are distinctive factors to promote technological commercialization in KAIST including: (i) strong and consistent government support for KAIST; (ii) research capabilities tightly linked with industries; (iii) balanced approaches to academic and industrial research; (iv) championing roles of faculty members; (v) commercialization hub of the research laboratory; and (vi) infrastructure for entrepreneurship. However compared with technological commercialization of leading universities in the USA, KAIST is still in the very early stages of generating value from commercialization. There remain substantial challenges for KAIST leadership to pursue new directions that move KAIST towards becoming a world-class university.

The chapters on Hong Kong and Singapore both utilize the Triple Helix framework (Etzkowitz et al. 2000) to examine the changing role of economies that are attempting to move towards a knowledge-based strategy for economic growth. The case studies of Hong Kong University of Science & Technology (HKUST) and the National University of Singapore (NUS) provide interesting contrasts in approaches, given the similarities in the underlying national contexts.

Singapore's policy makers have charted a course for the nation to transition from an investment-driven economy to an innovation-driven economy, emphasizing the building of intellectual capital and its commercialization to create value and jobs. Increasing prominence has been given to universities' role in stimulating economic growth through industrially relevant research, technology commercialization, high-tech spin-offs, attracting foreign talents and inculcating entrepreneurial mindsets. The authors examine the changing role of the NUS in the Singapore economy in this current transitional period. NUS has responded to the globalization of the knowledge economy by adopting a comprehensive, integrated approach to promoting university entrepreneurship and technology commercialization. This is part of a conscious and planned effort to shift towards an 'entrepreneurial university' model. In addition to traditional forms of technology commercialization, the NUS framework has several novel features such as the emphasis on experiential entrepreneurial education and a focused approach towards facilitating and supporting university spin-offs. The NUS enterprise framework has evolved over time, with increasing importance placed on the quality and impact of the university's innovations, rather than an over-reliance on quantitative performance indicators

In their chapter on the HKUST, Sharif and Baark set the scene by describing how Hong Kong's university system is changing in response to the attempts by the Hong Kong government to transform the commercial-business hub into a knowledge-based economy. The authors argue that Hong Kong universities face particular challenges to effectively transfer knowledge; firstly, the lack of capacity and demand for university knowledge in the private sector, and secondly, the continually changing economic landscape shaped by Mainland China's rapid growth. The authors' analysis suggests that HKUST is uniquely placed in the university system, having been envisioned from its inception as a progressive and entrepreneurial world-class research university. Through its Entrepreneurship Programme and other initiatives, the university has achieved progress in entrepreneurship promotion, technology commercialization and collaborative R&D, and contract research. The Technology Transfer Centre has proactively sought to bridge the gap between university and industry by modestly expanding its role to develop new forms of joint ventures for the take-up of university technology by industry. The authors emphasize that technology transfer should be seen as a means of creating or strengthening university-industry research collaborations, and not as an end in itself. They conclude that while HKUST's progress is promising, its transition to the entrepreneurial university model is a work in progress. There is stubborn resistance in transforming technology commercialization into a university-wide focus and it is difficult to predict success in light of heightened public expectations.

Liu et al. study two Taiwanese universities; the National Taiwan University (NTU), which is Taiwan's leading comprehensive university, and the National Taiwan University of Science and Technology (NTUST), which is a newer university with a more specific science and engineering focus. Taiwan's industry structure poses particular difficulties for university technology transfer: most industries in Taiwan are dominated by Small and Medium-sized Enterprises (SMEs) which prefer to have full ownership of their patents and prefer to engage universities through commissioned research rather than licensing arrangements. The authors examine how this industrial structure historically arose from the socio-economic development strategy implemented by the Taiwanese government, and provide a detailed account of how the government has devoted years of planning to promoting industry-academia collaboration and university technology commercialization. In particular the Taiwanese government in recent years has implemented a slew of new measures, including providing funds, enacting legislation, delegating responsibilities to government agencies and monitoring the impact made by universities. At NTU, human resource issues are at the forefront. Staffing of the TTO, named as the Office of Research and Development, has been difficult due to lack of qualified professionals in technology transfer and licensing, as well as technology marketing. Growth in patenting activities has also increased the need for manpower and escalated costs. In NTUST, the approach is positive, ambitious and business-like. The emphasis is on placing highly experienced and well-networked Tech Transfer Managers in R&D teams, where they will connect external resources with NTUST's R&D capabilities. Knowledge transfer is deemed as more important than technology transfer or patent licensing, hence NTUST researchers work closely with companies to evaluate technology portfolios and processes to identify potential synergies. Case studies from both NTU and NTUST highlight the importance of individual faculty members who are entrepreneurially minded and passionate about the commercialization potential of their inventions for industry.

### 1.3.3 Emerging Economies

The chapter on Tsinghua University provides insights into the rapid evolution of university technology transfer and commercialization in China. Although initially designed as institutions aiming at providing well-trained human resource to other sectors in the society, policy reform has led to Chinese universities being asked to contribute more actively to the national science and technology development effort. The authors describe Tsinghua University's connections to and cooperation with the industry sector, framed against a discussion on the broader role of universities in the Chinese NIS. The authors highlight five major forms of university-industry linkages being established by Tsinghua: (i) technology contracts, the most important source of research funding; (ii) IP transfer and licensing; (iii) joint research centres; (iv) university science parks; and (v) university-owned enterprises. To support these different forms of industry linkages, Tsinghua University has developed the Science and Research Institution to manage industry research collaboration and IP creation, while a company, Tsinghua Holdings Co. Ltd, has been established to manage the commercialization of university IP. This chapter concludes with a detailed case study of NUCTECH, a successful commercialization initiative at Tsinghua University. Several salient issues are highlighted as being key to the success or otherwise of technology transfer activities at Tsinghua: the environment for access to financial support; human capital management within the university, in particular the issue of balancing research performance with involvement in commercial projects; IPR protection and policies on sharing and division of rights; and the role of government in Chinese university technology transfer and commercialization.

In contrast to China, the place of universities in India's knowledge production and knowledge transfer system from the perspective of commercialization has not been emphasized until fairly recently, as the traditional values of teaching and research excellence remained the most important missions of universities. The chapter on two campuses of Indian Institutes of Technology (IITs), which are at the forefront of education excellence in India's higher education system, provides interesting insights on the process of introducing the third mission to universities in a national context where there has historically been little national policy emphasis on industry engagements by universities. The authors firstly describe India's NIS and follow this by introducing the IIT system and explaining its uniqueness within India's higher education and innovation systems. While they were established within a national system and receive public funding from a central budgeted allocation, the individual IIT campuses actually operate quite independently of each other and adopt different institutional arrangements for industry engagement as well as different IPR policies. The authors study in particular two of the IITs, IIT Madras and IIT Bombay. The approach to technology transfer and spin-off creation has evolved differently in the two institutes. IIT Bombay has institutionalized a formal incubation unit and a TTO that manages university-industry relations. This has led to several success stories. On the other hand, IIT Madras has for the most part adopted an unconventional mode of spin-off creation without formal institutional structures, although an incubation unit was set up in 2006 to promote ventures with a rural development focus. The authors describe the active formation of new firms by the Telecommunications and Computer Networking (TeNET) group at IIT Madras. TeNET was formed as a research grouping of faculty members from the faculties of electrical engineering and computer science. The group has incubated over 15 enterprises from the knowledge generated through TeNET projects. Additionally staff, students and partners from outside IIT Madras have also formed enterprises based on the institute's technology. From an overall policy perspective, the authors conclude that universities in India lack the appropriate 'innovation ecosystem' and venture funding support structure. Government schemes to promote university-industry relations are recent initiatives and their impact will not be felt for several years. The legal framework governing IPR in universities is also still in the process of being framed.

### 1.3.4 Developing ASEAN Economies

In her chapter on Malaysia's Multimedia University (MMU), Ming-Yu Cheng contributes the perspective of a private university, the only such case among the 13 focal universities covered in this volume. The case study of MMU is prefaced by a discussion on Malaysia's policies that have focused on science and technology as the main driver for economic growth. Among the policy reforms towards this end was the restructuring of the higher education system towards a more business-oriented approach. As part of this reform, public universities were given greater autonomy with several being corporatized and for the first time, private universities were granted approval to set up campuses and offer degree-granting programmes. MMU is the first private university in Malaysia, established in 1996 by Malaysia's largest telecommunications corporation, Telekom Malaysia Berhad. The pursuit of research excellence had been emphasized from inception, as had the importance of working with private industry for technology development. As a private university with bottom-line considerations, MMU has evolved its technology commercialization forms differently from public universities. Additionally, as a university specialized in information and communications technology (ICT) which has a relatively short shelf-life, speed to market is key. Rather than the more traditional forms of IPR and technology licensing which are costly and time-consuming, the focus is on staff consultancy and student technopreneurship. The University Consultancy Policy makes it compulsory for all academic staff to engage in consultancy work, ensuring that they are in touch with current industry requirements. MMU's entrepreneurship education programmes are a unique feature of both its undergraduate and postgraduate curricula. The author argues that financial considerations remain the key challenge for MMU to support the development of a more substantive research and technology commercialization programme. The answer appears to lie in closer collaborations with industry to provide the needed support such as laboratory facilities, consultancy opportunities and grants for R&D.

Thailand's Mahidol University is the subject of the final chapter authored by Thanaphol Virasa. This chapter is particularly instructive for universities in developing economies where the main mission of the higher education system has traditionally been human resource development. In recent years, Thai universities have been drawn into increasing their linkages to industry, the impetus being exerted top-down by the government and reinforced by interest on the part of firms. Among the reforms under this government push has been the transitioning of public universities to autonomous universities with less bureaucracy and greater capability for technology development. The government has also implemented schemes to encourage university incubation and instituted a fund for the setting up of Technology Licensing Offices in universities. The author examines how Mahidol University manages the technology transfer and commercialization process in order to play a more proactive role in knowledge diffusion and exploitation in line with the government's directive. The author shows that the management of intellectual capital portfolio in Mahidol University is still a relatively new phenomenon and that personal contacts and informal channels remain the principal mechanism for knowledge transfer to industry. Several key issues in the university's technology transfer process are identified: (i) the university faces pressure to generate income from commercialization but the linear model of innovation still dominates, with little awareness of the commercialization process; (ii) the system and structures for supporting technology transfer are not known or well understood by university stakeholders; (iii) substantial infrastructural gaps exist, especially in terms of finance; (iv) the lack of an entrepreneurial culture in the university as it has been governed by a research-oriented ethos for over a hundred years. The policy implications arising from these are the need to educate university stakeholders on commercialization and market matters, to implement changes with an understanding of the university's research/publication-oriented culture and to balance entrepreneurial activities, such as start-ups, with traditional forms of relations with industry.

## 1.4 COMPARATIVE ANALYSIS

### 1.4.1 Salient Differences and Common Themes

Given the diversity of the Asian economies covered in our study, it is not surprising that there are significant differences in the scale, composition and level of sophistication of technology commercialization activities across the universities covered in the study. In the three developing economies of Thailand, Malaysia and India, university technology commercialization is still largely at an emerging stage. The focus appears to remain at the stage of consulting and contract research, with less emphasis on joint/sponsored research, IP creation and commercialization. Enterprise creation is less systematically pursued, being hampered by gaps in the entrepreneurship support ecosystem, especially in terms of venture financing and other forms of early-stage funding. Although also a developing economy, government policy reforms of higher educational institutions in China appear to have been more ambitious, with a more strategic focus on creating a number of world-class universities that can contribute significantly to strengthening the national innovation system. This has enabled a number of the leading Chinese universities to adopt a much more commercialized and entrepreneurial approach, as evidenced by their ability to attract large investments in science and technology parks, and the creation of university-owned enterprises.

In contrast, most of the universities in the four Asian NIEs have implemented technology commercialization policies and organizational mechanisms that are similar to those practised at the leading US universities. In particular they have tended to adopt the transactional approach of North American universities, as typified by the use of benchmark metrics similar to those of the Association of University Technology Managers (AUTM), which emphasize patenting and their licensing to industry to generate returns to the university community. These returns are then shared among the faculty, department and central university administration through a well-defined royalty income-sharing policy to align incentives. In contrast, the Japanese universities appear to have pursued a more distinctive relational approach, where the focus is on building long-term cooperation and relationships with industry partners.

Despite the significant variations observed above, a number of common themes can nonetheless be discerned when these Asian universities are examined collectively. First and foremost, despite the diversities in the characteristics of the universities and the level of socio-economic development of the host economies in which they operate, it is clear that a key driver for involvement in technology commercialization has been increasing government emphasis on universities playing a greater role in the commercialization of technology and engagement with industry. In some ways this mirrors earlier developments in Europe (Wright et al. 2007); in the 1990s European policy makers became increasingly aware that economic growth would depend strongly on technology transfer from public research to industry. In Europe, public policy has been directly oriented at the stimulation of spin-offs from public research, with policy instruments and legal reform aimed at creating an innovation system conducive for the formation of new technology-based firms. In Asia the phenomenon is more recent and policy efforts have taken different forms. The impetus for greater university involvement in technology commercialization appears to have occurred largely in the 2000s, with many governments in the Asian region initiating major policy changes to reform higher education systems. These aim to make universities more responsive to the changing needs of economic development in general and play a greater role in the development of the NIS in particular.

Secondly, the initial response from many of the universities covered in our study appears to have been mostly a mimetic adoption of the established organizational mechanisms and performance indicators as practised by the leading North American universities. Indeed many of the initial approaches are similar to those being promoted by AUTM. These include the establishment of TTOs, invention disclosure systems, IP management policies and royalty income-sharing policies. While this mimetic approach appears to be particularly pronounced among those universities which are relatively new in embarking on technology commercialization activities (for example, the universities in Thailand and Malaysia), it is nonetheless heartening to note that universities with a longer history of involvement appear to have subsequently shifted to more refined approaches tailored to their own particular contexts and environment. This includes revising performance indicators to focus on quality and impact indicators rather than simplistic quantity measures such as basic patent counts (as witnessed particularly in NUS and the two Taiwanese universities), organizational adaptation and policy framework improvement to support joint research with industry (particularly evident among the Japanese universities), and enhanced TTO capabilities to provide more value-added technology marketing and packaging services instead of a more passive approach to licensing.

Thirdly, compared to the leading US universities, the technology commercialization role of Asian universities appears to have a broader scope than the conventional technology licensing function. There appears to be a greater emphasis placed on providing supporting services to facilitate spin-off formation. For example, not only have most of the Asian universities covered in our study established incubators to support start-ups by professors and students, most of these incubators have also expanded their functions over time beyond mere physical facilities to providing additional 'software' support services such as mentoring and networking with industry and investors. A number of the universities have also strongly emphasized entrepreneurship education to encourage faculty and students to engage in start-up activities. In addition, several universities (most notably Tsinghua) have also actively expanded their scope of activities to include encouragement of alumni to start companies in the vicinity of the universities. While the actual pace of shift towards such broader scope of engagement varies among the universities covered, taken collectively, such a broadening trend is evident.

It is true that this greater array of activities carried out by the Asian universities to directly or indirectly support technology commercialization may not yet have been uniformly translated into significantly university-related spin-off successes (with the notable exception of KAIST and Tsinghua). Nevertheless the growing policy emphasis on supporting spin-off formation does indicate a strategic awareness that the external environments within which these universities are located are still not wholly conducive for technology licensing; basically most of the Asian economies (with the exception of Japan and possibly Korea) still do not have sufficient number of sophisticated private high-tech firms that have the capabilities to commercialize upstream technologies from universities, unlike the situation in Japan or in high-tech hubs like Silicon Valley and Boston in North America. The lack of such sophisticated 'receiver-active' players in the NIS (Kodama and Suzuki 2007) makes it necessary for Asian universities to promote new spin-offs as an alternative route to commercialize university technology.

The above observed development trends over time – refinement of TTO performance metrics and organizational capacities, and implementation of a broader range of start-up supporting activities – suggest a healthy process of organizational learning among the Asian universities covered in the study. Not only are the universities acquiring more technology commercialization capacity through learning by doing, but the government ministries that oversee them also appear to be learning over time to adapt and refine their policy instruments for fostering university technology commercialization to achieve better fit with the unique context of their respective economies. For example, in Singapore, the government has recently introduced incentive schemes adapted from the SBIR scheme in the USA to channel more funding resources to specifically support technology spin-offs, based on inputs provided by the university itself. Likewise, the Taiwanese government has refined the performance indicators to be used to monitor the progress of university technology commercialization activities.

### 1.4.2 Emerging Policy Issues

The case studies of the focal universities have revealed several issues which would be of interest for further discussion and analysis. While it is beyond the scope of the present volume to answer all the questions raised here, it is instructive to note these issues as some of these concerns may become relevant to universities in the future as they further evolve and develop their third mission role.

1. *Is the mode of university-industry collaboration favouring large private companies too much?*

This imbalance was a key concern raised in the case of Tokyo University. Would this be unique to Japan, or is it also likely to affect other Asian economies? We can anticipate several cases where such a situation may arise:

- Foreign multinational corporations (MNCs) operating in China demanding exclusive patent rights to sponsored research with Chinese Universities.
- Korean chaebols benefiting more from university technology collaboration than SMEs.
- Private industry getting more out of Hong Kong universities under the government scheme to promote university-industry collaboration.
- Private firms in India and Malaysia preferring to engage university scientists in personal consulting contracts to capture the IP generated, instead of engaging in sponsored research with the university where IP may have to be jointly owned.

On the other hand, such imbalances may be temporary as universities undergo a transitional phase. As the universities' organizational capacity for technology commercialization develops, the bargaining power of universities may increase. In the long run, it is feasible that a fairer collaborative relationship may emerge.

2. *Is enough being done by the university and government to promote technology transfer to local SMEs?*

In most universities, technology licensing and industry collaboration involves engaging large corporations, with minimal participation of SMEs. This is due to the lack of absorptive capacity of local SMEs to internalize sophisticated technologies emerging from universities. There may therefore be a need for more active government support schemes to support technology transfer to local SMEs in most Asian economies. Indeed such support schemes have recently been introduced by the Singapore government in the form of a Proof of Concept (POC) and a Proof of Value (POV) grant which are modelled after the SBIR grant scheme in the USA. The Taiwanese government has also extended the joint R&D consortium approach, which has been successfully implemented by the state-owned Industrial Technology Research Institute (ITRI) to enable SMEs to pool resources to participate in state-sponsored R&D projects, to the university context as well.

3. *Is enough being done by the university and government to promote technology commercialization through spin-offs?*

More advanced university technologies may be left on the shelf if there is a lack of industry readiness or capability to commercialize these technologies. In such cases, formation of university spin-offs becomes the only alternative mechanism to bring advanced university technologies to market. The evidence from the case studies appears to suggest that, despite growing effort by the universities to nurture such spin-offs, this route has not been very fruitful yet, as the number of spin-offs from the universities is still relatively small, and with the exception of KAIST and Tsinghua, success rates have remained low. This may be attributable to a number of external and internal factors. The main external impediment factor is the lack of early-stage angel investor and venture capital to support early stage start-ups. The lack of sophisticated lead-users in the domestic economy who could serve as customers to such university start-ups is another factor. The key internal factor could be the lack of entrepreneurial drive and business skills among the university faculty and students, and the lack of experienced and successful high-technology entrepreneurs to serve as role models or to provide mentorship. This internal factor could be made worse by external socio-cultural factors such as cultural norms leading to fear of failure and hence preference of graduates to work in large corporations and government. While start-up support programmes such as incubators and entrepreneurship education are increasingly being implemented by many Asian universities to overcome the above internal impediment, such efforts are insufficient and need to be complemented by government policies to address the weaknesses in the external ecosystem.

4. *The role of entrepreneurship education/outreach programmes to inculcate entrepreneurial mindset among university faculty and students*

The case studies of MMU and the NUS highlighted entrepreneurship education as a key element in increasing the level of technology commercialization through spin-off formation in the long run. Such education programmes need to address not only the skills/knowledge gap of the university community, especially the faculty and students in science and engineering disciplines who traditionally do not receive training in business skills and knowledge, but also to play a change-agent role in terms of raising awareness or interest in entrepreneurial pursuits and inculcating a more entrepreneurial mindset. In particular, the experience of NUS suggests that an experiential learning approach involving internship in start-up companies or involvement of students in attempting to start their own ventures (for example, through participation in business plan competitions) is likely to be more effective than the traditional classroom academic approach.

5. *The need for universities to better leverage on informal networks such as alumni networks to facilitate technology commercialization*

There is a considerable resource base in the informal networks surrounding universities, such as alumni, angel investors and industry partners, which have not been fully tapped by the universities. As an example, Tsinghua University's alumni have founded several ventures housed in the Tsinghua University Science Park (TusPark), using Tsinghua University's technology and IP. Alumni networks have also been found to be effective in the Indian Institutes of Technology and KAIST in Korea. These examples of successful commercialization efforts suggest that there is a need for more universities to take an active role in facilitating informal interactions and exchange between university faculty, students and industry. These may be achieved through platforms such as regular enterprise forums, entrepreneur-in-residence programmes, angel investor networks and the development of science park or incubation facilities in close proximity with the university campus.

6. *Risk of using wrong technology commercialization performance indicators*

Many universities have begun their technology commercialization efforts by adopting the AUTM approaches and performance indicators. While these provide a sound starting point, slavish adherence to these metrics could lead to resource misallocation if the indicators are not aligned to the intended mission or objectives. For instance, using patent counts as a performance indicator has often led to a glut of patent applications in universities, directing undue amounts of resources to this exercise to the possible detriment of other activities such as technology marketing and licensing of non-patented technologies.

7. *Should faculty achievements in technology commercialization be recognized in promotion and tenure decisions, or other forms of rewards?*

An issue which surfaces in several case studies is the challenge of managing the allocation of faculty resources to commercialization activities. In many universities, faculty performance evaluation is primarily based on teaching and research, with entrepreneurial or commercialization achievements accorded little or no weight. A number of government agencies in Asia have suggested that such university policies could be revised to incentivize faculty participation in technology commercialization.

We believe that such proposals to amend the promotion and tenure policies to encourage faculty involvement in technology commercialization to be misplaced, at least in so far as the leading research-intensive universities are concerned. The primary role of the leading research-intensive universities in Asia remains teaching and research, and they need to focus on promoting excellence in both in order to achieve their aspirations to become competitive on a world scale. The experience of the leading universities in the USA and Europe like Stanford, MIT and Cambridge suggest that it is possible to achieve a high level of technology commercialization performance without compromising the pursuit of excellence in research and teaching.

## 1.5 POLICY IMPLICATIONS

Synthesizing the analysis and findings from our 13 university case studies, we highlight the following key policy implications for both government policy makers as well as senior administrators of universities in Asia to take into consideration when implementing programmes to promote university-industry linkages and technology commercialization:

- The government policy makers and the university senior administrators need to understand that the economic impact of technology commercialization activities involves long gestation periods, and despite the growing focus on generating royalty income from IP licensing, much of the spillover benefits generated from technology commercialization activities are not captured by the university itself. Consequently, university investment in such activities cannot be evaluated in terms of financial returns in the short or even medium term. Even in the USA where university technology commercialization activities have been pursued the longest, only a handful of universities have so far been able to achieve financial self-sustainability in their TTO operations, and the time taken by some of these TTOs to achieve financial autonomy is longer than two decades. For most Asian universities that have a shorter history in technology commercialization, the expectation of the government policy makers and the university leaders must therefore be realistic. The assessment of the economic impact of university technology commercialization must also be more holistic to take into account the larger spillover benefits to the economy, rather than just focusing on the licensing revenue captured by the university. At the same time in monitoring the performance of the TTO operations, the senior university administrators must avoid the pitfalls of using simplistic quantitative targets such as the number of patents granted and the number of spin-off companies created, without taking into account the quality of the underlying technology portfolio, or the size and growth potential of the employment and sales generated by the start-ups.

- The actual level of economic impact that can be derived from university technology commercialization activities depends not only on the efficacy of the university's policies and programmes, but also on the absorptive capacity of the existing industries in the host economy as well as the vibrancy of the larger external ecosystem that supports the growth of new ventures to commercialize university technology. Thus any government policy to induce the universities to invest more resources into technology commercialization activities must be complemented by appropriate public policies to enhance the absorptive capacity of industries as well as to plug the obvious gaps in the entrepreneurial support ecosystem such as availability of early-stage risk capital and reduction of regulatory costs for new start-up formation. Such external supporting environment policies may also need to take into account the characteristics of different technological fields, which may require different supporting environmental policies and financial resources (for example, life sciences versus ICT).
- For university technology commercialization through spin-off formation to occur on a larger scale, there must be a larger pool of faculty, researchers and students who are entrepreneurially minded and who possess relevant business skills. As such, university leaders must recognize that one of the enabling factors for greater university spin-off formation is the injection of a more entrepreneurial culture among the university community, which in turn may require a greater effort in implementing entrepreneurship education activities to raise awareness/interest and to impart relevant business skills. In addition the university may need to pay more attention to the software aspects of the technology commercialization infrastructure, including facilitation of networking with potential

investors and customers, provision of mentoring by experienced entrepreneurs and investors, and provision of seed funding to bridge the gap in funding for early-stage start-ups. In addition the university's policy framework should not impose excessive regulatory constraints on faculty and student involvement in technology start-up activities.

## NOTE

1. Mahidol University in Thailand has in the last decades expanded its academic scope to include the fields of social sciences, arts and humanities. However it still retains a strong focus, especially in terms of research and postgraduate programmes, on medical and science disciplines, reflecting its origins as the first medical school in Thailand, established in 1890.

## REFERENCES

Association of University Technology Managers (AUTM) (2001), *AUTM Licensing Survey 2000*, Northbrook, Il: AUTM.

Clark, B. (1998), *Creating Entrepreneurial Universities: Organizational Pathways and Transformation*, Oxford: Pergamon-Elsevier Science.

Etzkowitz, H. (2002), *MIT and the Rise of Entrepreneurial Science*, London: Routledge.

Etzkowitz, H., A. Webster, C. Gebhardt and B.R.C Terra (2000), 'The future of the university and the university of the future: evolution of ivory tower to entrepreneurial paradigm', *Research Policy*, **29** (2), 313–30.

Heher, A.D. (2006), 'Return on investment in innovation: implications for institutions and national agencies', *Journal of Technology Transfer*, **31** (4), 403–14.

Hershberg, E., K. Kabeshima and S. Yusuf (2007), 'Opening the ivory tower to business: university-industry linkages and the development of knowledge-intensive clusters in Asian cities', *World Development*, **35** (6), 931–40.

Kodama, F. and J. Suzuki (2007), 'How Japanese companies have used scientific advances to restructure their businesses: the receiver-active national system of innovation', *World Development*, **35** (6), 976–90.

Laredo, P. (2007), 'Revisiting the third mission of universities: towards a renewed categorization of university activities?', *Higher Education* Policy, **20** (4), 441–56.

Mowery, D.C., R.R. Nelson, N.S. Bhaven and A.A. Ziedonis (1999), 'The effects of the Bayh-Dole Act on US university research and technology transfer', in L.M. Branscomb, F. Kodama and R. Florida (eds), *Industrializing Knowledge: University-Industry Linkages in Japan and the United States*, Cambridge, MA: MIT Press, pp. 269–306

Mowery, D.C., R. Nelson, B. Sampat and A. Ziedonis (2004), *Ivory Tower and Industrial Innovation: University-Industry Technology Transfer Before and After*

*the Bayh-Dole Act in the United States*, Palo Alto, CA: Stanford University Press.
Paytas, J., R. Gradeck and L. Andrews (2004), *Universities and the Development of Industrial Clusters*, Pittsburgh, PA: US Department of Commerce and Carnegie Mellon University.
Shane, S. (2002), 'Selling university technology: patterns from MIT', *Management Science*, **48** (1), 122–37.
Shane, S. and T. Stuart (2002), 'Organizational endowments and the performance of university start-ups', *Management Science*, **48** (1), 154–70.
Thursby, J.G. and S. Kemp (2002), 'Growth and productive efficiency of university intellectual property licensing', *Research Policy*, **31** (1), 109–24.
Wright, M., B. Clarysse, P. Mustar and A. Lockett (2007), *Academic Entrepreneurship in Europe*, Cheltenham, UK and Northampton, MA, USA: Edward Elgar Publishing.

# 2. The role and impact of universities in national and regional innovation systems: Kyushu University's experience

**Toru Tanigawa, Hiromi Sasaki and Megumi Takata**

## 2.1 INTRODUCTION

### 2.1.1 Background and Purpose of Research

With the spread of the internet, the world economy has become increasingly close-knit, and the information and knowledge gap between countries and regions has narrowed markedly. As Thomas L. Friedman says in his book *The World Is Flat*, a newly competitive country or region can appear anywhere in the world. Under these conditions, it is commonly recognized that the competitiveness of a country or region depends largely on its power to bring about innovation, and that much of that power comes from universities.

In the discussion below, we will explain how university-industry collaboration (UIC) in Japan is contributing to the national and regional innovation systems, mainly focusing on the example of Kyushu University. University-industry collaboration is steadily expanding across Japan, primarily through joint research and sponsored research, but also through technology transfer and university start-ups. These contributions made by Japanese universities through UIC are being highlighted in various policies set out by the government, such as the Science and Technology Basic Plan. More and more achievements are being made. However, compared with the USA which has great university-industry collaboration, the level of UIC in Japan, and universities' awareness of UIC, are both low. In the next section we will discuss the contributions made through university-industry collaboration by Japanese universities as a whole to the national and regional innovation systems in Japan.

## 2.2 THE CONTRIBUTION OF JAPANESE UNIVERSITIES TO THE NATIONAL AND REGIONAL INNOVATION SYSTEMS (WITH A FOCUS ON UNIVERSITY-INDUSTRY COLLABORATION (UIC))

### 2.2.1 Roles of Companies and Universities in Japan's Innovation System

A Swiss-based international research organization, World Forum, reports that Japan was ranked ninth in international competitiveness in 2008 and 2009. It was ranked fourth in innovation. In other words, Japan is internationally evaluated as a country that is maintaining world-leading science and technology. Typically, research by universities and companies underlies science and technology. In Japan, which is ranked second overall in World Forum's evaluation, companies make great contributions to research. Figure 2.1 shows a comparison of the distribution of research and development (R&D) expenditure among government, industrial and academic organizations. The diagram indicates that companies in Japan account for slightly more than 70 per cent of total expenditure on R&D, revealing that they play an important role in R&D (universities account for approximately 20 per cent, equivalent to the level in Europe).

On the other hand, looking at the type of research as shown in Figure 2.2, almost all corporate R&D expenses (94 per cent) are spent on applied R&D; development in particular accounts for 74 per cent. This demon-

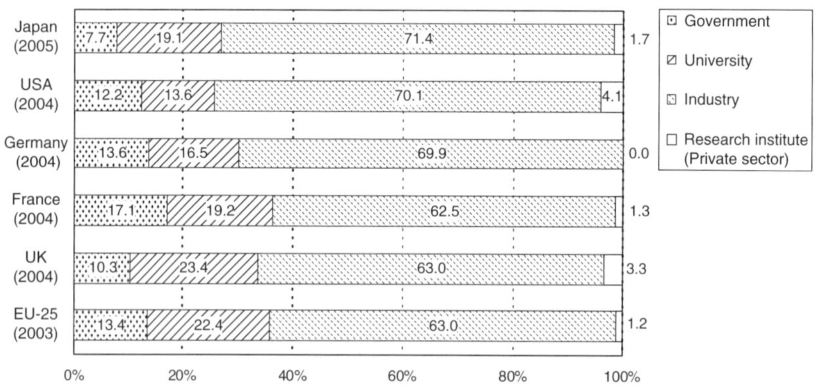


*Source:* White Paper on Science and Technology.

*Figure 2.1 Share of R&D expenditure in Western countries and Japan (2006)*

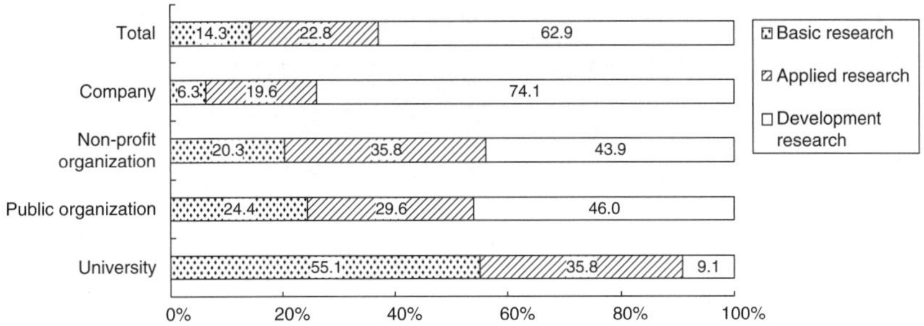


*Source:* White Paper on Science and Technology.

*Figure 2.2 Main type of research in Japanese organizations (2006)*

strates that the purpose of companies' R&D activities is development, targeting commercialization of technology. In contrast, for universities, the majority of research (55 per cent) is basic research, while development accounts for less than 10 per cent. It is obvious that there is a difference in the purpose of research between companies and universities. This is to be expected, because the supreme mission of companies is to bring products and services to the market in order to generate revenue, while that of universities is to clarify principles of various natural phenomena in order to create knowledge. However, for the innovation system of the country to function effectively, it is necessary for both companies (industry) and universities (academia) to be committed to the practical application and commercialization of knowledge. This requires cooperation by mutually complementing each other's functions and sharing their roles.

In reality, as shown in Figure 2.3, companies' need for collaboration with universities is high, including their need for basic research. In this sense, UIC is critical in the national innovation system. However Japan is only ranked 21st in the world in UIC, resulting in a decrease in the competitiveness of Japan as a whole. However, over the past decade, particularly since 2004 when Japanese universities started to be given corporate status, Japanese universities have reversed their traditional academia-oriented policies and are clearly attaching importance to UIC.

### 2.2.2 The Contribution of Japanese Universities to the National and Regional Innovation Systems

The contribution of Japanese universities to the innovations system in Japan takes the following forms: (1) research results from universities are diffused through presentations at academic society meetings and through

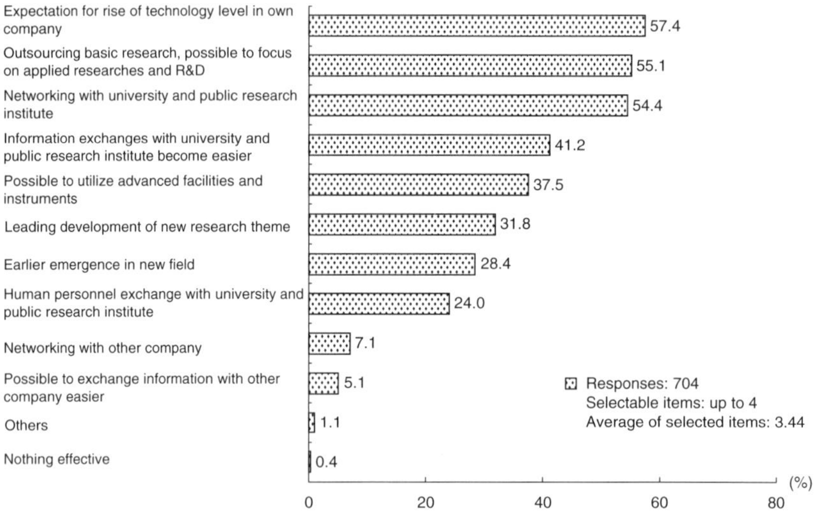


*Source:* Science and Technology/Academic Policy Bureau of the Ministry of Education, Culture, Sports, Science and Technology (MEXT) (2006).

*Figure 2.3 The reasons companies collaborate with universities*

publishing of papers, and so influence various quarters in society; (2) nurturing of innovators through higher education; and (3) various types of cooperative activities between university and society based on UIC, including joint research, sponsored research, technology transfer and creation of university start-ups. This chapter concerns mainly cooperative activities.

Universities' contribution to national and regional innovation systems can be subdivided into contribution in a broad sense (creating value for society as a whole), and contribution in a narrow sense (meeting needs within the industrial sector and contributing to innovation in that sector). This study emphasizes universities' contribution in the narrow sense, and their contribution in the broad sense shall be dealt with only briefly.

### 2.2.3 Comparison of University-Industry Collaboration in Japan and the USA

Like Japan, the USA boasts high levels of research in universities and in industry. In this section, we compare the two countries with respect to three types of UIC: joint/sponsored research, technology transfer and university start-ups. This will allow us to clarify the background behind

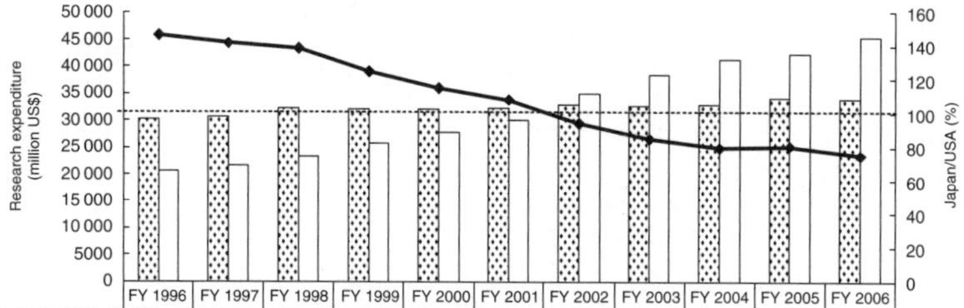


| | FY 1996 | FY 1997 | FY 1998 | FY 1999 | FY 2000 | FY 2001 | FY 2002 | FY 2003 | FY 2004 | FY 2005 | FY 2006 |
|---|---|---|---|---|---|---|---|---|---|---|---|
| Japan (million US$) | 30 131 | 30 592 | 32 229 | 32 091 | 32 084 | 32 334 | 32 823 | 32 631 | 32 740 | 34 074 | 33 824 |
| USA (million US$) | 20 560 | 21 630 | 23 250 | 25 670 | 27 870 | 29 960 | 34 960 | 38 500 | 41 200 | 42 300 | 45 300 |
| Japan/USA (%) | 146.6 | 141.4 | 138.6 | 125.0 | 115.1 | 107.9 | 93.9 | 84.8 | 79.5 | 80.6 | 74.7 |

*Source:* Data from Association of University Technology Managers (AUTM) (various years) and from Ministry of Economy, Trade and Industry (METI).

*Figure 2.4 Comparison of universities' total research expenditure (Japan–USA) (1996–2006)*

the characteristics of Japanese universities' contribution to the national innovation system.

### 2.2.3.1 Industry-funded research expenditure and joint research in universities in Japan and the USA

As is evident from Figures 2.4–2.6, the total amount of university research expenditure in recent years has been smaller in Japan than in the USA, with Japan's expenditure being approximately 80 per cent of that in the USA. Given the difference in size between the two economies, however, the difference in research is not as large as the figures suggest. The share of university R&D expenditure that is funded by industry (business enterprises) for joint research and so on is far smaller than the share that is funded by the government and other public institutions in both countries – about 7 per cent in the USA and about 1 per cent in Japan. It should be noted, however, that in Japan university R&D expenditure that is funded by businesses does not include personnel expenses, such as the manpower cost of university researchers. Administrative expenses (that is, indirect research costs) comprise 10 per cent of university R&D that is funded by industry for joint research and 30 per cent for sponsored research, as compared with 50 per cent or more in the USA. Therefore, the total amount of R&D funding that national universities in Japan receive from industry is only slightly more than 10 per cent of that in the USA. However, although it is estimated that the absolute amount of industry-funded R&D by universities in Japan is not very large, the actual scale of joint research between Japanese universities and industry (business enterprises), as well as the universities' contribution to national innovation, are quite close to that of the USA.

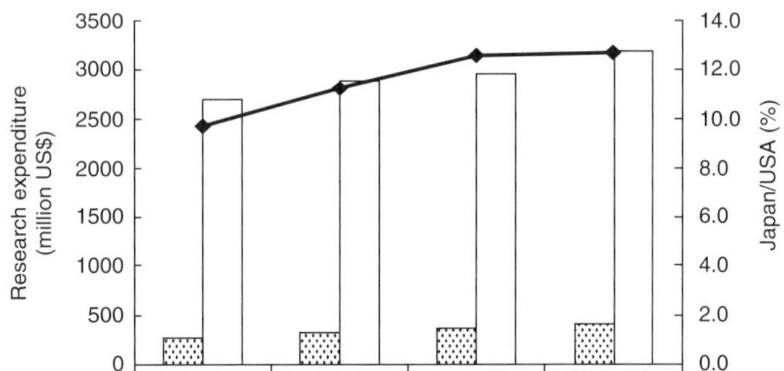


| | FY 2003 | FY 2004 | FY 2005 | FY 2006 |
|---|---|---|---|---|
| Japan (million US$) | 262 | 323 | 371 | 403 |
| USA (million US$) | 2700 | 2880 | 2960 | 3180 |
| Japan/USA (%) | 9.7 | 11.2 | 12.5 | 12.7 |

*Source:* Data from Association of University Technology Managers (AUTM) (various years) and from Ministry of Economy, Trade and Industry (METI).

*Figure 2.5 Comparison of university research expenditure funded by industry (Japan–USA) (2003–06)*

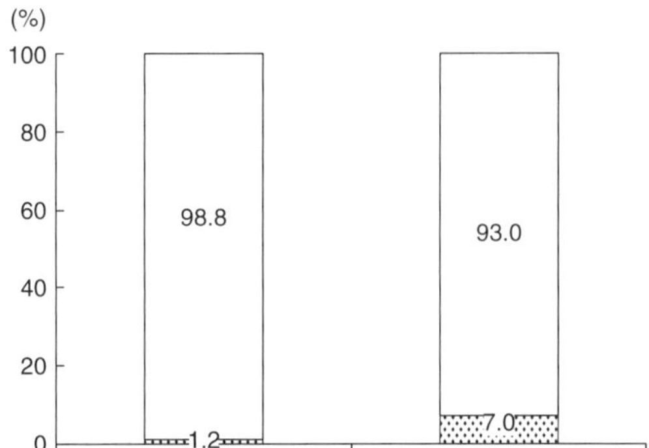


| | Japan | USA |
|---|---|---|
| From government and others | 98.8 | 93.0 |
| From industry | 1.2 | 7.0 |

*Figure 2.6 Comparison of sources of universities' research expenditure (Japan–USA) (2005)*

As shown in Figure 2.7, the majority of patent applications by the eight major research-oriented universities of Japan are made jointly with business enterprises. Also, technology transfer to industry is often the result of joint research or sponsored research between universities and businesses. These

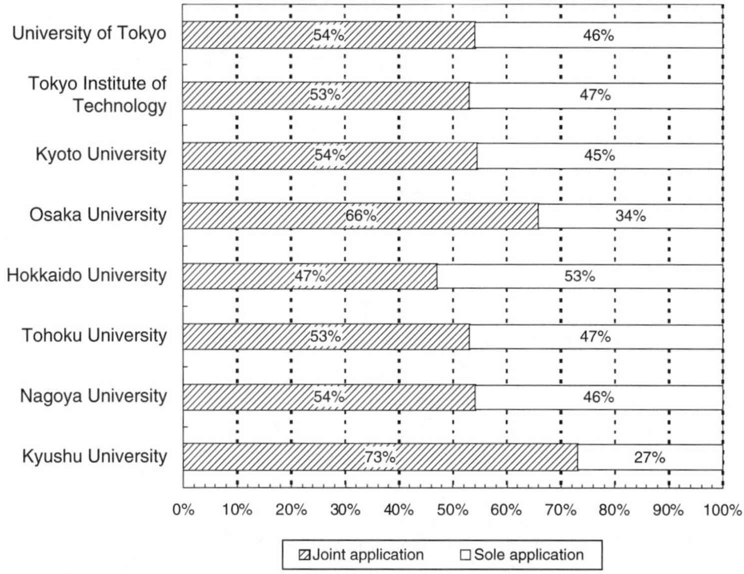


*Note:* Joint application means the application is the result of joint research with industry.

*Source:* Ijichi and Nagaoka (2007).

*Figure 2.7 Disclosed patent applications by main universities in Japan (2004–05)*

facts suggest that, as in the case in the USA, joint/sponsored research is important for universities' contribution to industry and innovation in Japan.

#### 2.2.3.2 University-industry technology transfer in Japan and the USA

Figure 2.8 shows that the number of inventions reported to universities by their researchers has been increasing rapidly in Japan in recent years. This is due, at least in part, to the fact that more and more Japanese universities are attaching importance to UIC, and are striving to increase their portfolio of intellectual property. The number of inventions reported in FY 2006 was 53.2 per cent of that in the USA. It should be noted, however, that as shown in Figure 2.9, the total amount of royalties received by Japanese universities for licensing their intellectual property (including patent rights) is only about 1/100 of that received by their USA counterparts.

The major reasons for this are as follows:

1. Before the National University Corporation Act was put into effect in 2004, the right to any intellectual property invented within a

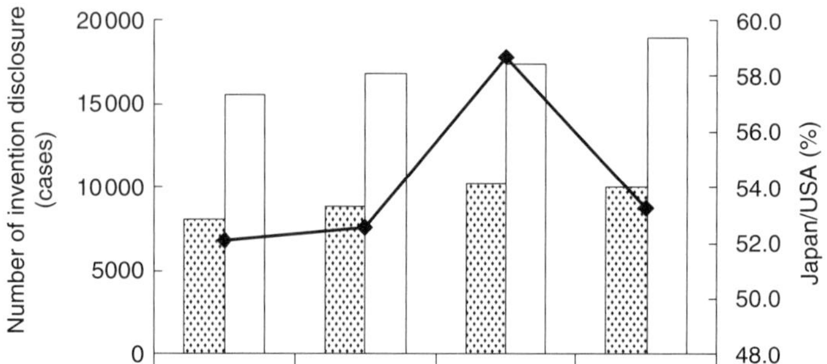


| | 2003 | 2004 | 2005 | 2006 |
|---|---|---|---|---|
| Japan (number of invention disclosure) | 8078 | 8833 | 10 202 | 10 048 |
| USA (number of disclosure) | 15 510 | 16 811 | 17 382 | 18 874 |
| Japan/USA (%) | 52.1 | 52.5 | 58.7 | 53.2 |

*Source:* Data from Association of University Technology Managers (AUTM) (various years) and from Ministry of Economy, Trade and Industry (METI).

*Figure 2.8 Comparison of number of invention disclosures (Japan–USA) (2003–06)*

university belonged to the inventor or the government, not the university, and therefore it was difficult for the university to secure royalties from the invention.

2. Japanese university researchers focus more on academia than those in the USA and other countries, showing less interest in creating intellectual property and utilizing research conducted by universities.
3. Protection of intellectual property and marketing are required for inventions created solely by university research. However, universities have not sufficiently developed or acquired university staff with expertise and experience in this field.
4. As shown in Figure 2.7, the majority of patent applications by universities are filed jointly with the business enterprises with which the universities have carried out the research. Under these conditions, it is difficult for the universities to secure revenues by licensing their intellectual property.

Recently however, amid the strong current of administrative and financial reforms in Japan, there is a mounting cry for activation of the economy through an effective use of the knowledge owned by universities, and for promotion of competition among universities and greater efforts

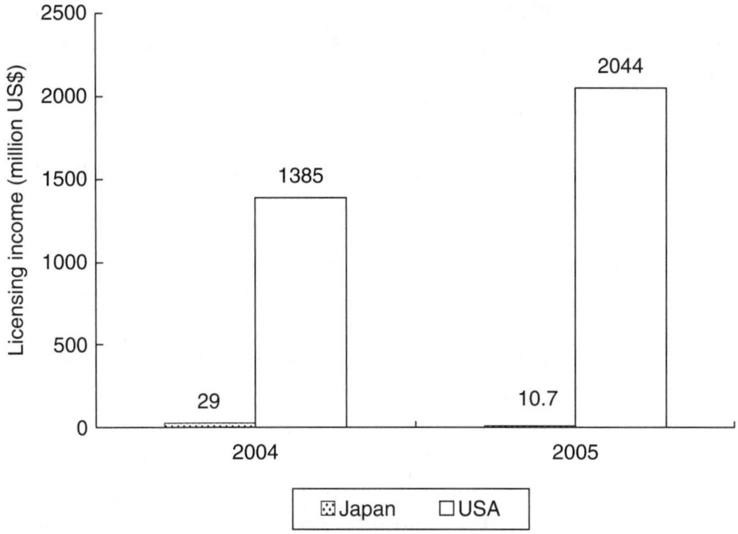


*Source:* Data from Association of University Technology Managers (AUTM) (various years) and from Ministry of Economy, Trade and Industry (METI).

*Figure 2.9 Comparison of licensing income (Japan–USA) (2004–05)*

by universities through transformation of the national universities into independent administrative corporations – that is, a cry for reduction of government aid to universities. In addition, it is increasingly being voiced that Japan should follow the example of USA universities, which have gained increasing revenue from technology transfer since enforcement of the Bayh-Dole Act. As a result, Japanese universities have started pressing ahead with the transfer of their knowledge and technology.

At the present stage, it must be said that universities' influence on Japanese society, and their contribution to innovation in Japan through technology transfer, are still minor as compared to those in the USA and so on.

#### 2.2.3.3 Creation of start-ups from Japanese universities

Commercializing university research results or inventions is not limited to transferring or licensing the intellectual property to a business enterprise. If it is difficult for the university to find a suitable transferee or licensee, or if it is not likely that the technology will be utilized effectively, the university may start up a new business to commercialize it. This is called the creation of a business venture by a university. In this respect, there is a marked difference between Japanese and USA universities. The number of business ventures established from intellectual property licensed from universities is more than ten times greater in the USA than in Japan. Although government and public support for the creation of business ventures by universities in Japan has been strengthened considerably, Japan is far behind the USA, where business ventures flourish remarkably. This trend in Japan is true of Kyushu University.

Thus, contributions of university-launched companies have not yet had significant influence on the Japanese innovation system. No university high-tech venture in Japan has had any social impact, as compared to Genentech and Google in the USA.

As is evident from the above comparison, the USA shows remarkable achievements in all modes of UIC, including joint/sponsored research, technology transfer and creation of university start-ups. It may be said that the contribution of USA universities to their national innovation system is large and well-balanced. This is in contrast to Japanese universities' contribution to their national and regional innovation systems through UIC; in Japan, the contribution from technology transfer and creation of university start-ups has just started and therefore is still insignificant, but the contribution from joint/sponsored research has been increasing steadily.

## 2.3 KYUSHU UNIVERSITY'S CONTRIBUTION TO THE NATIONAL AND REGIONAL INNOVATION SYSTEMS (WITH A FOCUS ON UNIVERSITY-INDUSTRY COLLABORATION)

### 2.3.1 Profile of Kyushu University

Kyushu University was founded as Japan's fourth imperial university in 1911. It is one of the most distinguished research-oriented national universities in Japan. Having some 100 years of history and tradition, the university is especially strong in natural sciences (material sciences and information sciences). When the university merged with the Kyushu Institute of Design in 2003, it came to have research and educational functions in the unique field of design. At present, it has 17 graduate schools, 16 faculties, 11 undergraduate schools and 3 research institutes. The university has 2224 teachers and 18 647 students (11 084 undergraduate students and 6843 graduate students) (Table 2.1). Since the university has comparatively close connections with East Asia and Southeast Asia both historically and geographically, it has a policy of focusing on Asia, and so

*Table 2.1 Profile of Kyushu University (2008)*

| | | |
|---|---|---|
| Chronological Outline | 1911 | Kyushu Imperial University was founded |
| | 1947 | Name of university changed from Kyushu Imperial University to Kyushu University |
| | 1994 | Advanced Science and Technology Center for Cooperative Research Kyushu University was founded. Currently, Art, Science and Technology Center of Kyushu University: KASTEC |
| | 1995 | Venture Business Laboratory (VBL) was founded |
| | 1998 | Business Liaison Office (BLO) was founded |
| | 2000 | TLO Kyushu was founded |
| | 2003 | IMAQ was founded |
| | 2003 | Kyushu University integrated with Kyushu Institute of Design |
| | 2004 | Kyushu University became a National University Corporation |
| Graduate Schools (17) | Humanities, Social and Cultural Studies, Human-Environment Studies, Law, Law School (Professional Graduate School), Economics, Sciences, Mathematics, Systems Life Sciences, Medical Sciences, Dental Science, Pharmaceutical Sciences, Engineering, Design, Information Science and Electrical Engineering, Interdisciplinary Graduate School of Engineering Sciences, Bioresource and Bioenvironmental Sciences | |
| Faculties (16) | Humanities, Social Cultural Studies, Human-Environment Studies, Law, Economics, Languages and Cultures, Sciences, Mathematics, Medical Science, Dental Science, Pharmaceutical Sciences, Engineering, Design, Information Science and Electrical Engineering, Engineering Science, Agriculture | |
| Undergraduate Schools (11) | Letters, Education, Law, Economics, Science, Medicine, Dentistry, Pharmaceutical Sciences, Engineering, Design, Agriculture | |
| Research Institutes (3) | Medical Institute for Bioregulation, Research Institute for Applied Mechanics, Institute for Materials Chemistry and Engineering | |
| Budget | ¥1 200 560 000 (October 10 2008)<br>(Rate 1US$=¥0.01) | |
| Staff (Research) | 2224 | |
| Staff (Administration) | 2607 | |
| Students | 18 647 (11 804 undergraduate students; 6843 graduate students) | |

*Source:* Data from Kyushu University (2008) and Kyushu University website.

has been active in the interchange of research, personnel and so on with countries in this region.

### 2.3.2 University-Industry Cooperative Activities of Kyushu University

#### 2.3.2.1 Outline and strategy of the intellectual property management center of Kyushu University

After universities were given corporate status in 2004 to further advance research and education – one of Kyushu University's aims – and to develop as an internationally competitive university, Kyushu University set two goals: development of new scientific areas and focusing on Asia. Based on these two important goals that Kyushu University aims to achieve in the future, the university has attached importance to four activity fields: research, education, social contribution and international contribution. In particular, with respect to social contribution through UIC, the Intellectual Property Management Center of Kyushu University (IMAQ) was established in 2003. IMAQ is a core and centralized organization for industry-academia collaboration in Kyushu University, vigorously promoting UIC activities.

IMAQ is the university's central organization for UIC, functioning as the interface between Kyushu University and the local community, society and industry. IMAQ was established on 1 October 2003, following the merger between Kyushu University and Kyushu Institute of Design, and the adoption by the Ministry of Education, Culture, Sports, Science and Technology (MEXT) of the university's proposal for the University Intellectual Property Management Center Development Project. IMAQ is a general one-stop service organization that not only creates, manages and utilizes intellectual property, but also supports all kinds of UIC, such as coordinating joint research, supporting university start-ups and promoting internships and international UIC. The university's objective in UIC is to maximize effective use of all its resources, including intellectual property, talented personnel, facilities, brand name, networks and its proximity to Asia; it also aims to return the results of its research efforts to society.

Strong leadership exerted by IMAQ has resulted in brisk UIC in Kyushu University. This indicates the importance of a core university organization promoting UIC in a country such as Japan which otherwise has insufficient UIC experience.

#### 2.3.2.2 Organization of IMAQ

IMAQ consists of eight groups staffed with a total of 59 persons (as of 1 October 2008). The policy of recruiting young staff members who have experience in industry has allowed for smooth cooperation between the university and industry (Figure 2.10). In addition, ten external technical

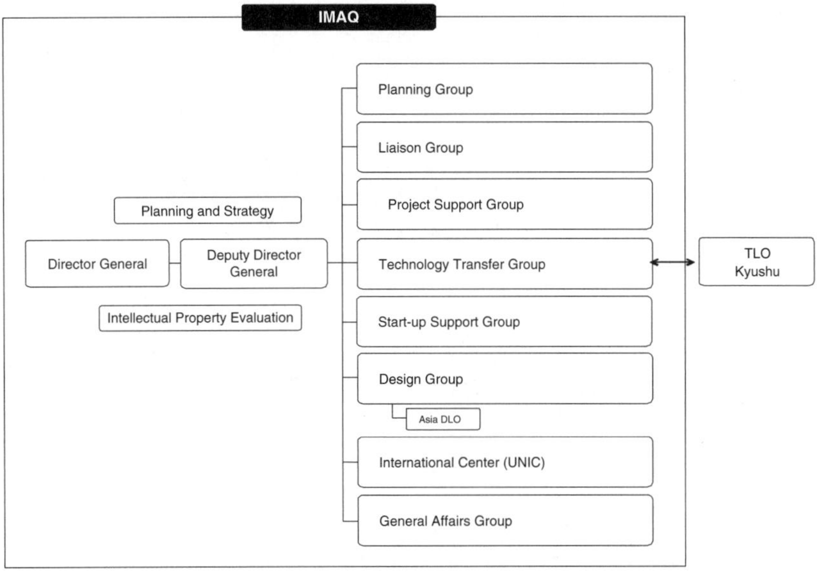


*Figure 2.10 Organization of IMAQ and operation of individual groups*

advisors from industry, academia and the government provide extensive support for the activities of IMAQ. TLO Kyushu and the Technology Transfer Group jointly take care of technology transfer activity.[1]

### 2.3.3 Kyushu University's Contribution to the National and Regional Innovation Systems

#### 2.3.3.1 Joint research and sponsored research in Kyushu University

The achievements in UIC (joint research, sponsored research, technology transfer, support for university start-ups and so on) at Kyushu University have been growing steadily over the years. Since IMAQ was established in 2004, the growth has been especially conspicuous.

The volume of joint research and sponsored research has been increasing each year. In FY 2007, the number of joint research projects was 579 and the total expenditure on these projects was $21.2 million (Figure. 2.11). In the same fiscal year, the number of sponsored research projects was 560 and the total expenditure was $79.4 million (Figure. 2.12). The growth of joint research has been particularly rapid. In terms of total expenditure on joint research in FY 2007, Kyushu University ranked fourth among the universities in Japan (Figure. 2.13).

Technology transfer by Kyushu University, measured by the number of

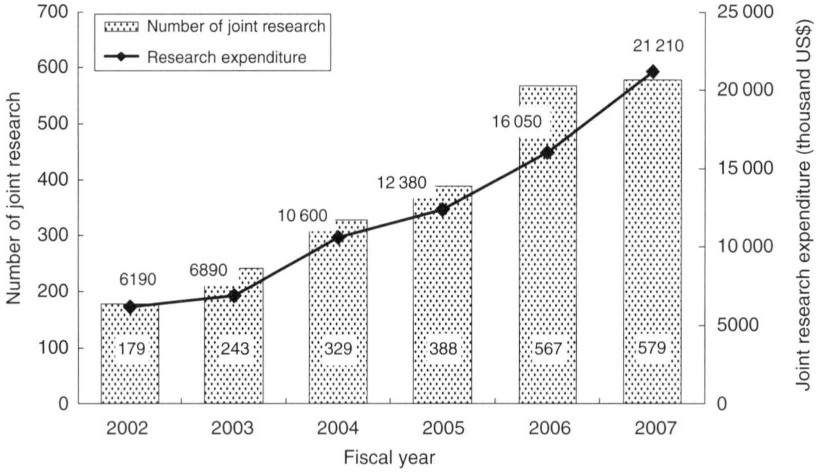


*Source:* Data from Kyushu University.

*Figure 2.11 Number of joint research projects and expenditure on joint research*

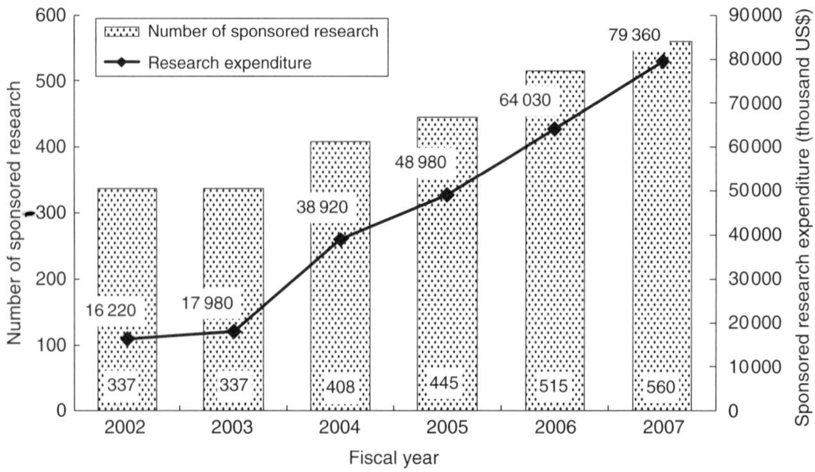


*Source:* Data from Kyushu University.

*Figure 2.12 Number of sponsored research projects and expenditure on sponsored research*

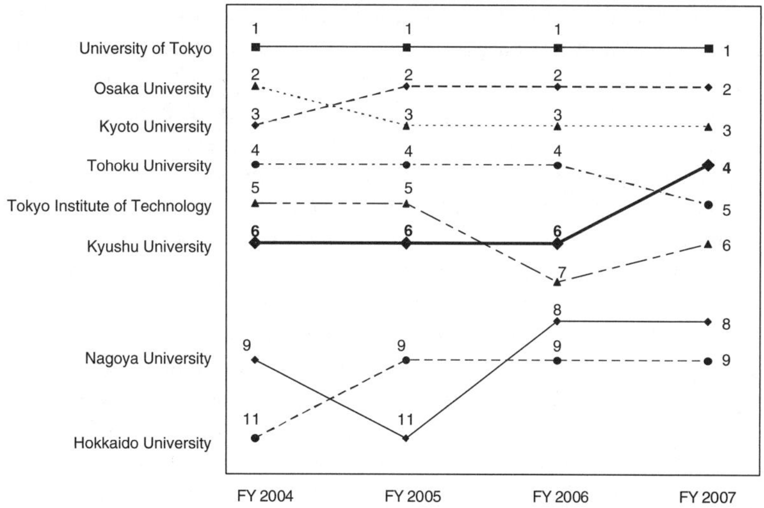


*Figure 2.13 Ranking of joint research expenditure in main universities*

invention disclosures, patent applications and licenses (285, 294 and 205, respectively in FY 2007) and by revenues from licensing ($194600 in FY 2007) has also been increasing.

In addition, the university created a cumulative total of 53 business ventures by 2007, ranking sixth among all Japanese universities.

As mentioned earlier, Kyushu University has a proud history as one of the seven imperial universities. Today it is one of Japan's major research-oriented universities. Therefore, as shown in preceding sections, Kyushu University has been among the highest-ranking universities in Japan in terms of the total amount of research funding received from the government, businesses and so on. In particular, in view of intensifying competition among universities due to the incorporation of the national universities and ever-expanding UIC in recent years, Kyushu University has made strenuous efforts to strengthen its ties with business enterprises (specifically, to win joint/sponsored research contracts). As a result, the university's joint/sponsored research has grown markedly in recent years, outpacing many of Japan's major universities. This suggests that the university's contribution to the innovation system owes largely to its growing volume of joint/sponsored research.

On the other hand, the volume of technology transfer by Kyushu University is relatively small. Although the university is striving to increase the numbers of inventions, patent applications and the amount of licensing income and so on, they are not very large as compared with other universities in Japan. This is due to Kyushu University's strategy of attaching special importance to increasing joint/sponsored research with business enterprises. As shown in Figure 2.7, joint applications with business enterprises account for more than 70 per cent of total patent applications by Kyushu University, the largest proportion among the major universities. The reason for this is that Kyushu University strictly limits patent applications for independent inventions because of cost considerations, while it gives top priority to obtaining joint research contracts within the framework of organization-level cooperation (see Section 2.3.4).

The results of the university's creation of start-ups have yet to be verified, because none of those businesses are fully fledged. Despite this, the number of start-ups from Kyushu University has been steadily increasing, so that the university is amongst the top ten Japanese universities on this dimension. In the past one to two years, not only the number but also the quality of university start-ups has improved. Several of the business ventures created by Kyushu University have won the grand prize at business plan competitions. In addition, with the support of venture capital, some have begun operating overseas.

#### 2.3.3.2 Implementation of questionnaire survey of Kyushu University's researchers on joint/sponsored research

*Purpose and outline of survey*

With the aim of understanding the condition of joint research with business enterprises, Kyushu University carried out a questionnaire survey of researchers in major postgraduate courses in natural science. The survey covered a total of 181 joint research and sponsored research projects carried out jointly by 155 researchers in the Faculty of Engineering/Faculty of Information Science and Electrical Engineering (ISEE)[2] and in business enterprises.[3]

*Survey results (excerpts of main items)*

Characteristics of industrial partners Approximately half of the companies that conduct joint research and sponsored research with the university have their head offices in Tokyo. Approximately three quarters of them are large companies, 16.8 per cent of the companies (20 companies) are located in Fukuoka Prefecture, in which Kyushu University is located, and 21 per cent of the companies are in the Kyushu area (Figures 2.14 and 2.15). Thus, local (that is, Kyushu-based) companies together account for approximately 40 per cent of collaborating partners. Of the partners headquartered in Tokyo, 86.8 per cent are large companies. By contrast, in

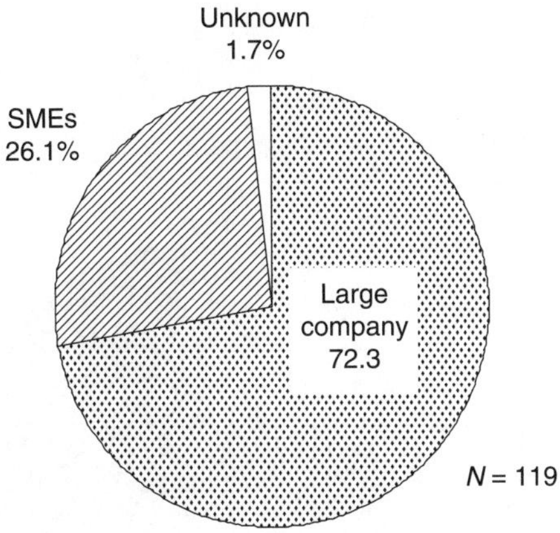


*Source:* Data from Kyushu University.

*Figure 2.14 Breakdown of companies by scale of company*

Fukuoka Prefecture, the proportion of small- and medium-sized companies is somewhat higher. With regard to the sectors in which the companies are operating, electrical equipment accounts for the highest percentage (13.4 per cent), followed by precision equipment (12.6 per cent). The total of electric/machinery companies (comprising machinery, transportation equipment, electrical equipment and precision equipment) comes to 47.8 per cent. In Fukuoka Prefecture, electric/gas companies account for the highest percentage.

It can be seen that as a result, Kyushu University mostly conducts joint research with Tokyo-based large companies in spite of the fact that the university is located far from Tokyo. In contrast, Kyushu University has less collaboration with local companies. Hence, it is possible to say that Kyushu University's social contribution is mainly to the national innovation system, subordinating its contribution to the local innovation system. It would be fair to say that this fact shows the following points: (1) Kyushu University is among the research universities in Japan that focus on basic research; (2) companies with capabilities to leverage such a high level of research are large companies in Japan; and (3) these types of companies converge on Tokyo. It may be conjectured that other Japanese research universities are in a similar situation.

Reasons for companies conducting joint research and sponsored research with universities Amongst the reasons business enterprises give for carrying out joint/sponsored research with the university, the top reason

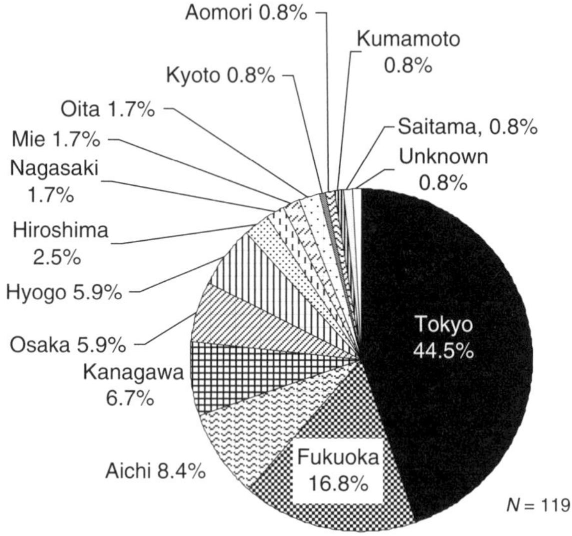


*Figure 2.15 Breakdown of companies by location of headquarters*

is 'To do more advanced research' (78.2 per cent), followed by 'To do applied research' (59.7 per cent) (Figure. 2.16).

This result closely matches the country-wide trend (Figure 2.3). In other words, facing an environment of fiercer competition, companies are being compelled to consolidate corporate resources. In this situation, companies choose universities focusing on basic research and advanced research which they do not have the resources to perform, rather than choosing universities specializing in applied and development research in which the companies are already strong. In short, they collaborate in a symbiotic relationship. Therefore it is not possible to immediately put university research results to practical use, even through technology transfer. Thus, the following points form the background for the above trend: in order to realize the commercialization and use of university research results, it is necessary for companies to promote commercialization through joint/sponsored research with universities by utilizing their own strength in applied and development research; meanwhile, companies position universities as the performers of basic research and advanced research, in which the companies themselves have a weakness, and aim at joint/sponsored research with universities to discover potentially beneficial business ideas and seeds for new businesses.

Results expected by university's researchers Amongst the expectations that university researchers have from cooperation with a business enter-

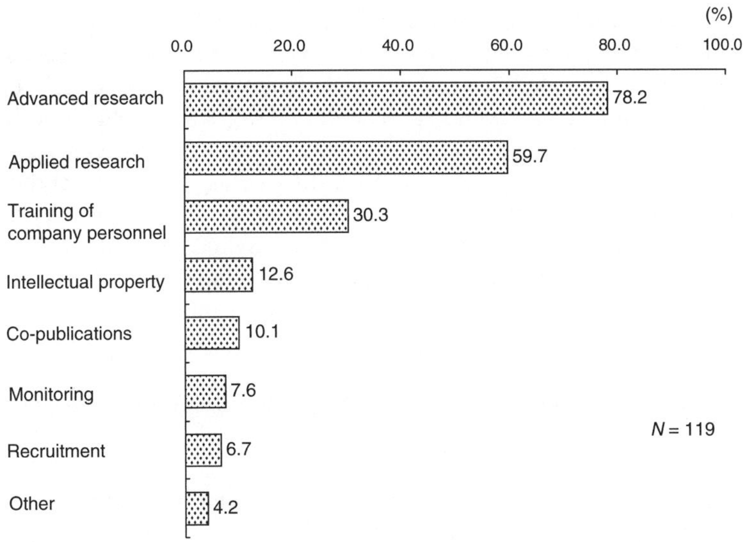


*Figure 2.16 The reasons for companies collaborating with the university*

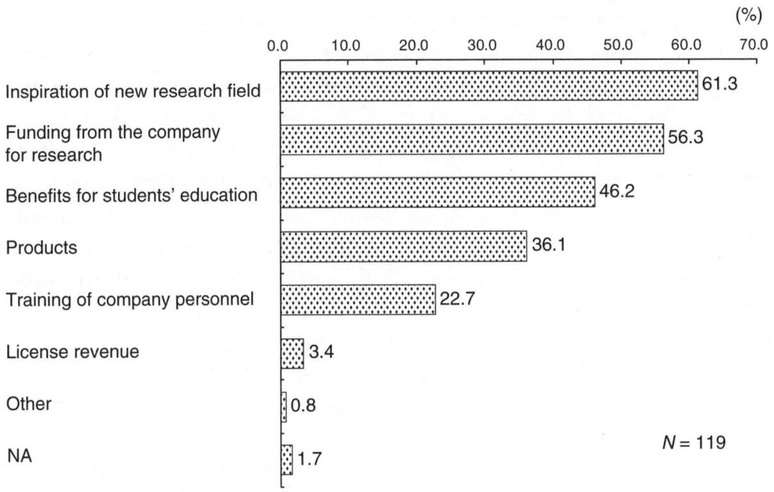


*Figure 2.17 Results expected by university's researchers*

prise, 'Inspiration of new research field' ranks top (61.3 per cent), followed by 'Funding from the company for research' (56.3 per cent) and 'Benefits for students' education' (46.2 per cent) (Figure 2.17).

It is evident that many university researchers consider working in close cooperation with business enterprises to be meaningful not only because it permits them to grasp society's needs and market trends, but also because it provides them with ideas for new research themes and with research funds.

In short, joint research and sponsored research are beneficial not only to companies, but also to university researchers from the viewpoint of revitalizing their research (for example, discovering new research fields). In other words, this type of research also contributes to university innovation. In Japan, joint/sponsored research creates mutual innovation and knowledge benefiting both companies and universities. It should be noted that this contribution is not distinct from the university bringing innovation to the industrial community.

### 2.3.4 Kyushu University's Attempt to Expand its Joint Research and Sponsored Research

#### 2.3.4.1 Joint research and sponsored research promoting system

IMAQ engages in various activities to expand joint research between researchers of Kyushu University and those in companies. In IMAQ, groups are classified based on functions. Among these groups, the Liaison Group is taking the initiative to increase the number of joint research projects. In addition, most IMAQ coordinators (approximately 20) are expected to support researchers in each department and research school within the university. Thus, IMAQ has also adopted a research support system based on disciplines. Increasing the number of joint research projects is a mission of IMAQ as a whole. Furthermore, administrative staff members who are in charge of joint research contracts have been segregated from the university's administration division and transferred to IMAQ. This is uncommon in Japanese universities, and has sped up contract procedures and allowed IMAQ to achieve efficient administration and prompt decision-making. Thus, IMAQ is implementing various innovative schemes to promote joint research.

In addition, with regard to the acquisition of large-scale joint research projects, the Liaison Group (14 persons) introduced a large-scale and interdisciplinary mechanism for organization-level cooperation in 2003, leading to a drastic increase in the number of joint research projects.

#### 2.3.4.2 Overview and significance of the 'organization-level cooperation' technique

Organization-level cooperation is a large-scale, interdisciplinary joint research system developed by Kyushu University. In organization-level cooperation, joint research which was formerly carried out between a business enterprise and a university researcher (organization – individual)

## BOX 2.1 CHARACTERISTICS OF ORGANIZATION-LEVEL COOPERATION

- Organized management by the university, rather than management by individual researchers.
- Speedy, accurate management of the coordination and progress of the project by a specialized organization within the university (observance of secrecy, disclosure of research results, progress management, intellectual property management and so on).
- Arrangement for exclusive use of the university's share of intellectual property created by cooperation.
- Compatibility with large-scale or interdisciplinary joint research.
- Compatibility with joint research which needs to be continually coordinated.
- Compatibility with cooperative activities other than R&D operations (internship in R&D for graduate students, interchange of personnel, joint projects and so on).

*Source:* Kyushu University.

is now carried out between the business enterprise and the university (organization – organization). We will introduce organization-level cooperation as a tool that is effective in the cooperation of Japanese universities with companies, particularly in the promotion of joint research.

Organization-level cooperation is a needs-oriented joint research system whereby IMAQ gathers and organizes university teaching staff who wish to respond to the needs of a client (for example, a business enterprise). It is also a management-oriented system that has introduced measures to eliminate clients' uneasiness about the management of intellectual property and the control of research progress (Box 2.1). This new scheme has dramatically improved clients' satisfaction.

As a result, the number of joint research projects based on this scheme is increasing markedly every year. The amount of funding per joint research project is also increasing. The main clients are large companies.

Another salient characteristic of organization-level cooperation is that it utilizes project management, in which management of progress and intellectual property of projects are implemented in a consistent manner. Also, various new ideas are incorporated to prevent the research from

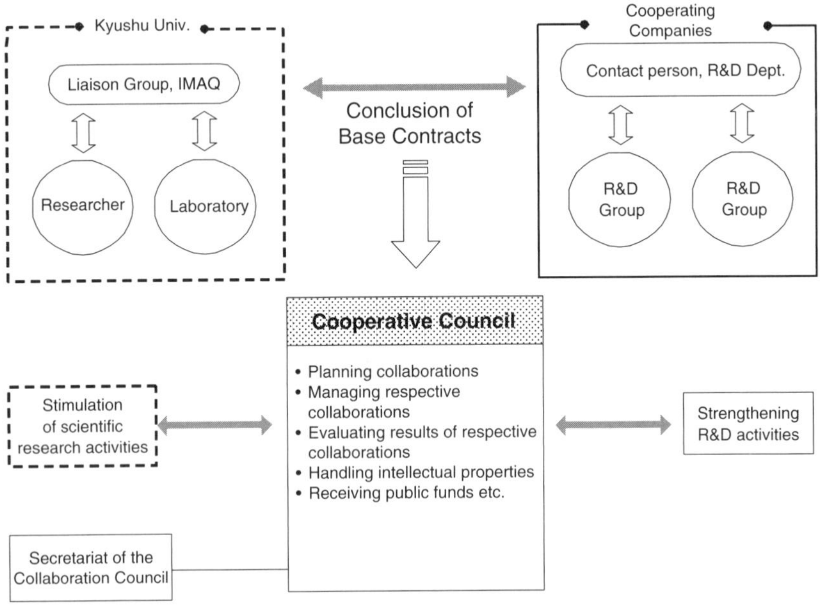


*Source:* Adapted from Art, Science and Technology Center for Cooperative Research (KASTEC) website.

*Figure 2.18 Structure of organization-level joint research*

becoming stereotyped. Since a window for cooperation is instituted between the university and the client, it is possible to efficiently coordinate the research and manage its progress in a consistent manner. In addition, a meeting of the Cooperative Council is held semi-annually, for the purpose of making decisions for promoting cooperation. Since attendees of the meeting are personnel from the university and companies who have the right to make final decisions, it is possible to solve various problems in the cooperation on the spot (Figure 2.18).

Furthermore, the university has recently introduced an internship that permits its students to experience R&D at a business enterprise. As a result of the above activities, the number of joint research agreements based on organization-level cooperation reached 90 in FY 2006, with total research expenditure amounting to $3 million.

### 2.3.4.3 Project scheme

In the operation and management of an organization-level joint research project, IMAQ functions as the secretariat, spearheading the project in cooperation with the department of the client that is in charge of UIC. In addition, the Cooperative Council serves to solve major problems that may occur between the two parties. The Cooperative Council consists of: the Trustee and Vice President (in charge of UIC) of Kyushu University; the officer in charge of R&D within the client; representative researchers from the university and the client; persons in charge of UIC and so on. The persons in charge of research within the two parties jointly formulate a draft research plan that shows the target results, time schedule and so on for each individual research project. The plan is discussed by the Cooperative Council and modified as required before it is finalized.

The progress of the research is periodically reported to the Cooperative Council, which speedily alters R&D policy, projects and so on where necessary. Through the management of the Cooperative Council, Kyushu University and the client strive to get rid of any discrepancies in their recognition of research, combine the preliminary research results of individual laboratories and promote the creation of innovative new technology that satisfies both parties.

#### 2.3.4.4 Achievements in organization-level cooperation

Kyushu University's achievements in organization-level cooperation are very highly rated in terms of both quality and quantity. In FY 2007, the number of joint research projects carried out at the university was 579. Of these, joint research projects based on individual contracts of organization-level cooperation number 91, or 15.7 per cent of the total number of joint research projects. Individual contracts also account for 16.1 per cent of total expenditure on joint research, indicating that organization-level cooperation occupies an important position at Kyushu University. The number of organizations and businesses that sign an organization-level cooperation agreement is increasing every year. As of September 2008, the number was 46.[4] It is expected that the ratio of organization-level joint research to all joint research will continue increasing in the future. In the five years since 2003, the number of patent applications from individual contracts of joint research based on organization-level cooperation was 76.

#### 2.3.4.5 Example of organization-level cooperation: cooperation with Nissan Chemical Industries, Ltd

The information below was obtained through an interview in October 2008 with Dr Misao Miyamoto, a board member who is in charge of new business at Nissan Chemical.

Nissan Chemical is a Japanese medium-sized chemical maker founded in 1887 (capital: US$189.4 million, number of employees: 1614). In order to emerge as a real value-creating enterprise having a world-class brand,

*Table 2.2 Profile of Nissan Chemical*

| Company name | Nissan Chemical Industries, Ltd |
|---|---|
| Location of headquarters | 7-1, Kanda Nishiki-cho 3-chome, Chiyoda-ku, Tokyo |
| Branches | Laboratories: Chemical Research Laboratories, Electronic Materials Research Laboratories, Special Materials Research Laboratories, Biological Research Laboratories.<br>Offices: Sapporo Sales Office, Sendai Sales Office, Nagoya Sales Office, Osaka Sales Office, Fukuoka Sales Office<br>Plants: Sodegaura Plant, Saitama Plant, Toyama Plant, Nagoya Plant, Onoda Plant |
| Establishment | April 1887 |
| Capital | US$189.4 million (US$1 = ¥100) |
| Employees | 1614 |

*Source:* Website of Nissan Chemical, http://www.nissanchem.co.jp/index.html. accessed May 2010.

Nissan Chemical launched 'Vista 2010' in April 2009. Vista 2010 is a medium-term business plan scheduled for completion in FY 2010. In the fields of organic materials, inorganic materials and electronic materials that form the company's core business/technology fields, Nissan Chemical is positively pressing ahead with the creation and development of new businesses and next-generation materials (Table 2.2).

*Background of the collaboration with Kyushu University*

Several years ago, in order to restructure the once-downsized company-wide research division, and to strengthen it into a division that would seek out and conduct research in new growth areas, the company set out to supplement its research capabilities through a basic strategy of collaboration with universities. At that time, the company investigated and discussed the possibility of collaboration with several universities and approached the universities. Kyushu University provided the most excellent proposal related to UIC management, incorporating a cross-sectional joint research system. Thus, the company and the university agreed to start joint research.

Nissan Chemical Industries requested the university to 'conduct joint research on the practical use and new application possibilities of 'new organic nanoparticles currently being investigated and at the initial development stage,' which would be provided by the company as seeds (materials)'. Kyushu University searched for researchers

## BOX 2.2 OUTLINE OF ORGANIZATION-LEVEL COOPERATION WITH NISSAN CHEMICAL

1. Plan

In December 2005, Kyushu University and Nissan Chemical signed a five-year agreement on organization-level cooperation starting from FY 2006, with the first two years dedicated to basic studies (research on common base/creation of functional materials) and the following three years to application-oriented research/applied research.

2. Purpose of cooperation/themes of joint research

- Research development collaboration for practical application of organic nanoparticles, which are expected to be materials for next-generation optical and electronic materials as well as medical materials.
- Individual joint research: 15 themes in four fields, including biotechnology/medicine and engineering electronic devices. Currently, the cumulative total of joint research projects is 54 (10 to 18 per fiscal year).
- Joint research expenditure: several hundreds of millions of yen in the past three and a half years. (The expenditure has been increasing sharply every year.)

3. Cooperative Council

- Registered members: Kyushu University – three members, Nissan Chemical – three members (board members and persons responsible for UIC).

4. Expanded Cooperative Council

- Registered members: about 60 persons of each party. (From among the following registered members, some 30 of each party attend each meeting of the Cooperative Council.)
  Kyushu University:
  Trustee/vice-president; liaison group leader and several persons of the department in charge of university-industry cooperation (IMAQ); 30+ researchers (from six departments, including the Engineering Faculty).

Nissan Chemical:
Chairman, president and board members; general manager and several persons of the department in charge of university-industry cooperation (Advanced Materials and Planning Division); several persons of indirect departments, including the Intellectual Property Department; about 30 researchers (from research laboratories, factory development departments and so on.)

- Meetings: the Cooperative Council meeting was held a total of eight times during the period from 2005 to September 2008 (about semi-annually).
- Main agenda: confirmation of requests from each party and of matters to be discussed by both parties; report on the progress of each party; presentation of the content of research (by Kyushu University, about seven to eight researchers at each presentation); special lectures (on the content of research in advanced fields by researchers of Kyushu University).

5. Results achieved (2005 – September 2008)

- Number of patent applications: 28
- Number of papers presented at academic meetings: 51.

6. Status of new developments of cooperation

- A Cooperative Research Department (for lectures on sponsored research) was established in KASTEC as a campus lab office to reinforce the joint research organization (in 2008).
- The interchange of personnel between the two parties has started (for example, reception of student interns by Nissan Chemical and reception of Nissan Chemical's researchers by Kyushu University).

within the university and collated research themes compatible with the topic requested by Nissan Chemical. The company was then presented with a proposal. Nissan Chemical judged that, relative to other universities, Kyushu University's response was the most appropriate for their needs, and the university's research level was the highest in the nano-technology field. Thus, the company decided to collaborate with Kyushu University.

*Circumstances that prompted Nissan Chemical to positively implement university-industry joint research*

- As a result of management selection and consolidation, the research department was incorporated into the business department, and the company-wide research department was abolished as the company began seeking and studying new business policies.
- As the company recognized that carrying out research in most advanced fields was vitally important, it discussed re-establishing the company-wide research department. The company saw that it would inevitably find the most advanced seeds outside the company, and also decided that collaboration with a university was required.
- In view of the necessity of hedging against risk, and the speed of change in the market, the company felt the need to cooperate with external organizations.
- The company received a proposal for organization-level joint research from Kyushu University.

*Advantages of carrying out organization-level joint research with Kyushu University*

- Since IMAQ has a well-established joint research management organization staffed with highly capable persons, the company can reasonably expect that the research will produce tangible results.
- The company understands that the level of the university's researchers in the field of materials is world-class.
- The proposed research is effective from the standpoint of nurturing young researchers within the company and securing (recruiting) talented researchers.
- The proposal allows for interdisciplinary joint research on a single subject between the company and the university (plane-plane), rather than between the company and a particular researcher (or laboratory) of the university (point-point).

*Contribution of the results of joint research with Kyushu University to product development at the company*

- The university conducts joint research for the future, not for the present. Although the university's research has resulted in many

patents, it takes years for these patents to be converted into marketable products.
- Nissan Chemical carries out joint research with the concept that it is acceptable if their results can be converted into marketable products for profit within 10 to 20 years.

*Direction of R&D based on joint research with external organizations, including universities*

- In this age of open innovation, in view of the need to provide some hedge against risk and to keep up with the speed of change in the market, it is difficult for the company to carry out effective research by itself.
- The company is seeking seeds of research locally and abroad. It is also considering not only UIC but also industry-industry collaboration.
- The company is aiming not only for product innovation (product development), but also process innovation through an alliance with an equipment manufacturer and so on.

*Perceptions of technology transfer from the university*

- Technology in patents obtained by the university are not ready to be commercialized. It is difficult for a company to judge which of these patents could lead to marketable products. The university also finds it difficult to put its patents into practical use by itself. Besides this, there are cases where the patents obtained by the university do not meet the needs of industry.
- Rather than trying to utilize the university's patents, understanding the content of research conducted by the university's researchers and carrying out joint research with the university is effective from the standpoint of matching research results with industry's R&D needs.
- Research carried out by a company must be scientific. Any R&D and product that is not based on science will not last long.

The cooperation between Kyushu University and Nissan Chemical described above is the most important and successful example of organization-level cooperation for the university. It may be said that Nissan Chemical, a technology-based company, has found a good partner in UIC, and this is indispensable for the company to rebuild its own research organization.

The organization-level cooperation has made it possible for Nissan

Chemical to entrust to Kyushu University research in fields in which the company does not have sufficient expertise. In the process leading up to the output of research results, the superb management system of the university has enabled the company to engage in joint research with a sense of security. The organization-level cooperation has also made it possible for the university to discover new research themes by becoming acquainted with the company's needs. In addition, some researchers of the university say that they have obtained new ideas through participation in joint research with researchers from other departments or in other fields.

Thus, the organization-level cooperation between Kyushu University and Nissan Chemical may be regarded as an advanced model of UIC through joint research which is ideal for both parties. Conventional joint research between a company and an individual university researcher is also a means by which the university can contribute to industry. In addition, the researcher can gain new ideas from the joint research activity. Therefore, it benefits both parties. However, more often than not, joint research conducted outside the framework of organization-level cooperation has posed a number of problems in the assessment of research results, the progress of research activity, the handling of intellectual property and so on, all of which differ with the individual researcher. Thus they have often left both the company and the university unsatisfied. It may be said that this situation has impeded the spread of joint research and the progress of UIC.

In the organization-level cooperation between Kyushu University and Nissan Chemical, the above-mentioned problems are treated as organizational matters, and the departments in charge of the UIC in both parties maintain a good liaison activity. This largely accounts for the success of the cooperation.

#### 2.3.4.6 Summary of organization-level cooperation

Organization-level cooperation is a joint research scheme that was started in early 2004 by Kyushu University – a research-oriented university with many researchers in diverse fields and a strength in basic research. Organization-level cooperation allows for large-scale, interdisciplinary joint research between the university and business enterprises. Some of the other major universities are also tackling large-scale joint research projects under the name 'comprehensive joint research'. Thus, organization-level cooperation is becoming a strong trend at prestigious research-oriented universities in Japan.

With the progress of information technology (IT) and the diversification and globalization of markets, more and more business enterprises feel that they can no longer confine themselves to their conventional business fields and that they must seek new profitable fields of business. Thus it has become increasingly difficult for large companies to carry out research in every phase of their business fields, from basic and exploratory research to applied research and development (that is, from upstream research through to downstream research). Under these conditions, many companies are downscaling or abolishing their central research institutes that were dedicated to basic research, and are turning to applied research and development in fields other than their core business. Collaboration with some external organization is considered a promising substitute for the central research institute. In particular, there is a growing expectation of UIC (joint/sponsored research). It may be said that we are entering an age of open innovation.

Needless to say, collaboration with universities in applied R & D is also common. However, for Japanese business enterprises which have strong capabilities in applied research for the development of new products and new businesses, expectations of universities primarily concern the universities' superior capacity to conduct basic research and explore new fields. This was made clear by the questionnaire survey described earlier; it has been reconfirmed through interviews with companies that have signed an organization-level cooperation agreement with Kyushu University.

On the other hand, there are two major concerns business enterprises have about universities in Japan: the ability of university researchers to manage the progress of their research; and the mismatch between the company's needs and the researchers' motivations. In addition, researchers in different fields of research and in different laboratories at Japanese universities have a strong sense of independence. Under these conditions, interdisciplinary research can hardly be hoped for. In other words, companies feel uneasy about whether university researchers will carry out the joint/sponsored research within the specified period, and whether researchers will bring about the results anticipated by the companies. Thus, companies' uneasiness is closely connected with researchers' abilities to manage research progress, and with the environment in universities which is not conducive to interdisciplinary research.

The framework of organization-level joint research of Kyushu University is one that removes companies' uneasiness. In fact, it is highly rated by business enterprises that have a cooperative relationship with the university. IMAQ coordinates the UIC to ensure that the needs of the company are met. In addition, IMAQ – which has made various arrangements to break through the barriers of conventional joint research – plays a role in increasing companies' satisfaction.

When it comes to assigning teachers/researchers of the university to a particular joint research project, IMAQ makes it a rule to select those who are interested in the need and topic proposed by the company. Thus, IMAQ pays due attention to the interests and volition of individual teachers/researchers. In addition, in the course of the project, the researchers have an opportunity to exchange views with other researchers in different fields. Some researchers say this is very enlightening. As a result of all this, the company will be much more satisfied with the research and at the same time, university researchers find that the research is meaningful to them.

As already mentioned, joint/sponsored research between companies and universities plays the major central role in UIC in Japan. The positive stance on the part of universities has significantly helped to promote UIC. As a result, the level of university-industry joint research is improving each year. Under these conditions, the organization-level cooperation developed by Kyushu University is positioned as a framework which removes the limitations of conventional joint research and sponsored research and which allows for a higher quality of UIC. Comprehensive cooperation implemented by other universities is oriented in the same direction. However, the organization-level joint research of Kyushu University is evidently at an advanced new stage. It may be said that, from the standpoint of contributing to innovation in the country (mainly in industry), it is a highly effective and efficient framework. Thus as observed in the example of Kyushu University, to promote UIC in an efficient and effective manner, universities in Japan have endeavored to implement various innovative schemes. These are mainly led by an organization for UIC. They have started to achieve excellent results.

## 2.4 KYUSHU UNIVERSITY'S ATTEMPT TO ACHIEVE INDUSTRY-UNIVERSITY COLLABORATION TO CONTRIBUTE TO THE REGIONAL INNOVATION SYSTEM

In addition to supporting joint research and implementing it through organization-level cooperation with large companies, Kyushu University carries out various activities including: supplying technological consultation to small and medium-sized enterprises (SMEs) in the local area; enhancing local residents' understanding of science and technology; engaging in science communication to deepen the understanding of the problems in the local region and the way they should be solved; and the building of local networks which support entrepreneurs in the local community. At the same time, Kyushu University supports industrial policies of the local government, such as obtaining direct investment from technological firms in the region.

### 2.4.1 Technological Consultation

#### 2.4.1.1 Outline of technological consultation

IMAQ provides technological consultation and coordination of university-industry joint research services, mainly for small and medium-sized companies in the local area. This includes chargeable technological counseling, joint/sponsored research, introduction of other organizations and so on. Requests for technological consultations accepted by IMAQ are put through a prescribed procedure (to be described later), and some lead to guidance, advice or counseling by researchers, or to joint research. Every year IMAQ accepts some 150 technological consultations, about 20 per cent of which lead up to specific university-industry cooperative projects such as joint research.

The requests for technological consultations accepted by IMAQ are divided into three categories: (1) technical problems which IMAQ considers possible to solve by supplying academic knowledge; (2) technical matters which are likely to lead to the signing of a joint/sponsored research agreement or the acceptance of external researchers; and (3) other matters which are considered to be of help in promoting local industries. Technological consultation is requested from the university by a company seeking guidance or advice, in order to solve a particular business problem. There are some cases in which solving a problem takes considerable time and cost as it calls for joint or sponsored research. In many cases, however, problems are solved in a relatively short period of time by making effective use of university researchers' knowledge.

#### 2.4.1.2 Achievements in technological consultation

There were a total of 198 technological consultations during the period 2005 to 2007, with the number increasing gradually. Small and medium-sized companies in the local area (Kyushu area) account for a majority of the applicants (Figures 2.19 and 2.20).

In many of the technological consultations, the problems are minor and call for relatively quick solutions. Because of this, the research-oriented Kyushu University was somewhat passive in responding to technological consultations. The university's strength is in basic research, which does not produce tangible results in the short term. The fact that there is not always a guarantee of an adequate supply of funds from the consulting company – and hence technological consultation does not offer much incentive to researchers – was another factor in the university's passive stance. However social conditions have changed, and Kyushu University – a university located in Kyushu – has come to feel strongly that it is important for it to contribute to local business enterprises, including SMEs. In

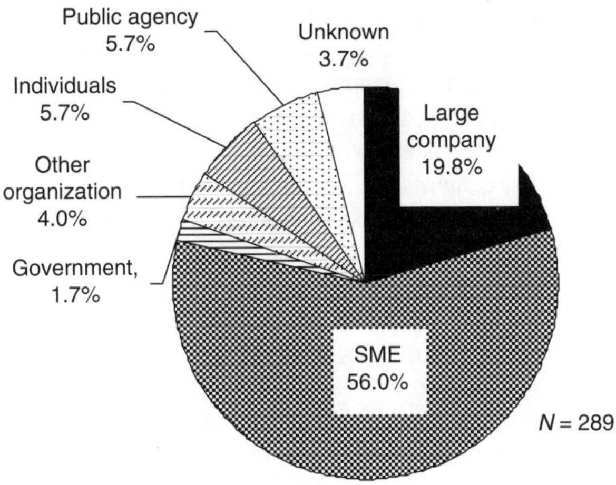


*Source:* Data from Kyushu University.

*Figure 2.19 Breakdown of consulting companies by type of company (2005–07)*

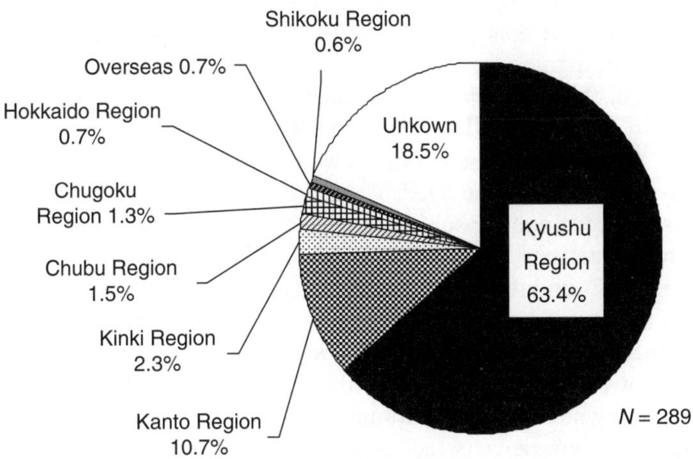


*Source:* Data from Kyushu University.

*Figure 2.20 Breakdown of consulting companies by location of headquarters (2005–07)*

this way it contributes to the innovation system in the Kyushu area. Thus the university has enhanced the functions of its technological consultations and has established an organization for utilizing university resources to meet the specific needs of consulting companies. In addition, the university has developed a system which permits utilizing external resources, and is making arrangements for improving its problem-solving ability to better respond to the technological consultations that it does accept (see Section 2.4.2).

### 2.4.2 Kyushu University-Steinbeis Japan Inc. Transfer Center (KSTC)

#### 2.4.2.1 Outline of KSTC

KSTC is a technological consulting project jointly established by the Steinbeis Foundation (a German technology transfer agency) and Kyushu University. The purpose of KSTC as a consulting scheme located within Kyushu University is to strive to solve companies' business problems. These problems should be of interest to university researchers, and no specific time limit is given. For this purpose, KSTC mediates between the companies and the university's researchers. Previously, Kyushu University accepted technological consultations brought in by SMEs as part of the university's contribution to society (other than academic research and education). Depending on the content and scale of technological consultation, however, it was difficult for the university to completely satisfy consulting companies.

Because of this, the university decided to devise a new system. The university also decided that the new system should be highly transparent. For example, technological consultations which require continual involvement of teachers should be managed by the university in such a way that the teachers can continue consulting for the company while engaging in their own research activity. Eventually, in July 2005, the university tied up with the Steinbeis Foundation (StW) of Germany (which has rich experience in research for business enterprises and extensive know-how in diverse fields) and with Steinbeis Japan (StWJ) to establish KSTC. For technological consultations which do not fall within the category of joint/sponsored research and which require continual involvement of the researchers concerned, the system permits the researchers to continue working on these while also working for KSTC. This is to enhance the level of satisfaction of the consulting companies.

If external parties consult Kyushu University to resolve an issue, the university responds through joint research, sponsored research and the scholarship donation system. However various kinds of requests are brought to the university from companies, and there are many cases that the university cannot handle through joint research or sponsored research, such as specific issue-resolution actions.

It is KSTC that facilitates researchers' involvement in response to these diverse requests from business enterprises (consulting, assessment of technology, support for R&D and so on) and permits introducing external know-how so as to meet the needs of local companies.

The above activity is an improved version of technological consultation, the majority of which are conducted for local companies. With the support of the Steinbeis Foundation, Kyushu University has built a new system in order to make a break from the limits of traditional universities in Japan. This allows the university to positively respond to the needs of local companies, and is in line with its aim of contributing to the regional innovation system. Japanese universities which emphasize fundamental research are not good at providing problem-solving consultations. As a result these universities – including Kyushu University – were not active in providing this type of consulting service. The KSTC project is popular and activity is increasing in both the number of, and revenue from, consultations every year. Requests for consultations from large companies are also increasing. It is believed, therefore, that the KSTC project of Kyushu University will enhance not only the university's contribution to innovation in its locality, but also to Japanese industry as a whole.

### 2.4.3 Other Projects in Cooperation with Local Government

Kyushu University, in cooperation with the Fukuoka Prefecture authorities, has been proactively engaged in forming the semiconductor cluster in the Fukuoka region, which is promoted by Fukuoka Prefecture. The university has also been involved with building a wide-ranging collaboration with Asia in the cluster. Kyushu is called a silicon island, dominating semiconductor production in Japan. (Approximately 30 per cent of the industry's production value comes from Kyushu.) Fukuoka Prefecture, located in Kyushu, is a base not only for semiconductor production, but also for system LSI (Large-scale integration) design, which will lead future semiconductor demands. In the intellectual cluster plan[5] led by the Japanese government, Fukuoka received the highest assessment in Japan as a cluster focusing on semiconductors.

The intellectual cluster plan encourages technology-oriented companies in each region to cluster together. This policy prompts research organizations such as universities to advance joint research and other types of research with companies while cooperating with the industrial community and administration. Since Fukuoka Prefecture was designated as a core research region in 2002, Kyushu University has contributed to the promotion of the Fukuoka Intellectual Cluster Plan in cooperation with Fukuoka Prefecture. For example, to cluster semiconductor design engineers in the Fukuoka region and enhance their skills, Kyushu University has cooperated with Fukuoka Prefecture in establishing an educational institution for LSI technology (College of System LSI, Fukuoka). Kyushu University took the initiative in the operation. In addition, Kyushu University proactively participated in a project organized by Fukuoka Prefecture for establishing the Fukuoka Institute of System LSI Design Industry, a research institute for system LSI and a semiconductor-related business incubator. The System LSI Research Center of Kyushu University has an office in the institution and is a core research organization of the institution. As a result of this collaboration, the number of system LSI design companies in Fukuoka Prefecture increased more than five-fold (from 21 companies to 110 companies) between 2000 and 2007. As a result, in the 2006 evaluation for the first stage of the intellectual cluster project, Fukuoka Prefecture received the highest assessment. In the 2007 evaluation of the second stage of the project, Fukuoka Prefecture also received the most positive evaluation, obtaining a budget of approximately nine billion yen for five years. In addition, Kyushu University has promoted inter-regional collaboration, playing a significant role in the formation of a network of semiconductor centers in East Asia coastal regions ranging from Kyushu and Korea to Singapore (the Silicon Sea Belt). Part of this collaboration includes participating in joint projects.

Thus, Kyushu University, in close collaboration with administrative authorities in the Fukuoka region and the industrial community, has played a great role in promoting the clustering of high-tech industry in the region. In other words, it has significantly contributed to the regional innovation system.

## 2.5 CONCLUSION

1. *University-industry collaboration has made a significant contribution to the innovation system in Japan. This is based on joint research and sponsored research between companies and universities. In addition, to quantitatively expand and qualitatively enhance joint/sponsored research, Japanese universities are implementing various new systems such as organization-level cooperation (comprehensive cooperation).*

In short, from the perspective of universities' contribution to the national innovation system in Japan, it is most important that universities respond to needs from the industrial community by: setting a research theme;

conducting joint/sponsored research; and realizing the development of the industrial community and national innovation system through the processes employed and the results achieved. This has been implemented widely. On the other hand, for technology transfer and the creation of university start-ups, there have been no remarkable results.

Out of several UICs initiated by universities, the three main research-oriented collaboration activities are as follows: joint/sponsored research; technology transfer; and creation of university start-ups. In other Asian countries, the primary contributions of universities to the national and regional innovation systems are the transfer of intellectual property from university research to the industrial community (technology transfer), and the creation of new businesses based on technologies developed in universities (creation of university start-ups). This is because in these Asian countries, R&D and technology commercialization capabilities of companies are weaker than in Japan and the Western countries, and so industry has not been equipped to create new value through joint research with universities. Hence, it is inferred that in these countries, universities generally look for parties to whom they can transfer research results (technology transfer), and attempt to commercialize research results in the market as business ventures (creation of university start-ups), thus achieving UIC (contribution to the innovation system).

In contrast, the level of technology development capabilities and technology absorption capabilities in the Japanese industrial community is high. Hence, the general trend of UIC is through joint/sponsored research. In addition, joint/sponsored research has provided universities, which are ignorant of social needs, with excellent stimuli to review their research direction and ideal research style. In other words, universities have also undergone innovation. Thus, joint/sponsored research has great benefits both to the industrial community and to universities. It is possible to say that joint/sponsored research is both the mainstream and the desirable form of Japanese universities' contribution to the national innovation system.

Furthermore, most Japanese research universities have recently introduced large-scale interdisciplinary joint research projects with companies in the form of organization-level collaboration (comprehensive collaboration). This not only expands the scale of joint research but also realizes cross-sectional and interdisciplinary research, which was not common before. This mechanism actualizes the potential of innovation available in universities. There are similar mechanisms in the USA. In Japan, this mechanism has been generalized as evolved joint/sponsored research, which is a desirable trend.

2. *With respect to Japanese universities' contribution to their regional innovation systems, research universities focusing on basic research have contributed in the form of providing technology consulting, and in cooperating with the government in attracting and clustering companies in the region. These efforts need to be reinforced.*

Mainly led by Japanese national universities, which were given corporate status in 2004, most Japanese universities have started to attach importance to integrating themselves to the region and contributing to their regional innovation systems as well as to the national innovation system. Several efforts have been made along this path. Universities have also been motivated by the Japanese government's policy of vitalizing regional economies through the utilization of university strengths (for example, the intellectual cluster policy); consequently, Japanese universities are strengthening various attempts to contribute to their locality. Many universities provide technical advice and technology consulting to local companies. The intellectual cluster policy primarily utilizes knowledge owned by universities and research organizations. It aims to create university start-ups and to realize company clustering and new business creation in regions where universities and research organizations are located. As observed from the example of Kyushu University, in some cases this regional vitalization and innovation policy takes effect through universities' proactive cooperation. It is still far from a great success when compared to the relationship between Stanford University and Silicon Valley in the USA. However, it can be said that there is certainly a drive to contribute to the regional innovation system.

In addition, Japanese universities are not good at technology consulting because they focus on research with a long-term perspective, such as basic research and advanced research. However, universities in each region have started to address regional needs wherever possible. To improve the efficiency and outcome of technology consulting, Kyushu University collaborates with specialized foreign institutions. By doing this, the university supplements its insufficient know-how, and thus achieves results through technology consulting.

In Japan, the economy has continued to be sluggish a long time after the collapse of the economic bubble around 1990. In this situation universities have contributed to the national and regional innovation systems through UIC, particularly over the past few years when there has been strong demand from the government and the industrial community. The corporate status given to national universities arose partially from this background. We have discussed Japanese universities' contributions to their regional innovation systems through UIC in the light of this background, focusing on the example of Kyushu University. We will be happy if this discussion aids understanding of the current situation of UIC and the future direction in Japan.

## NOTES

1. The Director General of IMAQ serves as President of TLO Kyushu.
2. The Faculty of Engineering of Kyushu University consists of three department groups: Material and Chemical Engineering; Global Environment Engineering; and Mechanical and Aviation Engineering. The Material and Chemical Engineering department group consists of four departments: Chemical Engineering; Applied Chemistry (applied chemical functions); Applied Chemistry (applied chemical molecules); and Materials Engineering. The Global Environment Engineering department group consists of five departments: Construction and Design; Environment and Urban; Ocean System Engineering; Earth Resources and System Engineering; and Energy and Quantum Engineering. The Mechanical and Aviation Engineering department group consists of three departments: Mechanical Science; Intelligent Machinery and Systems; and Aerospace Engineering. The faculty has a total of 314 teaching staff (96 professors, 89 associate professors, 116 assistant professors, six instructors and seven associate assistant professors). The Faculty of Systems and Information Science consists of six departments: Information Science; Intelligent Systems; Information Engineering; Electric and Electronic System Engineering; Electronic Device Engineering; and Superconducting Science. This faculty has a total of 104 teaching staff (34 professors, 38 associate professors, 30 assistant professors and 2 associate assistant professors).
3. The number of projects was sampled on a contract basis using the information registered in the database of IMAQ. The projects surveyed are based on the data collected during the period 2004 (after enforcement of the National University Corporation Act) to 28 December 2007.
4. The organizations that have signed an organization-level joint research agreement comprise 31 business enterprises, five local self-governing bodies and ten other organizations (as of the end of August 2008).
5. http://www2.lab-ist.jp/english/, last accessed May 2010.

## REFERENCES


Art, Science and Technology Center for Cooperative Research, Kyushu University (TLO Kyushu) (2004), *Bulletin 'LINK'*, **12**, September.

Art, Science and Technology Center for Cooperative Research, Kyushu University (2007), *Yearbook No. 13*.

Art, Science and Technology Center for Cooperative Research (KASTEC) (n.d.), Kyushu University website: http://www.astec.kyushu-u.ac.jp/eng/html/liaison construction.html, last accessed May 2010.

Association of University Technology Managers (AUTM) (various years), *Licensing Activity Survey*, Deerfield, IL, USA: AUTM. Intellectual Property Management Center of Kyushu University, homepage.

Ijichi, T. and S. Nagaoka (2007), 'Management of IPRs for upstream research inventions in universities and national R&D institutes in Japan: results from the IIR IP Survey 2006, Hitosobashi University, Institute of Innovation Research.

Itoshima Academic Research Community Promotion Council Maebara City, Nijo Town, Shima Town, Fukuoka Prefecture, *FY 2006 Itoshima Academic Research Community Promotion Council Activity Report*.
Japan Association of University Intellectual Property and Technology Management (2006), *Survey of Technology Transfer by Universities: University Intellectual Property Management Yearbook 2006*.
Japan Association of University Intellectual Property and Technology Management (2007), *Survey of Technology Transfer by Universities: University Intellectual Property Management Yearbook 2007*.
Japan Economic Research Institute (2006), *Economic Effects of University-Industry-Government Cooperation*, University Intellectual Property Management Center Development Project, Sponsored Research on Model Program for Building 21st Century University-Industry-Government Cooperation Method of the Ministry of Education, Culture, Sports, Science and Technology.
Kyushu University 2008, *Outline of Kyushu University 2008*, Fukuoka: Kyushu University.
Kyushu University website, http://www.kyushu-u.ac.jp/index.php, last accessed May 2010.
Ministry of Education, Culture, Sports, Science and Technology, homepage.
Osaka University Intellectual Property Management Center (2005), *Type of Organization-level Collaboration and the Way it Should be at the University*.
Science & Technology/Academic Council, Science & Research Foundation Subcommittee, Committee for Promotion of University-Industry-Government Collaboration, University Intellectual Property Center Examination/Evaluation Subcommittee (2008), 'University intellectual property management center development project', *Report on Business Evaluation Results*, 4 August.
Science & Technology/Academic Policy Bureau of the Ministry of Education, Culture, Sports, Science and Technology (MEXT) (2006), *Survey Report on Research Activities of Private Enterprises 2006*.
Venture Business Laboratory of Kyushu University, homepage.

# 3. Invention management in Japanese universities and its implications for innovation: insights from the University of Tokyo

**Robert Kneller**

## 3.1 INTRODUCTION

This chapter describes the system of technology management in the University of Tokyo and the dominance of sponsored joint (collaborative) research as the main mechanism of technology transfer. As is probably the case in other major national universities in Japan, this system has resulted in large, established companies controlling a large proportion of university discoveries, while opportunities and incentives for start-up formation have been constrained. The University of Tokyo is Japan's most important university, accounting for approximately 12 percent of all Japanese university research and development (R&D).[1] Its faculty and administrators have played important roles in shaping changes in Japan's system of university-industry collaboration, and it serves as a model for other universities. Nevertheless practices do vary and some of these instances are noted.

## 3.2 PAST AND PRESENT LEGAL FRAMEWORK FOR TECHNOLOGY TRANSFER

Prior to 2004, ownership of inventions arising in Japanese national universities, which account for approximately 75 percent of all university research expenditures and include most of Japan's major academic research centers, depended upon the source of funding that gave rise to the inventions. Inventions arising under government research grants as well as all contractual sponsored research were supposed to be owned by the nation – usually in effect by the Ministry of Education, Culture, Sports,

Science and Technology (MEXT, or its predecessor, Monbusho). This was true even in the case of inventions arising under contractual company-sponsored research. Government ownership meant that inventions either were free for anyone to use or would be licensed non-exclusively for modest royalties, except that in the case of contractual company-sponsored research, the sponsors could obtain exclusive licenses.

On the other hand, inventors could retain ownership of inventions arising under donations (in theory, charitable gifts, but prior to 2000, the principal mechanism for companies to fund university R&D) or the standard research allowance allocated to each faculty member (usually only several thousand dollars annually after deductions for infrastructure and other fixed laboratory expenses). Considering the proportions of the various sources of funding, inventions ought to have been roughly equally distributed between government and inventor ownership. In fact, almost all commercially relevant university inventions were attributed to donations. As quid pro quo for receiving donations, the donor companies expected faculty to pass their inventions and related intellectual property (IP) rights to them and also to recommend those companies to their capable students as places to work (Kneller 2003).

Thus prior to 1998, Japan had a system of de facto faculty ownership of university inventions, shaped in large part by direct links between faculty and companies. Donations were among the most important of these links. University researchers did invent, but companies usually applied for the patents.[2] The average number of patent applications per year was about 50–70 percent of the average following incorporation of national universities in 2004 (Kanama and Okudawa 2008; Takahashi and Carraz 2009).[3] But it was nevertheless respectable and represented a first significant step in technology transfer, in that companies were sufficiently interested to apply for patents.

There were other barriers to university-industry cooperation. As government employees, faculty members in national universities were barred from compensated consulting for outside organizations (although donations facilitated extensive informal consulting). Also, research funds generally could not be used for personnel expenses, the exception being use of donations to pay secretaries and occasionally technicians (Kneller 2003). Using research funds to supplement salaries of permanent faculty was (and remains) prohibited. In other words, unlike the USA, Japan still does not have a soft money system of funding tenured faculty, although steps in this direction are under discussion.

The following four laws implemented between 1998 and 2004 changed the legal framework governing IP management and university-industry cooperation, so that it resembles on the surface the US system after passage of the 1980 Bayh-Dole Act:

- The 1998 Law to Promote the Transfer of University Technologies (the TLO Law) established a system for the government to approve and subsidize university technology licensing offices (TLOs). Starting with five approved in 1998, the number of approved TLOs had increased to 39 by 2005 (UTTAJ 2009). Even more important than approval and subsidies, this law legitimized the transparent, negotiated, contractual transfers of university discoveries to industry and the channeling of royalties back to the inventors, their laboratories and the university. Although the system of IP ownership was left unchanged, careful inquiries as to the sources of invention funding were avoided, and faculty in universities with competent TLOs began to let those TLOs manage a considerable number of inventions even though they may have arisen at least in part from government grants or contracts.
- The 1999 Law of Special Measures to Revive Industry (the Japan Bayh-Dole Act) has the same effect as the US Bayh-Dole Act, except that it did not apply to national universities until they obtained legal status as semi-autonomous administrative entities in 2004.[4]
- The 2000 Law to Strengthen Industrial Technology established procedures permitting university researchers to consult for, establish and even manage companies. It also streamlined the procedures for company-sponsored commissioned and joint research. Thus it opened the door for university researchers to found companies. At the same time, it eliminated many bureaucratic obstacles that had previously dissuaded companies from using contractual sponsored research, rather than donations, to fund university research. Among other things, it opened the door to the use of sponsored research funds to pay personnel expenses, although not to the extent of covering salaries of permanent administrative and teaching staff.
- The University Incorporation Law gave national universities independent legal status when it went into effect in April 2004. Previously they were merely branches of MEXT. By gaining status as independent legal entities, Article 35 of Japan's Patent Law, which enables employers to require assignment to them of employee inventions, became applicable. So too did the Japan Bayh-Dole Act. MEXT has urged the incorporated national universities to assert ownership over commercially valuable inventions (MEXT 2002).

## 3.3 KEY STEPS AND POLICIES IN THE UNIVERSITY OF TOKYO'S INVENTION MANAGEMENT PROCESS

### 3.3.1 Invention Reporting and Determining Whether an Invention is Work-related

Under Article 35 of Japan's Patent Law, universities can require their employees to assign work-related inventions to the university. Thus determining which inventions are work-related is an important issue. University of Tokyo guidelines set forth a broad definition, saying that any invention that relied substantially on university facilities or whose conception was related to an employee's research in the university is work-related.

All inventions should be reported using a set form to departmental-level invention committees, whose main responsibility is to make a preliminary determination whether an invention is work-related. The standard form can be supplemented with diagrams, articles and so on. In addition to information about the invention and the identity and affiliation of the inventors, it asks:

- whether the invention relied upon university funds or facilities
- whether it arose under joint research (and the identity of any joint research partner)
- whether the invention is work-related ('no' is possible only if the answers to the two preceding is 'no'. Also the inventors must explain why the invention is not work-related) and
- even if the inventors claim their invention is not work-related, whether they nevertheless want to assign it to the university.

It also asks the inventors to provide their perspectives on:

- how their invention solves unresolved problems in its technical field
- practical uses and likely important markets
- ability to produce as (or incorporate in) a commercial product
- antecedent technologies/discoveries
- background IP (for example, patent applications on prior related discoveries)
- plans to disclose the invention in publications or conferences
- whether patent applications ought to be filed overseas
- development plans and
- whether they have a desired development/commercialization partner.

One notable feature is that the lead university inventor designates the percentage contributions of each individual inventor. In the case of inventions arising under joint research contracts with private companies, corporate researchers are almost always listed as co-inventors (Kneller 2006). These percentage allocations are usually decided by discussions between the lead university inventor and the company collaborators. They are rarely, if ever, questioned at the university/TLO level, and outside attorneys who file patent applications seldom scrutinize them.

Co-inventorship automatically gives the sponsor co-ownership rights, whereas if all inventors were university researchers, the sponsor would have to negotiate with the university for a license. Because Japanese patent law (Article 73) requires the consent of all patent co-owners even for a non-exclusive license to a third party, co-ownership by a sponsor means the sponsor has a de facto perpetual, royalty-free, non-transferable exclusive license to the invention. In other words, a co-owning sponsor has unlimited use rights and the right to veto any transfer by other co-owners. It cannot license the invention without the university's permission, but this limitation rarely is a concern for large companies. The University of Tokyo has tried harder than any other university to circumvent these provisions and to require that companies that want unified exclusive rights negotiate an exclusive license for the university's ownership share. However, companies that oppose the university licensing its rights to a third party (essentially giving the third party a non-exclusive license) are able to do so by insisting on their Article 73 rights. Thus in all universities, if the sponsor wants exclusive control over a co-owned joint research invention, all it usually has to do is to pay the patent application and maintenance costs.[5]

The ease with which university inventors can designate company employees as co-inventors and thereby obtain exclusive control has been a key factor in the resurrection of the pre-1998 donation system, the main difference now being the requirement for a formal joint research contract.[6] As in the past, companies sponsor research (however such support is still small in proportion to total university R&D funding and also in relation to the proportion of US university research funded by industry[7]), they employ students and they receive non-transferable exclusive rights at low cost with no development obligations. Most university researchers seem comfortable with this. Joint research and joint patent applications have become the dominant mechanism of technology transfer in the University of Tokyo and probably most other major universities. The scope and implications of joint research partners having so much control over a large proportion of publicly funded university research are discussed in Section 3.4 below.

### 3.3.2 Deciding Whether to Require Assignment and to Apply for Patents

The departmental-level invention committees in the University of Tokyo should decide within two weeks of receipt of an invention report whether the invention is probably work-related and forward the report and their assessment to departmental-level administrators. These then forward the report to central university administrators and then to the TLO. In practice, many inventors first send their reports directly to the TLO so as to receive early feedback on patentability and commercialization prospects.

The Intellectual Property Management Office in the central university administration has final authority whether to require assignment from the inventor and to apply for patents, copyrights and trademarks. In practice, in the University of Tokyo, the TLO makes these decisions, although the IP Management Office (also known as the IP Headquarters, or IPHq) must ratify them.

MEXT authorized the formation of over 30 national university IPHqs in 2003 in anticipation of the incorporation of national universities the following year. MEXT felt that major national universities needed IP management capabilities within their official structures, even though they may also have a TLO. (Most of the existing TLOs had been formed as foundations affiliated with their universities or as nominally independent for-profit corporations. They were so organized in order to be able to collect royalty payments and to recruit staff from the outside at competitive salaries. Before incorporation it would have been administratively difficult for any organization within a national university to perform these tasks.) This often overlapping dual authority sometimes led to friction (Kneller 2004). But now these have been largely resolved in favor of the IPHqs which have absorbed the TLOs and confined their responsibility to licensing. The outstanding exception is the University of Tokyo, whose TLO had sufficient license revenue to hire competent staff who could take the lead in all aspects of invention management. Even in this case, however, the vetting of decisions by the university's IPHq is becoming more pervasive.

As a matter of practice, the University of Tokyo does not require assignment of inventions by company researchers engaged in joint research or by students. In the case of the remaining inventions, the TLO assesses patentability and market demand. Approximately 12 licensing associates carry out this work. Most are relatively young and have university science or engineering training and industry experience. Inventors are usually required to assign to the university rights to inventions deemed patentable and for which there will likely be market demand. Thereupon the same licensing associate assumes responsibility for licensing.

In principle, the TLO and IP Headquarters should decide within two weeks whether the university will require assignment. Thus within a month of submission of an invention report, the inventor ought to know whether the university will require assignment and apply for a patent or let the inventor retain title. If the university does not decide within a month, in principle, the inventor should retain title unless the university provides reasons for its delay.

The ease with which universities can let inventors retain ownership stands in marked contrast to the situation in the USA. There the Bayh-Dole law requires that the right to apply for a patent reverts back to the government funding agency if the university declines to patent. The university, on behalf of the inventor, can request that the government waive this right in favor of the inventor. However this process is not streamlined. If the inventor has a development plan (including companies or investors who can assist in company formation), easy transfer of rights to the inventor can provide an important safety valve in case the TLO lacks resources or expertise to manage the invention. Several Canadian universities (for example, Queens University and the University of Waterloo) go even further and give inventors the choice of assigning their inventions to the TLO or managing the inventions themselves. Such flexibility is absent from the US system in the case of government-funded inventions.

### 3.3.3 Patent Application, Marketing and Licensing

In 2006 the University of Tokyo applied for 432 Japanese patents, and about 170 international patents (that is, Patent Cooperation Treaty applications). Overall it files patent applications on about half the inventions reported to it. The proportion for some other major universities is higher indicating that the University of Tokyo is more discriminating concerning which inventions it decides to patent. This probably reflects a better capability to assess marketability and to find licensees.[8]

Royalty income between 2005 and 2007 ranged from $1.7 million in 2006 to $0.9 million in 2007 (Cabinet Office and University of Tokyo data). The TLO license associates use a variety of databases and personal contacts, as well as cold calls to search for potential licensees. They also advertise some inventions on the Internet, but this method rarely produces licenses. Close communication with the inventors is an important part of the licensing associates' work. Although some faculty members are still skeptical about invention ownership by the university and management by the TLO, those faculty members who have actually worked with the TLO usually seem satisfied. In other words, the staff seem to be doing a competent job. However, many other TLOs lack such competence. Staff of the University of Tokyo's TLO frequently take part in training sessions for staff of other TLOs. The future of Japan's new system of technology management depends largely upon efforts to increase the capabilities of TLOs in other universities.

## 3.4 JOINT RESEARCH AS THE DOMINANT MECHANISM OF TECHNOLOGY TRANSFER AND THE DEARTH OF VIABLE START-UPS

The above notwithstanding, in the University of Tokyo and probably most other major universities, licensing accounts for a minority of IP-based technology transfer. Instead, joint research is the dominant mechanism of technology transfer. The first year following incorporation, 2004, the number of license agreements by the University of Tokyo exceeded the number of joint patent applications for joint research inventions. However since then the number of licenses has increased only modestly, while the number of joint patent applications for joint research inventions has grown considerably. In recent years (2006–08) over 70 percent of transfers to industry of university inventions have occurred via joint patent applications by the university and the joint research partner, while less than 30 percent have occurred via licenses unrelated to joint research.[9]

Such precise data are not available for other universities. However conversations with technology transfer officials in other major national universities indicate that the ratios are similar. Approximately half of all patented inventions are attributed to joint research and three-quarters of all inventions that are actually transferred to companies are transferred by joint patent applications associated with joint research.

The dominance of joint research by large companies is especially strong outside of the life sciences. An analysis of a sample of invention reports from the University of Tokyo shows the differential impact on start-ups versus established corporations by technology field. Only about 20 percent of life science inventions are attributed to joint research. Thus the TLO is free to license 80 percent of the life science discoveries to the companies most able and willing to develop them. In fact, these are licensed to a wide variety of companies including start-ups. Of the 20 percent that are attributed to joint research, over three-quarters of the sponsors are start-ups or other small companies. In other words, in the life sciences, established companies do not receive the lion's share of university discoveries, and start-ups have access to a large proportion of university discoveries.

However in non-life science fields (engineering, chemistry/materials science and software) 40 percent of reported inventions are attributed to

*Table 3.1 Prevalence of industry co-ownership of university patents: a comparison of five major high technology economies*

| | Approx. no. US patents issued 2008/4–2009/3 where one assignee is a university in this country (A) | Univ. patents per $1 billion univ. R&D spend ((A)/ (2004 or 2006 R&D spend by higher education institutions in country)) (B) | % of (A) where at least one co-assignee is a private co. (C) | % of (C) where co-owner is a foreign co. (D) |
|---|---|---|---|---|
| Canada | 92 | 11 | 10 | 50 |
| Germany | 38 | 4 | 16 | 33 |
| Japan | 125 | 6 | 61 | 0 |
| UK | 70 | 9 | 6 | 100 |
| USA | 2249 | 48 | 3 | 0 |

joint research, and among those on which patent applications are filed, well over half are joint research inventions. More than 80 percent of the sponsors of the inventions attributed to joint research are large, established companies. So in non-life science fields, which account for two-thirds of the University of Tokyo inventions, most of the inventions are transferred automatically to large, established joint research partners (Kneller 2007c, 2006).

More macro-level data confirm the dominance of large Japanese companies over Japanese university inventions. Among technologically advanced countries, Japan appears to be unique in the high proportion of university discoveries that are automatically exclusively controlled by companies by virtue of joint patent applications. Approximately 60 percent of all US patents assigned to Japanese universities are co-assigned to companies. In other words about 60 percent of all issued/registered US patents on which at least one applicant is a Japanese university also have a corporate co-applicant. Almost all these co-applicants are large Japanese companies. The country with the next highest rate of co-ownership is Germany, but only about 15 percent of its university patents are co-owned by companies, about one-third of which are foreign. Propensity of universities to patent as measured by US patents issued per $1 billion R&D spend does not seem to be able to explain the difference, as Japanese universities' output of US patents per unit of academic research expenditure is equivalent to that of Germany and the UK.

This analysis is based upon a sample of 40 patents issued April 2008 to March 2009 from each of the countries listed in Table 3.1. In the case of Germany, all 38 US patents with a German university listed as assignee were reviewed. The time period is late enough so that there are few applications by Japanese universities made before Japan changed to a system of university ownership of IP in 2004 or by German universities before Germany changed to a system of university ownership in 2002. Data for academic R&D expenditures are from NSB (2008).

One interesting aspect of these findings is that patent law concerning the rights of joint patent owners is similar in Japan and almost all other countries, except the USA. Except in the USA, agreement of all co-owners is necessary for any transfer of patent rights (even a non-exclusive license), and thus if a company co-owns a patent with a university, it essentially has a perpetual, royalty free, non-transferable, exclusive license. This would seem to make co-ownership of university discoveries by Canadian and European companies participating in joint research just as desirable as it is for Japanese companies. Yet rates of company co-ownership of German, UK and Canadian university patents are low compared with those of Japanese university patents. This may indicate that cooperation between industry and universities is indeed closer in Japan compared to other countries. However it may also indicate that Japanese universities are letting companies assume control over a large proportion of university discoveries without diligently assessing whether these inventions really ought to be attributed to company funding or trying to ensure that the companies are committed to developing them. In other words, Japanese universities are guarding the public interest (cynics might argue, the universities' interests) less carefully than their European and Canadian counterparts. A recent report to the Kauffman foundation comparing industry perspectives on cooperation with universities in Japan, Canada, the UK and the USA suggests that both factors are at work (Ternouth et al. 2009).

Under the Japanese system, engagement between companies and universities is primarily based on the relationship between companies and professors. TLOs, which are sometimes accused of being obstructionist, bureaucratic or simply greedy, generally do not get in the way. Expectations between professors and companies are generally well aligned, with professors willing to let companies control IP in exchange for research support.[10] Many Japanese TLOs simply are not capable of effectively managing inventions, and the surrounding environment is not supportive of startups. For these universities letting joint research partners manage inventions is the best option for technology commercialization. Even under more favorable circumstances, having a large company that is interested in a discovery is an important first step to commercialization.

However there are also reasons for concern regarding the Japanese system. First, a majority of inventions from major universities are monopolized by large companies under the assertion that they are joint research inventions. Since joint research accounts for less than 5 percent of total R&D funding, the fruits of much publicly funded research is being exclusively appropriated. This may not be a problem if the transferees make concerted efforts to develop the discoveries, but this is probably not the case. The normal licensing mechanisms (including the inclusion in the license contract of development incentives, such as development milestones and obligations to pay royalties) have been bypassed and the joint research contracts impose, at most, weak development incentives. Also there is no system for tracking the fate of these discoveries. Interviews with large companies that have collaborated with universities suggest a real problem in transferring discoveries from the basic research to the development laboratories in big companies (Ternouth et al. 2009). Joint research partners, which control prosecution of most joint research patent applications, request the Japan Patent Office to review only about half of these. Most of the non-reviewed applications are, in effect, dedicated to the public. This mitigates the fact that the companies have chosen not to develop them, except that this decision is usually made three years after patent filing. Only then can other companies feel free to use them. The fate of the inventions on which patent applications are pursued is also uncertain. In some cases, the joint research partner makes concerted efforts to develop. In others, the patents are held to deny access to rivals, or as ammunition in case of patent litigation.

Second, interviews with the founders of university start-ups indicate that faculty inventors do perceive an either/or choice between forming a new company and entering into joint research with a large company in order to develop their discoveries. They usually choose the latter, not only because it is so much easier and the rewards in terms of research support much more immediate, but also because much of the IP covering their own or related university research is already controlled by large companies engaging in joint research. In other words, their energy and time and sometimes relevant IP are pre-empted by large joint research partners. This constriction of opportunities for faculty entrepreneurship might not be a problem, provided a large company is willing to develop a promising discovery in most of its various applications, but again this usually is not the case.

After a high point of start-up formation in 2000 and 2001, the number of start-ups formed annually has declined.[11] The University of Tokyo has established incubator space for start-ups and it has launched entrepreneurship training programs. With cooperation from main line investors it has also established an early-stage venture capital fund, which has become one of just a small number of sources of early-stage funding for science and technology-based ventures throughout Japan. However it seems that these efforts are unique. Except for incubator space which is still provided close to several universities, other universities are scaling back on their efforts to promote start-ups. A major university in Western Japan that has approximately 60 persons working in various fields of coordination with industry and technology commercialization devotes few personnel to fostering start-ups. Rather, in most universities, including the University of Tokyo, the emphasis has switched to obtaining more joint research support from large companies. Probably one reason is the immediate financial support that such support affords (including overheads of about 30 percent), since all universities are now facing steady annual cutbacks in core support from the government (Kneller 2010).

A careful assessment of the technologies, sales and employment trends of all the start-ups established before 2005 from the University of Tokyo, Keio University and the Agency for Industrial Science and Technology (the umbrella organization for METI's main research laboratories) showed that almost all those with growth potential are in the life sciences. (About one-third of all University of Tokyo start-ups are in the life sciences.) Only a few start-ups in engineering, materials science, chemistry and software seem to have unique or high-demand technologies and the basic elements of a business development strategy in place (Kneller 2007a). Yet it is clear from invention reports and government research grants that first-class research is occurring in non-life science fields. But with just a few exceptions, only large, established companies are developing these discoveries. Outside of the life sciences, university entrepreneurship is weak. Currently there are one or two promising start-ups from the University of Tokyo each in material science and communications, but these seem to be the exceptions that prove the rule. Except perhaps in the life sciences, all of Japan's future innovation potential depends upon its large companies. There is no viable alternative mechanism to develop science and engineering discoveries that big companies are not willing to develop.

Third, standard joint research contracts (for example, Articles 25 and 30 of the University of Tokyo's model agreement) give companies considerable authority to restrict academic publications. How often they exercise this authority is still unclear, but interviews with companies and students show it is exercised occasionally and that professors and students are widely aware of this authority. More broadly, there is the yet unanswered question of the extent to which joint research narrows the education and the professional opportunities of graduate students working on joint research projects. Of course, participation in joint research projects can also benefit graduate students tremendously – scientifically and in terms of their careers.

## 3.5 CONCLUDING OBSERVATIONS

These findings show that under current conditions, transfer of a substantial proportion of university technologies via joint research, whatever its other merits, is not conducive to the formation of vibrant new companies. In light of Japan's experience under the pre-2004 donation system, it also shows that academic inventors retaining ownership does not necessarily promote entrepreneurship and the formation of new companies. This is a point that should be born in mind by critics of the so-called Bayh-Dole system of university ownership, who often advocate inventor ownership as an alternative. Japan had, and still by and large has, a system where inventors effectively control their discoveries because of the way they and corporate sponsors can manipulate the reporting system and thus determine the recipients of their discoveries. But this freedom usually results in the direct transfer of discoveries to research sponsors which, except in the life sciences, are almost always large, established companies.

There are other factors that contribute to the weakness of new high-technology companies in Japan besides the system of university-industry cooperation. These include the continued prevalence of lifetime employment in high-technology industries, a strong social preference for employment in large companies, low levels of immigration, low levels of smart investment capital for start-ups and a continuing preference by large Japanese companies to rely on their own laboratories for innovation.

In this latter regard, Japanese companies are willing to partner with overseas universities and start-ups because they perceive the need for expertise from overseas. They are increasingly willing to partner with Japanese universities, in part because the cost of research is relatively low and they can obtain exclusive control over emerging IP. But in what appears to be a self-fulfilling prophecy, there is still widespread skepticism towards domestic (Japanese) start-ups. Some deals are made. Also, fault often lies with the start-ups for not communicating effectively with potential large customers or not anticipating their needs and how they can help respond to those needs. Nevertheless numerous interviews suggest Japanese start-ups must try exceedingly hard to get the attention of potential large Japanese customers (Kneller 2007a). Also there is a complete absence on the part of large Japanese companies of the approach of some major US companies (such as Qualcom, Cisco Systems, Apple and even IBM) that supporting new entrepreneurial start-ups will benefit the major companies in the long term.[12] Innovation still remains largely closed in the established Japanese companies, at least from a domestic perspective. The system of technology transfer which gives large companies so much exclusive control over publicly funded discoveries helps perpetuate this closed innovation system – probably to the long-term detriment of Japan's industry and economy.

## NOTES

1. Based upon the University of Tokyo's share of funding from various project-specific funding programs over recent years. Chief among these funding programs are Grants-in-aid from the Ministry of Education, Culture, Sports, Science and Technology (MEXT) and various commissioned research programs from government agencies. See Cabinet Office Data (various years), Kneller 2007b, 2010.
2. Occasionally the inventors themselves would apply for patents, but this was common probably only in the case of a few well-known serial entrepreneurs.
3. That is, comparing patent applications published in 2001–02 with those published in 2005–06 from Tohoku, Tsukabo and Hiroshima Universities.
4. The Japanese Bayh-Dole Act authorizes, but does not require, government funding agencies to let contractors and grantees own inventions. However at the urging of the Ministry of Economy, Trade, and Industry (METI), most funding agencies now let universities claim ownership, the principal exception being inventions arising under the Explanatory Research for Advanced Technology (ERATO) projects funded by the Japan Science and Technology Agency (JST).
5. The standard joint research contract of the University of Tokyo as well as those of some other prominent public research institutes contain provisions that attempt to limit the effect of Article 73 of Japan's Patent Law. For example, Article 21.3 of the University of Tokyo's standard contract stipulates that, if the sponsor has not taken an exclusive license to the university's ownership share, the university can request permission from the sponsor to license its share to a third party, and the sponsor *should not refuse this request without justification*. Some sponsors demand up-front that even this clause be stricken from the agreement. When these clauses are retained and sponsors exclusively license the university's ownership rights (thus obtaining unified exclusive rights) their royalty obligations are usually no more than total patent application and maintenance costs. Based upon close communication with the university's TLO and understanding of its operating procedures, the amount of additional royalties beyond patent prosecution and maintenance costs required by the university for an exclusive license of its ownership share is modest. The TLO's main concern seems to be that companies that obtain exclusive control are committed to development. If a company foregoes an exclusive license to the university's share, the TLO will push for permission to license its share non-exclusively to interested companies. Sometimes the sponsor agrees. Sometimes a long process of negotiation ensues. Two licensing professionals in the TLO are responsible for managing these negotiations and the licensing of joint research inventions. Among companies, the University of Tokyo's TLO has the reputation of being the hardest bargainer among all university TLOs (Ternouth et al. 2009). Most other universities with fewer human resources probably simply accede to a sponsor's request (backed up by Article 73) that, even without a license from the university to the sponsor, the university will not license its rights to a third party.

   By 2007, some other leading national universities had shifted to letting university inventors retain their co-ownership rights in inventions arising under joint research contracts, thus stepping out entirely from the management of co-invented joint research inventions. Just as under the old donation system, these inventors can manage their co-ownership rights as they wish. Most will let the sponsors have unified exclusive ownership in return for research support and providing training and employment opportunities for students.

6. Commissioned research contracts, which do not permit collaboration between company and university researchers, are relatively rare (Cabinet Office; Kneller 2010, 2007a). This is partly because through actual collaboration, companies are able to claim their researchers are co-inventors which automatically leads to co-ownership and exclusively control over any emerging IP.
7. In 2007 joint research sponsored by private companies accounted for less than 6 percent of all project-oriented research funding in the University of Tokyo, that is, government grants and contracts, and corporate-sponsored research and donations, *but excluding* salaries for permanent staff, infrastructure and most overhead costs (University of Tokyo *2008 Data Book*). For Japanese national universities as a whole, joint research accounted for less than 8 percent of such project-oriented funding (Cabinet Office). According to Organization for Economic Cooperation and Development (OECD) statistics, which attempt to account for salaries and infrastructure attributable to research (as opposed to education and administration), industry funded only 2.8 percent of Japanese university R&D in 2004, compared with 5.3 percent of US university R&D in 2006 (NSB 2008).
8. This is based in part on discussions with various TLOs suggesting that many, aside from the University of Tokyo's, have not yet developed strong marketing capabilities but instead patent many inventions in the hope they can be licensed later. It is supported by the fact that only the University of Tokyo and Nagoya University are close to breaking even on revenues versus expenditures (that is, most of the universities that file patents on a high proportion of their discoveries but so far have earned meager license revenue). The University of Tokyo is in a stronger position financially to apply for patents, on the strength of its respectable royalties and also windfall earnings from a successful 2003 Initial Public Offering (IPO) by a company to which it had given a license in return for stock in the company. Also the number of invention reports, about 600 annually, is the largest among Japanese universities (Cabinet Office).
9. For example, in 2008 the university entered into 169 contracts with joint research partners to apply jointly for patents. Some of these contracts covered more than one invention, so 169 underestimates the number of inventions transferred in this manner. In the same year, the university executed 65 independent licenses. Some of these were non-exclusive licenses covering the same invention, so 65 overestimates the number of inventions transferred by licenses (University of Tokyo data available at http://www.casti.co.jp., see also Chapter 3 in Kneller 2007a.) Overall, a respectable two-thirds of patented University of Tokyo inventions are transferred to industry, but as the above indicates the vast majority of these transfers are by joint patent applications, not licenses.
10. However, direct and overhead costs are both quite modest compared to the USA (Kneller 2007a).
11. In the case of the University of Tokyo, seven start-ups actually based on university discoveries were formed in each of 2000 and 2001. By 2004 this number had fallen to two or three (Kneller 2007a). Using the more restrictive definition of start-ups actually *based upon licenses* from a university, no University of Tokyo start-ups were formed in 2005 or 2006, one or two were formed in 2007 and for all Japanese universities only 25–45 have been formed annually since 2005 (UTTAJ 2009).

    Readers who are familiar with Japan may recall official government statistics showing over 200 university start-ups formed annually, at least during the first half of the decade. However these include subsidiaries of established companies and other companies whose only tie with the university is the existence of a joint research contract or board membership by a university researcher or recent graduate. Defining start-ups as companies based upon university discoveries would require reducing the overall figure by 40 percent to 50 percent (Kneller 2007a). The definition used by the University Technology Transfer Association, Japan (UTTAJ) is probably too strict, however, and probably does miss some legitimate start-ups based upon university discoveries.
12. For example, by developing applications for their products, developing new technologies they can later buy and turn into new products, becoming customers, and developing workers with unique skill sets that may eventually be useful to them.

# REFERENCES


Cabinet Office (various years), *Understanding the Scientific and Technical Activities of GRIs and National University Corporations (Dokuritsu gyousei houjin, kokuritsu daigaku houjin nado no kagaku gijutsu katsudou no haaku, shoken torimatome)*, available at http://www8.cao.go.jp/cstp/budget/trimatome.html (accessed 6 June 2011).

Kanama, D. and K. Okuwada (2008), 'A study on university patent portfolios (2):the impact of intellectual property related policies and the change into corporation of national university', *National Institute for Science and Technology Policy (NISTEP) Report No. 154* , English summary available at http://www.nistep.go.jp/achiev/abs/eng/mat154e/pdf/mat154ae.pdf (accessed 6 June 2011).

Kneller, R.W. (2003), 'University-industry cooperation and technology transfer in Japan compared with the U.S.: another reason for Japan's economic malaise?', *University of Pennsylvania Journal of International Economic Law*, **24**, 329–449.

Kneller, R.W. (2004), 'The new Japanese system of technology transfer: concerns related to the role of university IP centers', *Les Nouvelles*, **39** (2), 69–72.

Kneller, R.W. (2006), 'Japan's new technology transfer system and the preemption of university discoveries by sponsored research and co-inventorship', *Journal of the Association of University Technology Managers*, **18**, 15–35. Republished with permission in *Industry and Higher Education*, **21**, 211–20.

Kneller, R.W. (2007a), *Bridging Islands: Venture Companies and the Future of Japanese and American Industry*, Oxford: Oxford University Press, pp. 110–45.

Kneller, R.W. (2007b), 'Prospective and retrospective evaluation systems in context: insights from Japan', in R. Whitley and J. Glaser (eds), *Changing Governance of the Sciences (Sociology of the Sciences Yearbook)*, New York: Springer, pp. 51–73.

Kneller, R.W. (2007c), 'The beginning of university entrepreneurship in Japan: TLOs and bioventures lead the way', *Journal of Technology Transfer*, **32,** 435–56.

Kneller, R.W. (2010), 'The changing governance of Japanese public science', in R. Whitley, J. Gläser and L. Engvall (eds), *Reconfiguring Knowledge Production: Changing Authority Relations in the Sciences and Their Consequences for Intellectual Innovation,* Oxford: Oxford University Press, pp. 110–45.

MEXT Academic Advisory Committee, IP Working Group (2002), *Intellectual Property Working Group Report* (Kagaku gijutsu gakujutsu shingikai, gijutsu kenkyuu kiban bukai, sangakukan renkei suishin i-inkai, chiteki zaisan wa-kingu gu-rupu, Chiteki zaisan wa-kingu gu-rupu houkokushou) (issued 1 November, in Japanese).

National Science Board (NSB) (2008), *Science and Engineering Indicators (NSB 08-01),* Arlington, VA: National Science Foundation, appendix table 4–37.

Takahashi, M. and R. Carraz (2009), 'Academic patenting in Japan: illustration from a leading Japanese university', University of Strasbourg, Department of Theoretical and Applied Economics, Working Paper Beta no. *2009-07*, available at http://cournot.u-strasbg.fr/users/beta/publications/WP/article.php?SELECTWP=2009-07 (accessed 6 June 2011).

Ternouth, P., C. Garner, M. Mongeon, J. Cope and R.W. Kneller (2009), 'What industry wants from universities (Kauffman Seminar White Paper) university-business interaction: a comparative study of mechanisms and incentives in four countries. Issued by Council for Industry and Higher Education (UK)', Paper presented at the seminar of the same title, sponsored by the Kauffman Foundation and the University of California San Diego in San Diego, 12–14 March. Now under review by the Kauffman Foundation for publication as a Kauffman Report.

University Technology Transfer Association, Japan (UTTAJ) Nihon gijutsu iten kyougi kai (2009), *University Technology Transfer Survey, Fiscal Year 2008 (Daigaku gijutsu iten saa bei, 2008 nendou ban),* Tokyo: Japan Institute of Invention and Innovation (Zaidan houjin, hatsumei kyoukai).

# 4. Academic patenting in Japan: illustration from a leading Japanese university

**Makiko Takahashi and René Carraz***

## 4.1 INTRODUCTION

Recent work on universities has led many scholars to investigate the incentives behind academic patenting. This stream of literature began with the enactment of the Bayh-Dole Patent and Trademark Amendments Act of 1980, which allowed American universities to receive patents and grant licenses from research funded by the federal government. Patent grants to American universities peaked in 2002 at just under 3300, compared to 300 in the 1970s. The biomedical-related patent classes dominate these awards (National Science Board 2008). Most observers attribute this tendency to the legislative change, but it is worth noting that the trend preceded the Act: based on case studies Colyvas et al. (2002) argue that two other factors could explain the surge. First, the period saw the rise of important new areas of university research, namely, molecular biology and computer science, both of which are of particular interest to industry. Second, over the same time, various Patent Offices increased the range of research results that were patentable. In their view, these two elements were leading the increase in patenting and licensing, the principal effect of the Act being to accelerate these trends.

The increasing reliance on patenting has raised many questions in the literature. The enthusiasts spoke with emphasis of the increasing role of universities in economic development. The 'Triple Helix' concept (Etzkowitz 2003) sees patenting by universities as an indicator of their involvement in commercialization activities, beyond the traditional role of research and teaching. In the same vein, Jensen and Thursby (2004) show that direct involvement of scholars has proven to be a determinant in the success of technology transfer. Skeptics consider that the increase in patenting and commercialization by universities could lead to some caveats. It is possible that industry uses its growing relative importance to shape research agendas, inducing a redistribution of resources from basic to applied research. Other possibly negative effects of academic patenting include potential conflicts of interest, secrecy issues, delays in the publication process and increased costs of research (Heller and Eisenberg 1998). A growing 'anti-commons' perspective highlights the negative role of Intellectual Property Rights (IPR) over scientific knowledge. Academic inventors may have to use patents to protect and exchange their new knowledge. In that respect, patenting is seen as a defensive mechanism to enable the diffusion of knowledge. This new situation may be less efficient than the previous reliance on pure 'open science' because of the transaction and maintenance costs associated with patenting. A large number of studies have examined the impact of patenting activity on academic research; while the majority of the research has been centered on the USA and Europe, very little has been said about Asia. The prime aim of this chapter is to start filling this gap. We will look closely at the Japanese case and provide an analysis of a leading Japanese research university, Tohoku University. To our knowledge there is no study available in English on this topic centered on the recent Japanese context. We will investigate the mechanisms at stake, and meticulously whittle the salient figures.

Our analysis is based on data in English and Japanese for the Japanese context and on internal documents for Tohoku University. The main objective of the work presented here is to analyse the implications of four years of drastic changes resulting from the Incorporation of national universities in 2004, as this marks the moment when Japanese national universities started gaining control of their strategic development. The national universities are no longer a homogeneous entity; they can develop their own distinctive strategies. The legal system in which they evolved has changed dramatically, and each university now has to adapt to this new environment. In this context, it is very interesting to study the changes in terms of IPR policies for the universities and to determine how these changes have modified the behavior of faculty members. For these reasons, we believe scholars would gain from an increased scrutiny of the Japanese experience: because of the difficulty of the Japanese language, many findings and documents have not yet been publicized to a wider audience. We intend to partly fill this gap.

This chapter is organized as follows: Section 4.2 introduces the different channels of university technology transfer. Section 4.3 presents a description of the Japanese case in terms of institutional reforms and links them to academic patenting. Section 4.4 then moves to a description of the Tohoku University case. Section 4.5 discusses the implications of the increasing use of IPR by Japanese universities.

## 4.2 THE DIFFERENT TYPES OF UNIVERSITY-BASED TECHNOLOGY TRANSFERS

It is valuable to note that universities have always been involved in technological transfer. For instance, with the role of German academicians in the development of the nineteenth-century chemical industry, or as the seminal work of Rosenberg and Nelson (1994) has documented, there is a long tradition of universities transferring knowledge and technology to firms, though this has mainly occurred through channels like publishing, consulting, conferences and so on. In the case of Japan, in 1886 the government launched the first Japanese Department of Engineering in the University of Tokyo. It was the first Engineering Department within a university (not in a college, institute and so on) in the world. This department has a long history of transferring knowledge and technology, as well as collaborating with industry (Murakami 1994). These three examples stress the importance and long record of university-based technology transfers. Nowadays, the difference is that the political and economical spheres have put increasing pressure on universities to encourage a more active transfer of their knowledge to society, and particularly to contribute to economic development. As noted by Geuna and Muscio (2008), there is an institutionalization of university-industry linkages through the direct involvement of the university bureaucracy. As we will see in the next section, this process is prevailing in Japan too.

In terms of technology transfer mechanisms, many different channels characterize university-industry relationships. First, there are the traditional publication and conference mediums. Within this first channel are personal contacts; on the one hand, among faculty members and students, on the other, between academic researchers, industry and government. The second way is the professionalization of knowledge transfer activities, the so-called third mission of the university. In order to fulfill this mission, Technology Transfer Offices (TTOs) or Technology Liaison Offices (TLOs) have been created within universities. TTOs tend to be focused mainly on the exploitation of IPR via licensing or the creation of spin-offs, but are also active in supporting university-industry interactions via contractual research or consultancy.

The majority of studies focused on knowledge transfer use three kinds of data: patents and invention disclosures, licenses and spin-offs. Such data is increasingly employed as it becomes widely available; the first two are now standardized information. This information is used as a proxy to measure technological transfer by universities. In this chapter, we intend to focus on the academic patenting of Japanese institutions, as it becomes a widely used mechanism by research universities. Nevertheless it is important to keep in mind that IPR-related activities are only a small part of the knowledge transfer activity of universities. Agrawal and Henderson (2002, p.45) argue that 'too great a focus on patenting may seriously misrepresent the nature of the impact of the university on the private sector'.

## 4.3 THE CASE OF JAPAN

In this section we introduce the legislative changes that occurred in Japan concerning the university-industry setting. We particularly focus on their influence on the IPR regime, and stress the importance of contractual research as a means of transferring knowledge and technology.

### 4.3.1 University Reform

In order to understand the development of academic patenting activities in Japan, it is necessary to restate some key institutional reforms that have led to a dramatic increase in university-owned patents. We will investigate the changes in the legal status of national universities, where the majority of academic research takes place, as well as the modifications that occurred in the university-industry legal framework.

University-industry collaboration has evolved recently in order to facilitate interaction between the two institutions. Until 1980 restrictive government regulations caused levels of university-industry collaboration to remain low. In 1983 the Ministry of Education relaxed its regulations, and notably allowed national universities to cooperate with industry. However it is only after the introduction of the 1995 Science and Technology Basic Law and the TLO Law that the real changes began.

We can step into the argument by briefly stating the main laws that are structuring the technology transfer activities within universities. Below are the three main laws shaping the legal framework:

1. The 1998 Law to Promote the Transfer of University Technologies (the TLO Law) legitimized and facilitated transparent and contractual transfers of university discoveries to industry.
2. The 1999 Law of Special Measures to Revive Industry (the Japanese Bayh-Dole Law).
3. The 2000 Law to Strengthen Industrial Technology established procedures through which university researchers can obtain permission to consult for, establish and even manage companies. It also streamlined the procedures for company-sponsored commissioned and joint research.

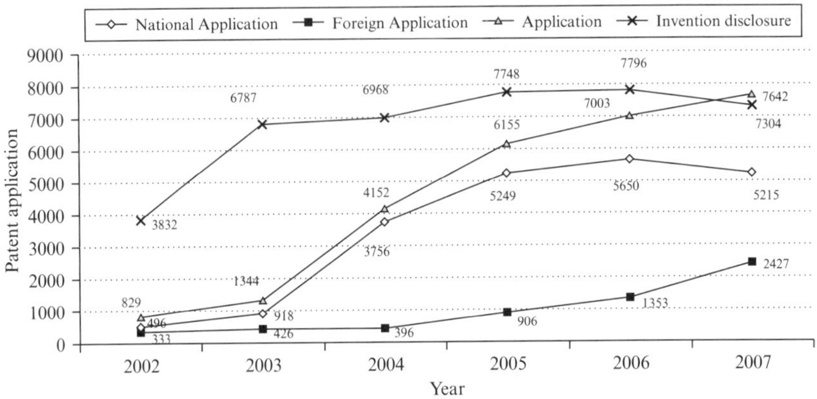


*Figure 4.1 Invention disclosures and patent applications by national universities*

All these legislative changes have been listed here to illustrate the increasing importance that the Japanese authorities place on university-industry collaboration and one of its corollaries, patenting.

Another important factor is the change in status of national universities and its influence on patenting patterns. The anchoring point of the university reform is the 2001 Toyama Plan named after the Minister for Education, Culture, Sports, Science and Technology Atsuko Toyama. This plan proposed three major reforms: (1) the reorganization and incorporation of national universities; (2) the development of universities that conform to the highest international standards by using third party evaluation; and (3) increasing the proportion of competitive funding. The plan recommended that national universities should be transformed into National University Corporations (NUCs), institutions that are legally separate from the government. In accordance with these proposals, in April 2004 the Japanese government incorporated the national universities as 'independent administrative entities'.[1] Since 2004 the universities have gained greater autonomy. They can more easily recruit academic and non-academic staff. Moreover they can maintain ownership of their inventions, which was seldom the case before the Incorporation. Consequently there has been a surge in the number and value of cooperative contracts, as well as in patents.

Since 2004, national universities have been solely managing their intellectual property. Figure 4.1 shows the influence of these changes of universities' status on invention disclosures and patent applications; invention disclosures had started to rise before the Incorporation, with a strong hike from 2002 and 2003, preceding the increase of patenting. Thereafter the number of invention disclosures increased only slightly, reaching a plateau around 7500. For patent applications, the number sky-rocketed in 2004 and increased steadily thereafter. In 2007 the number of national patent applications decreased for the first time, while the number of foreign applications intensified. These figures indicate two tendencies: first, the Incorporation entailed a huge increase in IPR activities; second, in 2007–08 it seems to have reached a peak. Furthermore universities seem to have gained expertise and quality, as the number of national applications decreased and foreign applications increased in 2007. Foreign applications are often judged to be of more value to the applicant as they cost more to start and maintain.[2]

Universities all over the world are increasingly patenting the outcome of their research (Mowery et al. 2001; Geuna and Nesta 2006). Our data shows that Japan is also following this upward trend. Together with research and teaching, universities are being asked to be the generators of future economic growth. Nowadays Japanese universities are directly managing their IPR, and are thus more prone to maximize the number of patents they can produce. New rules have been enacted about the invention disclosure process in order to facilitate patenting. Second, with time passing, the administrative staff have complied with the new regulations, and have accepted the notion of patentability of university inventions, and market orientation of some research. At the same time they have increased their expertise, which had the consequence of a slight drop in the number of applications as the inventions are now scrutinized more thoroughly.

On a more sociological note we can introduce the notion developed by Sine et al. (2003) on institutional prestige: in our case we could argue that the prestige of the university is linked to the level of the patenting activity together with the publication output. Recently information has started to be compiled in order to rank universities by technological strength, based on the number and relevance of patents issued (Staedter 2003). As these indicators are often used in rankings, more is necessarily better. Hence, legal changes have put more emphasis on patenting in the decision-making processes of universities. Figure 4.2 shows the number of patent applications by Japanese universities in terms of patents per year: in 2003, 61 universities had applied for one to nine patents, a number that rose to 115 universities in 2005. The tendency is the same for the highest bracket: in 2003 only one university applied for more than 200 patents, in 2005 there were seven. This illustrates the fact that universities at both ends of the spectrum quickly embraced the use of patents. However we should remain cautious about the total increase of patents applied for by Japanese universities; universities not previously active in patenting account for a

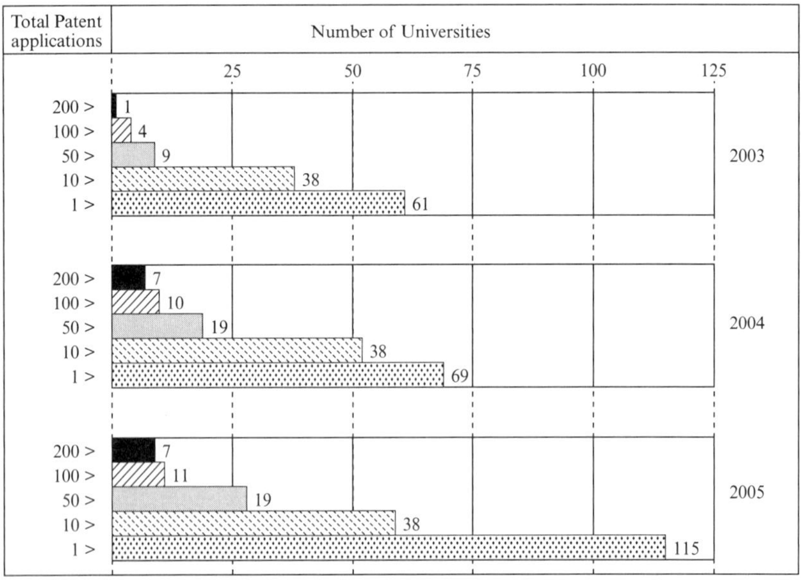


*Figure 4.2 Patent applications of Japanese universities by frequency of application*

significant part of the growth in overall university patenting. A similar phenomenon was observed in the USA in the 1970s, as noticed by Mowery et al. (2001).

### 4.3.2 Organizational Structure of Intellectual Property Management Institutions

In 1997 the Ministry of International Trade and Industry (MITI),[3] in coordination with the Ministry of Education, Culture, Sports, Science and Technology (MEXT), proposed to extend the support of university-industry cooperation. An important part of this initiative was the creation of TLOs. In 1998 national universities had no independent legal standing, and thus it was difficult to apply for patents on inventions by their faculty and to license such inventions. The Technology Transfer Law authorized universities to establish independent or semi-independent TLOs that could sell or license inventions and distribute royalties to inventors and universities. However academic inventors were not obliged to assign their inventions to the TLOs and could continue to transfer them directly to companies. Kneller (2003) suggests that inventors often turn to the TLOs when an invention has no takers.

In order to establish a comprehensive intellectual property (IP) management procedure, from the creation and evaluation of IPR to their management and licensing, MEXT established a program to support the creation of in-house IP management offices (hereinafter referred to as 'IP offices') within universities. By August 2003, just before the Incorporation, 43 universities had launched an IP office to develop their own technology transfer management system. Their responsibilities partially overlap those of TLOs. In general IP offices manage the whole IPR procedure from invention disclosure to patent application, and they have final authority over patenting and licensing decisions as the patent owner through the university. Parts of the procedure which need professional skills such as marketing, patent surveys and licensing are outsourced to TLOs. In some universities, relations between the IP offices and TLOs have been managed smoothly, while in others there has been friction. The issue is that they have different decision-making structures in terms of IP management and dealing with research contracts, particularly for managing license earnings, patent costs and contract specifications.

The inventor has different people to deal with, and the delimitation of power is not always clear between the TLOs and IP management offices. An important issue for the newly incorporated universities to address is how to resolve the inherent complexity and coordination of a dual structure system. In terms of manpower, IP offices and TLOs employ a relatively similar number of staff for licensing and IP management-related tasks, the difference being that IP offices have more personnel dedicated to the development of sponsored research and industrial liaison. Their job is not only to create IP but to find industrial partners to cooperate with the university.

On a more practical level, there are three types of IP management organizational configurations. While the IP office is always within the university, TLOs have three alternative settings. The first is the 'internal model', in which IP offices and TLOs are merged within the university. This is usually seen in private universities and in some national universities that have launched their TLOs within the last two or three years. In this configuration the two structures coexist within the university, but there is a division of labor. IP offices handle the invention disclosure process, the management of IPR resulting from contractual research, and they deal with the Japan Patent Office. TLOs are in charge of the practical use of the IPR, such as finding licensing partners, and supporting the creation of university spin-offs.

The second type of organizational configuration for TLOs is 'External & Exclusive',[4] and it is used in most national universities. The TLOs are outside the university structure, although they deal with only a single university. Recently some universities have dispensed with this type, for instance, the Tokyo Institute of Technology. Instead they have wholly integrated their TLOs to strengthen the coordination of their activities.

The third type is 'External & non-Exclusive'. In this case one university has alliances or cooperation agreements with more than two TLOs depending on its policy and/or characteristics. This organization has the merit of fostering specialization of TLOs in terms of technologies and/or activities; for instance, some specialize in start-up support. One of its main disadvantages is the resulting complexity of the arrangements between multiple universities and TLOs. This type of TLO configuration is concentrated in the Kansai area (Kyoto-Osaka-Kobe Region) where a large number of universities are concentrated in a relatively small territory; it is rarely seen in the rest of Japan.

Hence Japanese universities have two entities/offices simultaneously promoting technology transfer activities, especially IP-related activities. From the standpoint of IP management, they pursue three goals: (1) creation of IP; (2) protection of IP; and (3) exploitation of IP. Practically, these goals are achieved by a number of activities: implementation of commissioned and joint research; examination of new inventions; patent application and management; technology transfer through licensing and marketing; support to start-up companies; managing consulting activities of faculty members; and dealing with conflicts of interest. A recent MEXT survey of industry-academia technology transfer specialists has defined a total of 44 missions that are conducted by either IP offices or TLOs.[5]

Figure 4.3 shows the different sources of income of TLOs and IP offices for the fiscal year (FY) 2005. The first point to notice is that the income generated by licensing revenues is rather sleek. At best it accounts for 26.9 percent in the TLOs structures, and drops to 13.6 percent for the internal IP offices. These figures show that the lion's share of these institutions' revenue comes from subsidies. IP offices are heavily subsidized by the universities and by MEXT, which runs a five-year plan (2003–08) to help in fostering them. The TLOs receive important subsidies from METI. The main point here is that these structures have been promoted by the government as a way to develop university-industry relationships, but the figures showing their different sources of income reveal a financial imbalance in the activity. They have not been able to cover their operating costs after ten years of operations for the oldest TLOs, and nearly five years for the IP offices.[6] This phenomenon is quite common; indeed not all universities are wining in the patent casino. Few patents actually bring in income, and many universities are paying more to run their technological transfer activities than they are bringing in via license revenues.

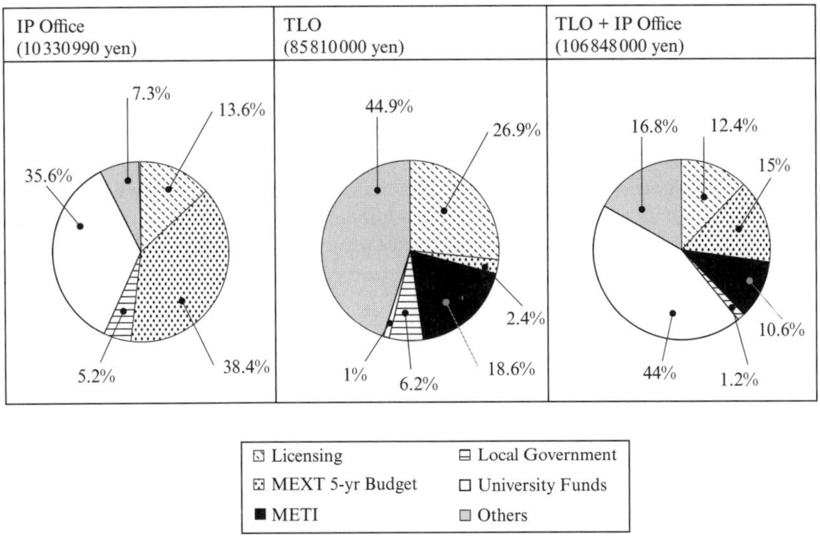


*Figure 4.3 IP management in Japanese universities: incomes*

### 4.3.3 Contractual Research

Contractual research is an important aspect of Japanese universities' technology transfer policies. It is one of the main channels of university-industry collaboration and IP-related activities. In this respect it has two important characteristics. First, research resulting from such contracts are likely to be licensed smoothly because there is already an industrial partner; second, IPR are at the center of the negotiation process while finalizing a contract. Thus we will explain concisely how it operates.

If a company, or any organization, wants to have a formal research agreement with a national university, it has to enter into either a commissioned[7] or joint research contract with the university. In the case of joint research, the university receives funds and researchers mainly from private firms to conduct research on common projects.[8] Under commissioned research, researchers in universities are appointed by firms, research institutes or governmental agencies to carry out research by contract. The principal difference between these two types of contract is that in joint research, company researchers can work in the university laboratories, while this option is not available under commissioned research. Generally more than 80 percent of commissioned research projects are conducted with the national government, or with a private company under a national

*Table 4.1 Contractual research by national universities*

| Category | Fiscal Year 1985 | Fiscal Year 1985 | Fiscal Year 1985 |
|---|---|---|---|
| Joint Research | 216 | 1704 | 12405 |
| Av. amount per contract | ¥5157405 | ¥2413730 | ¥2442790 |
| Commissioned research | 1700 | 3027 | 10082 |
| Av. amount per contract | ¥2051765 | ¥4662370 | ¥10926640 |

*Note:* The numbers are compiled from the MEXT website available at http://www.mext.go.jp/english/; $1=¥120.

project scheme. In contrast, the bulk of joint research contracts are carried out directly with private companies.

Joint and commissioned research is an important means of technology transfer for Japanese universities. The legal framework was enacted in 1983, and such transfers have grown in number and yen value ever since. Table 4.1 shows the trend over a 20-year period. In just the last ten years, joint research has multiplied by a factor of seven and commissioned research by a factor of three. For joint research, the average amount spent on one contract has been stable – if we exclude the first year's hike – whereas for commissioned research the amount has been increasing steadily.

On company use of joint and commissioned research, we could cite a MEXT (2007) survey which was sent to Japanese companies having capital of more than ¥1 billion, with the aim of understanding their research strategies better.[9] One part of the questionnaire was related to their outside partners, especially universities. In their response to where they intended to spend more money on external providers of research, Japanese universities came first. When asked to provide an evaluation of their joint activities with Japanese universities, 36.8 percent companies judged contractual research with universities to be a positive experience, second only to the ability of the universities to solve complex problems (49.9 percent). More specifically, a question was formulated to evaluate the pros and cons of conducting contractual research with universities.[10] On the positive side, the three most common answers were: the enhancement of the firm's research capabilities; the outsourcing of basic research; and the creation of a research network. On the negative side, the top three were: the non-applicability of some university research; the lack of secrecy; and the difficulty of gaining a monopoly on the IP resulting from common research. This shows that the companies value collaboration with the university, as it expands their knowledge capabilities; however they have some concerns about the openness of the relationship, which might be seen as a possible threat to the traditional open environment of university research.

## 4.4 TOHOKU UNIVERSITY

This section will present the case of Tohoku University, a leading Japanese research university. The aim is to show how it adapts itself to a new environment favorable to patents.

### 4.4.1 A Short Presentation

In this section we briefly present our unit of analysis, Tohoku University, and give figures about the recent trends in its patenting activity.

Tohoku University was founded in Sendai in 1907 as Tohoku Imperial University. It was the third Imperial university in Japan. It is located in Sendai, the most important city of the Tohoku Region (North-East of Japan). It is known as a strong research university; the 2008 Shanghai academic ranking put it in fourth place among Japanese universities and 79th in the world. The Thomson ISI list of the most cited papers in the world ranked Tohoku University second in the field of material science, 13th for physics and 22nd for chemistry. In the national context it is widely known as one of the flagship universities.[11]

In order to define its patenting activity, we must make a distinction between university-owned patents and university-invented patents. University-owned patents are patents where the university owns the IPR. Unfortunately data on university-owned patents are only a reliable indicator for universities in the USA and Canada. For European data, in the 1980s and 1990s there is a downward bias if you only look at university-owned patents. This is due to the tendency of European academic researchers to leave the property rights of their invention to the firm that financed the project and yet be included in the list of inventors. There is no official record of these transactions by the university administration, and we must use a different indicator to encompass this phenomenon: university-invented patents. These are patents that have at least one faculty member listed as inventor, regardless of whether the university is assigned the patent. In recent years there have been a few studies on university-invented patents to take this fact into account, notably in Belgium, Finland, France, Germany and Italy.[12]

In the case of Japan there was a similar phenomenon until the Incorporation. Kneller (2003) illustrated how a majority of university

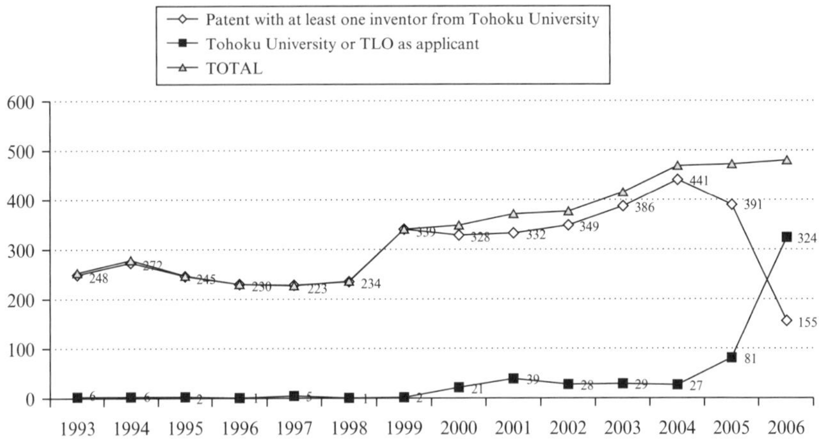


*Figure 4.4 Tohoku University patents*

discoveries were transferred directly from inventors to companies under the guise of donations, with the researcher being listed on the patent application as an inventor. Recent reports from the National Institute of Science and Technology Policy (NISTEP) (Kanama and Okuwada 2007, 2008) investigate this issue. They use the concept of university-invented patent. They listed the names of all researchers that have filed an invention disclosure during the period 1993–2004 at Tohoku University, then looked up all these names in the inventor section of the Japanese Patent database. In the case of a university-invented patent, the university does not have any ownership rights over the patent despite the fact that at least one faculty member is listed as an inventor; however university-owned patents are the property of the university. One of the many interesting results of this NISTEP study is presented in Figure 4.4. It illustrates Tohoku University-owned and invented patents. We see that up to 1999 university-owned patents were quite inconsequential. Their number started to rise in 2000 as a consequence of the TLO Law and Japanese Bayh-Dole Act, but really show a dramatic increase after Incorporation. Until 2000 only a minority of the invention disclosures led to a patent application by the university. In contrast university-invented patents were quite high throughout the period, with an increase in 1999–2000 and a decrease after 2004 when the university started to manage its IPR more aggressively. Overall we see from Figure 4.4 that university members have been active in the IPR business for a long time – although changes of ownership have occurred over time, moving from the research partner to the university – and that the trend is upward. This result can be used to give a better understanding of Figure 4.2. The rise of patenting activity that occurred in 2004 did not come from a vacuum; the potential was there before but diffused through more informal channels. In particular, faculty members were transferring their IPR to the companies with whom they were collaborating. We will see in the next section how this informal cooperation changed to a more formal arrangement, especially through cooperative research.

### 4.4.2 Patenting

The aim of this section is to explain the trend of Tohoku University patenting activity from the Incorporation to 2007. We will take into account: the total number of applications by the university; and the distinction between single application (that is, the university is the sole applicant, and so exercises full ownership) and co-application, in which the university shares ownership with a third party (third parties may be companies, research organizations, government agencies and so on). The following analysis focuses on the above-mentioned elements. To begin with, it is important to restate that one of the principal elements of university-industry cooperation in Japan is contractual research; its importance has been increasing since 1995 (1st Basic Plan on Science and Technology) until the year 2008, and the movement preceded and sustained itself after the Incorporation of 2004. By the same token, the proportion of co-applications has increased as well, most notably after the Incorporation. In 2005 national universities, TLOs and public research organizations applied for 5878 national patents, among which 28 percent were linked to joint research and 15 percent to commissioned research; collectively research contracts are the source of 47 percent of co-applications (Ijichi and Nagaoka 2007).

When a university files a domestic patent in Japan there are two options: a co-application or a single application. Co-applications may or may not be linked to a research contract. In cases of collaborative contracts between a university and a company, the name of the university inventor is listed as well. For the technological classification of the patent, the recommendation of the second Science and Technology Basic Plan (30 March 2001) is to divide the technologies into four categories as follows:

1. Life Sciences
2. Information and Telecommunications (IT)
3. Environmental Sciences
4. Nanotechnology/materials.

When the invention disclosure is filed, the inventor classifies the technology according to one of the above-mentioned categories. Practically, in the case of contractual research, the university manages the IPR in the following manner:[13]

1. Tohoku University (TU) owns all IPR arising from the results of its contractual research.
2. TU may license such IPR to the company for a fee or may transfer it wholly or partially to the company.
3. If TU and the company jointly hold any IPR, TU and the company shall enter into a joint application agreement or joint ownership agreement, which shall include the three following clauses:

a. TU may not exploit the IPR in any area outside of the research area;
b. The company may exploit the IPR itself without the consent of TU;
c. The company shall pay to TU all expenses incurred by TU to file an application and maintain such IPR until the company enters into an agreement with TU, and shall pay all necessary expenses after the execution of such agreement.

During 2004–07, 735 professors were listed as inventors on at least one patent; in 2008 there was an increase of 27.8 percent of the faculty involved in patents. As a matter of comparison, data for the Massachusetts Institute of Technology (MIT) shows that from 1983 to 1997 approximately half of the teaching staff were involved in at least one patent; in comparison, in three years about a quarter of Tohoku University staff filed an invention disclosure, many of which will lead to a patent.

In the case of co-applications there are two scenarios: a research contract that leads to a co-application; and a co-application without prior research contract. Unfortunately we do not have data for the first year (2004), but since 2005 the university administration has kept statistics on this subject. Figure 4.5 displays the proportion of single versus co-applications on the abscissa and the proportion of patent applications linked to contractual research on the ordinate. The data is grouped according to technological fields. The data clearly reveals some tendencies: there is an increase in the proportion of co-applications, and it is linked significantly with contractual research. There are also differences among the technological categories: IT is increasing its share in the total number of co-applications but is relatively less used in contractual research. In nanotechnology, a high proportion of applications are linked to prior contracts. On the other hand, life sciences have a high share of single applications by the university and fewer associations with contracts.

The overall trend arises from the fact that the administration of the university puts much emphasis on collaborative research contracts, as they are

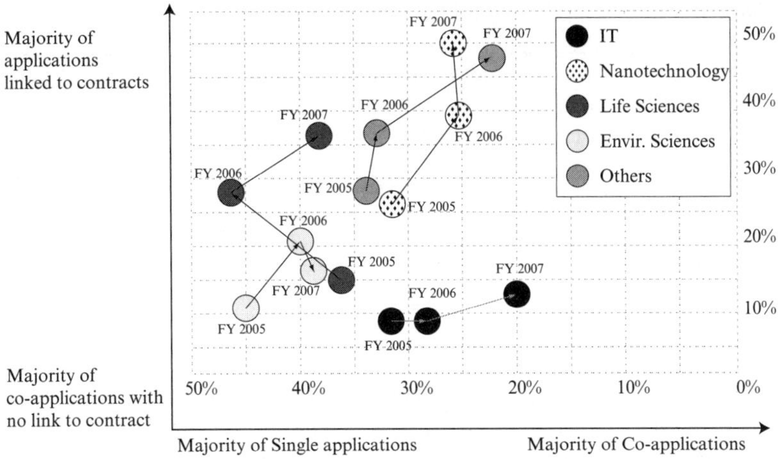


*Figure 4.5 Tohoku University patent applications by technological field and type of contract*

the source of research funding, future inventions and resulting patents. In addition, various measures have been taken to support university-industry relations in Tohoku University: the university-industry liaison office has been strengthened; each department now has administrative employees in charge of liaising with industry; and furthermore some professionals from outside the university have been hired to facilitate university-industry collaboration.

## 4.5 DISCUSSION

The increasing reliance on IPR by Japanese universities is definitely a widespread phenomenon. Furthermore the speed with which these changes took place is quite impressive, in only a few years they have greatly expanded their usage of IPR alongside an increased reliance on contractual research. The relations seem to have become more contractual since the Incorporation. Table 4.2 shows the speed and scope of IPR activities by Japanese and American universities. In only a few years Japan has increased its IPR-related activities. The amount spent per invention disclosure is on par with the American universities. In terms of volume, and considering their relative size, the number of invention disclosures and patent applications in the universities of the two countries are quite similar. The ratio of patent applications to invention disclosures is higher

*Table 4.2 Use of IPR by Japanese and US universities in 2005*

| | Research cost per invention disclosure | Invention disclosure | Patent application | Patent application/ inv. disclosure (%) | Patent registration | License | License income | University spin-offs |
|---|---|---|---|---|---|---|---|---|
| Japan | ¥200 m | 9924 | 7187 | 72 | 295 | 1056 | ¥ 107 b | 46 |
| USA | $2 m | 17 382 | 10 270 | 59 | 3278 | 4932 | $ 1395 b | 462 |

*Source:* Adapted from UTTA (2007); $1=¥120.

in Japan, which can be partly explained by the pressure on universities to develop their patenting activity, a relative lack of experience by the university administration in evaluating inventions and strong incentives given to researchers. The number of patents granted is relatively low, which is expected given that it is quite a new phenomenon. The numbers for licensing and university spin-offs in Japan are still quite low compared with the USA.

From a more theoretical perspective, an important question remains: what are the implications of the increasing use of IPR by Japanese universities? As we have illustrated, much emphasis has been put on this particular channel of technology transfer by the university. The question then is whether this is detrimental to the research done by faculty members.

First, we can argue that this is not the most important channel of technology transfer. Agrawal and Henderson (2002), based on both qualitative and quantitative data from MIT, have shown that patenting is a minority activity. Indeed a majority of the faculty in their sample had never patented. Furthermore publication rates far outstrip patenting rates. From a more qualitative point of view, most faculty members estimate that patents account for less than 10 percent of the knowledge that is being transferred from their labs. Additionally, considerable concern has been expressed in recent years regarding a possible negative impact of patenting on the 'public good' nature of academic work. This is called the crowding-out hypothesis. Two factors are at stake here: first, the risk of a possible switch towards more applied research by academic scientists as opposed to basic research; second, a change in the reward structure encouraging scientists to seek IP protection for their research results, thereby possibly postponing scientific publications.

As a consequence there is a need for better understanding of the influences that these policy changes have on the research activity of faculty members. One related question, often used as a proxy to this problem, is the link between patenting and publication activities of academic researchers. Empirical research tends to indicate that these activities are complementary rather than substitutive. Many studies of academic inventors have shown a positive correlation between patenting and publishing, suggesting that commercialization does not come at the expense of public disclosure of scientific discoveries by scientific members (Breschi et al. 2005; Van Looy et al. 2006; Carayol 2007; Stephan et al. 2007).

In the case of Japan, we must take into account its long history of university-industry relationships. For instance, the Department of Engineering of Tokyo University has an extensive relationship with industry; 32 percent of the department members have been an inventor on at least one industrial patent, 46 percent of them had no less than a paper collaboration with firms and 22 percent did both (Baba and Goto 2007). The figures are very high for both channels. The next step would be to gain a better understanding of the different mechanisms at stake in the Japanese context when deciding how to transfer a technology from the university. To start with, we have already shown (Figure 4.5) that depending on the technological field, patenting strategies differ in terms of contractual research and patent application considerations. Then, it is important to understand what incentives are at stake for faculty members. Further research is needed.

In the case of Tohoku University, Carraz and Takahashi (2009) have conducted an econometric analysis on the determinants of academic patenting by Tohoku University staff since the Incorporation. Their paper intends to explain which factors are at stake in the patenting behavior of faculty members. Three types of variables were tested: publications, funding and individual characteristics. We used a Zero Inflated Negative Binomial model as it fitted the characteristics of the variables. The first result relates to the publication level. We find a positive and significant relationship between patenting and publishing. Accordingly these two activities seem to be complementary, which is in agreement with many empirical studies. At the department level though, the quality of the department publications affects the probability of a researcher staying in the non-patenting regime. We can see in this pattern a kind of specialization, with some research units putting more focus on publishing for high-quality journals while others do research that is more related to patenting outcomes.

We also found a positive correlation between patenting and contractual funding, whether public or private. Moreover receiving public contractual funding significantly affects the probability of reaching the patenting regime. This appears to be counterintuitive, but it may be justified by the fact that the amount of public funding is a signal of the quality of the research. At the departmental level, public funding negatively affects the dependent variable. This is very interesting, as it shows again a kind of specialization, the amount of public contractual funding being negatively correlated with the patenting activity.

Let us now move to the individual characteristics. One of the main criteria of evaluation for a scientist and their parent academic institution is the quality of their publication portfolio. In the academic sphere, the game is defined by the priority of the discovery and the underlying peer-review system that validates the discovery (David 2007). There is a reinforcing mechanism in place, which provides more opportunities to those who are already successful – the 'Matthew Effect' as it is often called in the literature. Therefore young researchers may not consider patenting if it affects negatively their publication outcome, while older researchers may have more incentives to patent as they may place more value on social wealth (Carayol 2007). We therefore assumed an inverted U-shape relationship between age and patenting activity. In our data set, the *Age*1 dummy variable (researchers from 26 to 35 years old) correlates positively with patenting. Therefore the youngest scholars differ from the older ones in their patenting decision patterns. This could be linked to the fact that career advancement and patenting seem to be related; there is a positive link between the dummy *promotion* and patenting. It signals a possible change of reward structure as IP activity and promotion seems to be positively correlated, along with the fact that the younger generation is relatively more engaged in IP activities compared with the older generation.

## 4.6 CONCLUDING REMARKS

Japanese universities have embraced academic patenting enthusiastically since the Incorporation of national universities. They have increased their reliance on IPR strategies. At the governmental level, many regulations have been enacted since the mid-1990s in order to promote university-industry relations. At the university level, the supporting structures have been strengthened, but one should be cautious in interpreting these trends. First, they were not built from scratch. We have showed that IPR was already being transferred to companies on a large scale; the difference now is that these relationships are more formalized in the form of contracts. Second, the Japanese system is not exempt from cultural idiosyncrasies; for instance, one of the main mechanisms used for technology transfers is contractual research, whereas licensing and university spin-offs are quite rare. By developing the use of IPR, the relationship with contractual partners becomes more formal and consequently easier to monitor. However, if one of the tacit goals is to develop entrepreneurship based on IPR, there could be questions regarding the efficiency of the maneuver.

# NOTES

* The authors would like to thank Francis Gosselin and Yuko Harayama for their help in editing this chapter. The usual disclaimers apply.

1. For a more detailed account of the process that led to the Incorporation refer to Carraz and Harayama (2008).
2. Generally, in the case of an international application, Japanese universities use the unified procedure provided by the Patent Cooperation Treaty. They employ the same documents as for the national application, with a delay of one year.
3. In 2001 MITI was reorganized to the Ministry of Economy, Trade and Industry (METI).
4. Or 'Almost Exclusive' to be more precise. For instance, the TLO of Tohoku University (Tohoku Technoarch) deals with other universities of the region; however around 90 percent of its workload is centered on Tohoku University.
5. See Takahashi and Watanabe (2007) for an in-depth analysis of the subject.
6. IP offices have their own operating funds, so their accounting can be traced.
7. It is sometimes referred to by the term 'sponsored research'.
8. For FY 2007, 62.1 percent of the funds came from large companies and 14.4 percent from small and medium-sized enterprises (SMEs).
9. Of the 1791 initially contacted, 941 companies replied.
10. The companies could choose up to four responses in a panel of 12.
11. 'Flagship universities' in Japan are defined as top national and private research universities – namely, the seven former Imperial universities (Tokyo, Kyoto, Hokkaido, Tohoku, Nagoya, Osaka and Kyushu), the Tokyo Institute of Technology (a top national university in engineering) and three leading private universities (Keio, Waseda and Ritsumeikan) (Yonezawa 2007).
12. See Guena and Nesta (2006) for an overview of these studies.
13. To be precise, there are some minor differences in terms of IPRs between commissioned research and joint research contracts.

# REFERENCES

Agrawal, A. and R. Henderson (2002), 'Putting patents in context: exploring knowledge transfer from MIT', *Management Science*, **48** (1), 44–60.

Baba, Y. and A. Goto (eds) (2007), *Empirical Research on University-Industry Linkages in Japan* (in Japanese), Tokyo: University of Tokyo Press.

Breschi, S., F. Lissoni and F. Montobbio (2005), 'From publishing to patenting: do productive scientists turn into academic inventors?', *Revue d'Economie Industrielle*, **110**, 75–102.

Carayol, N. (2007), 'Academic incentives, research organization and patenting at a large French university', *Economics of Innovation and New Technology*, **16** (2), 71–99.

Carraz, R. and Y. Harayama (2008), 'Japanese university reform seen through bureaucratic reform and changes in patterns of scientific collaboration', in

Luc E. Weber and James J. Duderstadt (eds), *The Globalization of Higher Education*, pp. 93–107.
Carraz, R. and M. Takahashi (2009), 'Incentives to patent in a leading Japanese university', Mimeo, Strasbourg.
Colyvas, J., M. Gelijns, A. Mazzoleni, R.R. Nelson, N. Rosenberg and B. Sampat (2002), 'How do university inventions get into practice?', *Management Science*, **48** (1), 61–72.
David, P. (2007), 'The historical origins of 'open science': An essay on patronage, reputation and common agency contracting in the scientific revolution', Stanford University Working Paper, 1–82.
Etzkowitz, H. (2003), 'Innovation in innovation: the triple helix of university-industry-government relations', *Social Science Information*, **42** (3), 293–337.
Geuna, A. and A. Muscio (2008), 'The governance of university knowledge transfer', *SPRU Electronic Working Paper Series*, **173**, 1–39.
Geuna, A. and L. Nesta (2006), 'University patenting and its effects on academic research: the emerging European evidence', *Research Policy*, **35** (6), 790–807.
Heller, M. and R. Eisenberg (1998), 'Can patent deter innovation? The anticommons in biomedical research', *Science*, **280** (5364), 698–701.
Ijichi, T. and S. Nagaoka (2007), *Management of IPR for Upstream Research Inventions in Universities and National R&D Institutes in Japan: Results from the IIR IP Survey 2006*, Hitotsubashi University, Institute of Innovation Research (in Japanese).
Jensen, J. and M. Thursby (2004), 'Proofs and prototypes for sale: the licensing of university inventions', *American Economic Review*, **91**, 240–59.
Kanama, D. and K. Okuwada (2007), 'A study on university patent portfolios: portfolio of patent application from Tohoku university' (in Japanese), *Technical Report* 147, National Institute of Science and Technology Policy.
Kanama, D. and K. Okuwada (2008), 'A study on university patent portfolios (2): The impact of intellectual property related policies and the change into corporation of national university' (in Japanese). *Technical Report* **154**, National Institute of Science and Technology Policy.
Kneller, R. (2003), 'University-industry cooperation and technology transfer in Japan compared with the US: another reason for Japan's economic malaise?', *University of Pennsylvania Journal of International Economic Law*, **34** (2), 329–449.
MEXT (2007), *Survey on the Research Activity of Private Companies* (in Japanese), Ministry of Education, Culture, Sport, Science and Technology.
Mowery, D.C., R.R. Nelson, B. Sampat and A. Ziedonis (2001), 'The growth of patenting and licensing by US universities: an assessment of the effects of the Bayh-Dole Act of 1980', *Research Policy*, **30** (1), 99–119.
Murakami, Y. (1994), *The Cultural Aspects of Science* (in Japanese), Tokyo: Seidosha.
National Science Board (2008), *Science and Engineering Indicators*, Arlington, VA: National Science Foundation.
Rosenberg, N. and R.R. Nelson (1994), 'American universities and technical advance in industry', *Research Policy*, **23** (3), 323–48.
Sine, W., S. Scott and D. Di Gregorio (2003), 'The halo effect and technology licensing: the influence of institutional prestige on the licensing of university inventions', *Management Science*, **49** (4), 478–96.

Staedter, T. (2003), 'Academic patent binge', *Technology Review*, December/January.
Stephan, P., S. Gurmu, S. Albert and G. Black (2007), 'Who's patenting in the university? Evidence from the survey of doctorate recipients', *Economics of Innovation and New Technology*, **16** (2), 71–99.
Takahashi, M. and Watanabe, T. (2007), *Research and Analysis on Industry-University Collaboration: Activity of IP Management Centers of Universities and Technology Licensing Organizations* (in Japanese), Technical Report, 5th Annual Meeting, Intellectual Property Association of Japan.
UTTA (2007), *University Licensing Survey* (in Japanese), University Technology Transfer Association.
Van Looy, B., J. Callaert and K. Debackere (2006), 'Publication and patent behavior of academic researchers: conflicting, reinforcing or merely co-existing', *Research Policy*, **35** (4), 596–608.
Yonezawa, A. (2007), 'Japanese flagship universities at a crossroads', *Higher Education*, **54**, 483–99.

# 5. University technology commercialization and academic entrepreneurship in Korea: the KAIST experience

**Zong-Tae Bae and Sangmoon Park**

## 5.1 INTRODUCTION

Traditionally academic universities have been regarded as the source of basic research and important suppliers of high-caliber graduates for industry and society. Under these expectations of universities, academic researchers in universities have paid little attention to building strong collaborations with industry and commercializing their technological assets. Currently a group of leading universities in the USA has earned sizable economic returns from their technology commercialization and entrepreneurial activities (Siegel et al. 2007; Etzkowitz et al. 2000). These changing roles of universities are influencing university commercialization programs and the Korean government's policy (KRF 2007).

Among the catching-up countries, Korea has shown very successful economic growth based on science and technology (S&T) activities and research and development (R&D) investment (Lee et al. 1988; Kim 1997). In the era of technological catch-up, a major role of Korean universities has been to appropriately supply educated S&T engineers who can understand advanced technologies and modify existing knowledge into upgraded knowledge (Lee 2005). Since the 1990s Korea has been struggling to sustain its global competitiveness among advanced countries and emerging economies. There are increasing demands for new roles of universities, to generate world-class research output and commercialize their own intellectual property. Under much deregulation of university commercialization and entrepreneurial activities, Korean universities are now much more interested in how to commercialize their technologies and knowledge for economic results. However most Korean universities have little experience in, or means of, generating advanced knowledge for industry and commercializing their technological assets in many ways.

The experience of Korea Advanced Institute of Science and Technology (KAIST), a leading research university in Korea, is particularly unique in terms of the origin of research-oriented S&T universities in Korea, having a well-known nationwide heritage of entrepreneurship, and being active in technology commercialization. The case of KAIST's technology commercialization and entrepreneurship can be of interest to other Asian universities and policy-makers who want to know how they can promote university commercialization and entrepreneurship. This chapter focuses on the historical development and technology commercialization of KAIST. With a brief review of the general status of the Korean national innovation system and government policies, this chapter analyses key characteristics, performance and critical success factors of R&D commercialization and new venture creation. Finally, this chapter draws some lessons and implications from KAIST's case to promote university commercialization and entrepreneurship.

## 5.2 NATIONAL SOCIO-ECONOMIC AND POLICY CONTEXT

### 5.2.1 Socio-economic Development of Korea

The Korean economy has developed dramatically during the last 60 years, from being one of the poorest countries to one of the developing countries in the forefront of becoming a developed country, as shown in Table 5.1.

*Table 5.1 Overview of Korean economy and R&D investment*

| | | 1970 | 1980 | 1990 | 1995 | 2000 | 2005 | 2006 |
|---|---|---|---|---|---|---|---|---|
| GNI per capita (USD) | | 254 | 1645 | 6147 | 11432 | 10841 | 16413 | 18401 |
| GDP (KRW trillion) | | 2.76 | 38.77 | 186.69 | 398.84 | 578.66 | 810.52 | 847.88 |
| Total national R&D expenditure (KRW trillion) | | 0.01 | 0.21 | 3.21 | 9.94 | 13.85 | 24.15 | 27.37 |
| Ratio of R&D to GDP (%) | | 0.36 | 0.54 | 1.72 | 2.49 | 2.39 | 2.98 | 3.23 |
| Ratio of national R&D expenditure (%) | Public institutions | 83.9 | 49.3 | 18.4 | 18.7 | 14.7 | 13.2 | 12.8 |
| | Universities | 3.5 | 12.2 | 7.6 | 8.2 | 11.3 | 9.9 | 10.0 |
| | Private companies | 12.6 | 38.4 | 74.0 | 73.1 | 74.0 | 76.9 | 77.3 |

*Source:* Ministry of Science and Technology, Statistics on National R&D Activities (2007).

The development and growth history of the Korean economy makes it a successful role model of how technologically catching-up countries can leapfrog their economic development and build technological capabilities. The Korean economy has been characterized by strong involvement of the government and aggressive investments of large conglomerate firms known as *Chaebol* groups, such as Samsung and Hyundai. In the growth of Korea's economy, the Korean government and industrial firms have tremendously invested their resources into R&D activities to absorb advanced technologies and build their own technological capabilities. With the exception of the Asian Financial Crisis, the Korean government and key managers of firms have increased their R&D expenditure. Korea's R&D expenditure now exceeds 3 percent of the national GDP level, among the highest levels among Organization of Economic Cooperation and Development (OECD) members.

As the economy has developed over decades, the major technological contributors within the Korean innovation system have changed from government-supported public research institutes to globally competitive private companies. During the early 1960s and 1970s when there were few resources in the industrial sector, public research institutes accounted for most of the national R&D spending and played a key role in technology development through partially substituting for industrial laboratories and helping industrial R&D activities. Since the mid-1980s, when industrial firms competed in the global markets, the private sector has emerged and been positioned as a central player in Korea's innovation system. Korean universities have spent between 5 percent to 10 percent of national R&D expenditure over the years, thus lying between the public research institutes and the private sector.

Recently the Korean economy has encountered fierce global competition from catching-up countries like China and advanced countries with technological innovation. Under these environmental changes, the Korean government is changing its S&T policies from fast catching-up to creativity-oriented innovation. More investments of resources are needed for basic research and new knowledge to explore new innovations with high impacts on global society. Innovations are very risky and uncertain processes, and take a long time to commercialize in markets. New innovations require more advanced scientific knowledge and technologies. So the role of universities in generating innovations is very important for competing with global competitors. These social and economic needs encourage the Korean government to invest more resources to build overall university research capabilities and explore new scientific and technological fields.

### 5.2.2 Government Policy for Technology Commercialization

The growing investment in S&T is grounded on strong government support with long-term strategic planning, and R&D resource allocation aligning with economic development and social needs. Overall Korean S&T policies have been developed and coordinated by the Ministry of Science and Technology, and other government ministries have developed and initiated industry-oriented or technology-specific R&D programs. Most R&D programs are planned with the aim of acquiring and exploring advanced technologies and knowledge by providing competitive research grants to research institutions. Many of these R&D programs focused on administrative management including strategic selection of target technologies, allocation of R&D budgets, evaluation of R&D proposals and final R&D results. There was little concern for disseminating new knowledge from research projects and commercializing technological output.

During the Asian Financial Crisis, the Korean government initiated new policies to promote entrepreneurship in universities and public research institutes. At that time creating new jobs was critical for economic recovery from shocks in the Asian countries. Under these new policies, faculty members and researchers from government research institutes could take leave from their jobs to start new ventures in certain periods. As the information technology (IT) sector was booming it was relatively easy to attract considerable funding from venture capital and investment funds for start-ups. After the burst of the IT bubble, the number of new ventures founded and spun-off from universities and public research institutes reduced significantly.

To promote technology commercialization in universities and government research institutes, the Technology Transfer Promotion Act (Korean Bayh-Dole Act) was enacted in 2000. Under this Act universities and public research organizations can establish a formal organization to deal with technology licenses and intellectual property. Shortly after the enactment of the Act, most of Korea's universities organized new managing teams and established technology licensing offices (TLOs). This law was slightly modified in 2005 and explicitly indicates that the rewards to inventors should be more than half of the technology royalties, one of the key motivators for inventors to transfer and commercialize new technologies. Despite this new law and government support, the performance of technology commercialization is still below expectations. In Korea's new venture ecosystem, it is very difficult to attract angel investors and venture capitalists willing to invest in the early stages of technology development. In addition there is an increasing need to extend new commercialization channels, such as the establishment of holding

*Table 5.2 Patents and technology licensing of Korean universities*

| | | 2003 | 2004 | 2005 | 2006 |
|---|---|---|---|---|---|
| Domestic patents | Total number of applications | 1832 | 2151 | 2841 | 4635 |
| | Total number of grants | 926 | 1087 | 1326 | 2973 |
| International patents | Total number of applications | 398 | 597 | 626 | 731 |
| | Total number of grants | 166 | 155 | 122 | 183 |
| Technology licensing | Total number of licenses | 210 | 243 | 587 | 563 |
| | Total revenue from licensing | 1973 | 3184 | 6323 | 9033 |

*Source:* KRF (2007).

companies and joint ventures with industrial firms. The Technology Transfer Promotion Act was amended and became the Technology Transfer and Commercialization Promotion Act in December 2006. This new law includes most of the policies regarding technology commercialization of universities and public research institutes. As technology transfer and commercialization is inactive, the Korean government supports a group of TLO organizations in building a consortium to overcome limited stock and experience within each TLO organization.

The combination of government policies and intense competition among universities for government research grants has influenced the intellectual property and technology commercialization of Korean universities. The changes in university research activities have resulted in a rapid increase of domestic and international patents and technology licenses with royalties. The total number of Korean universities' applications for domestic patents increased from 1832 in 2003 to 4635 in 2006, and international patent applications almost doubled from 398 to 731 during the same period. The performance of technology licensing of university technologies also reported high growth rates: the licensing number tripled and licensing royalties quadrupled within only four years (Table 5.2).

Despite the rapid increase of technology commercialization, there are large performance differences among universities in Korea. According to a recent survey of 134 Korean universities, 87 universities (64.9 percent of the 134 universities) have more than one domestic patent and 58 universities (43.3 percent) have recorded at least one revenue-generating license with an industry licensee (KRF 2007). This indicates that Korean universities are still under-performing in technology commercialization. Compared with other Korean universities, however, KAIST has recorded successful performance in technology commercialization. Its cumulative revenues from technology licensing amount to more than 80 KRW million, almost 50 percent of total combined revenues of all Korea universities up to 2005. This performance also surpasses technology commercialization in other government research institutes except the Electronics and Telecommunications Research Institute (ETRI), a patent holder of parts of Code Division Multiple Access (CDMA) technologies that are commercialized by Qualcomm and a group of partners in Korea.

## 5.3 FOCAL UNIVERSITY: KAIST

### 5.3.1 Overview of KAIST

KAIST was established under strong support from the government and a long-term plan to supply talented S&T employees to industry which lacked skilled scientists and engineers. There had been increasing demands for highly qualified S&T human resources to accelerate Korea's industrialization since the 1960s. During the early days of industrialization in the 1970s and 1980s, Korean graduate schools were still unable to systematically and effectively educate S&T students. It was difficult to educate talented S&T individuals within existing university systems due to a lack of research capabilities and qualified faculty members. This was a key reason for establishing KAIST, the first research-oriented S&T graduate school in Korea. KAIST has two objectives. One is to nurture highly qualified S&T individuals with profound theoretical knowledge and application capabilities which are critical for the Korean economy. Another is to perform mid- and long-term basic and applied research for domestic development and cultivation of S&T, and to provide research platforms for other research and industrial institutions.

Soon after its establishment in 1971 KAIST substantially contributed to the technological development and economic growth of Korea. The first research-oriented university in Korea has been financially supported by the Ministry of Science and Technology (MOST) under a special enactment for its establishment and other benefits for faculties and students. It is quite different from other Korean universities that mostly focus on undergraduate programs under the Ministry of Education (MOE).

KAIST has been regarded as a deep well-spring of talented S&T individuals and technological entrepreneurs in Korea's economy (Table 5.3). One distinguishing feature is that the number of graduate students, accounting for 59 percent of all enrolled students, significantly exceeds the number of undergraduates. As KAIST was established as a graduate university, the number of graduate students is larger than that of undergraduate students. The latter first enrolled in 1986, 15 years after KAIST's inception. As of 2004, more than 20 percent of all Korean scientists and engineers

*Table 5.3 Profile of KAIST (FY 2006)*

| Category | Indicator | FY 2006 |
|---|---|---|
| People | Total no. academic staff | 745 |
| | Total no. of tenure-track faculty members | 421 |
| | Total no. of undergraduate students | 3021 |
| | Total no. of graduate students | 4315 |
| | Ratio of graduate students (%) | 58.8 |
| Patents | Domestic patents issued | 403 |
| | Cumulative no. of domestic patents issued | 1954 |
| | International patents issued | 186 |
| | Cumulative no. of international patents issued | 604 |
| Licensing | No. of licenses | 39 |
| | Cumulative no. of licenses | 191 |
| Publications | No. of research publications (SCI only) | 1407 |
| | Total no. of research publications | 2346 |
| Research funds | Total R&D expenditure (KRW billion) | 117.6 |
| | Ratio of research sponsored by industry (%) | 13.6 |

with doctoral degrees and 11 percent of S&T faculty members in Korean universities throughout the country were graduates of KAIST. Alumni have been successful in their careers and have ranked in key positions across multiple industries, especially in R&D-related areas. For instance, Samsung Electronics has been hiring numerous KAIST graduates for its R&D functions and businesses; approximately one in five Samsung R&D employees is a KAIST graduate. Furthermore, KAIST is well known as the incubating university for more than 400 new ventures and spin-offs since 1981 when the first new venture was spun-off without the support of any government program. A group of graduates of KAIST have founded new ventures to commercialize technologies and research results developed during their academic programs. Beyond its current performance, KAIST now boosts a new vision of becoming a world-class research university, by challenging more high-risk and high-return problems in global society and attracting top class faculty and talented students.

### 5.3.2 Knowledge Creation and Commercialization at KAIST

#### 5.3.2.1 Mission and key activities

Traditionally accepted roles of universities include education, research and service to society. Education is the most important area for all universities: to teach and train high-caliber personnel and supply qualified individuals to industry and society. Another role is to create new knowledge valuable for humanity and society. University research has created new knowledge to expand human intelligence over unknown fields and developed innovations to transform the world. The other role is to contribute to social welfare by disseminating new knowledge and innovations into society. This includes diverse outreach programs, interactive university-industry collaboration and being a source of new ideas and wisdom (Bae 2008).

More recently another new role for universities to adopt into their critical missions has emerged. With a changing environment, universities are required to play a more active role in creating new high-tech ventures and promoting entrepreneurship. First, universities can educate and cultivate potential entrepreneurs, and train entrepreneurship through diverse entrepreneurship education. Universities can also provide new technologies to be commercialized by students and faculty members through spin-offs and licensing arrangements. In addition universities can develop a wide range of supporting programs and services for new ventures such as business incubators, technology and business consulting, and investment funds with other investors.

Along with changing social needs and government support, KAIST has followed and maintained three initial key missions: education; research; and service to society. The first is the education and training of highly qualified scientists and engineers armed with theoretical and practical expertise. The second mission is the participation in mid- to long-term government research projects, and basic and applied research for the accumulation of Korea's competitiveness in S&T. The last is the provision of research platforms to other research institutes and enterprises. Since its inception, KAIST has sustained these three critical missions and has focused on key activities and outcomes to contribute to the Korean economy and society as shown in Figure 5.1. The organizational structure in KAIST has changed frequently to adapt to changing environments. One of the most recent initiatives in line with becoming a world-class research institute is to engage in more high-risk and high-returns (HRHR) research and interdisciplinary research collaboration over various departments and academic fields.

#### 5.3.2.2 R&D manpower

Research and education at KAIST are based in six colleges (natural science, life science and bioengineering, engineering, information science and technology, cultural science and business) with 27 departments and a number of research institutes and research centers. As of January 2007 there were a total of 7336 students comprising 3021 (42 percent) in the

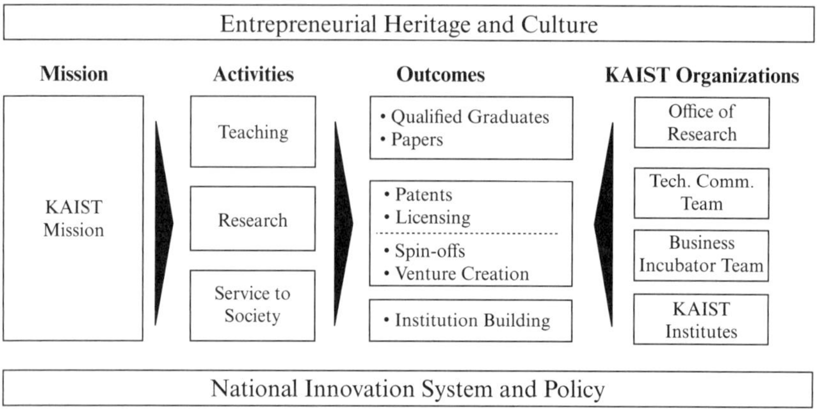


*Figure 5.1 Mission and key activities of KAIST*

bachelor programs, 2089 (29 percent) in the masters programs and 2226 (30 percent) in the doctorate programs. The number of graduate students, about 59 percent of enrolled students, is larger than that of enrolled undergraduate students. This student composition is distinctively different from other comprehensive Korean universities and reconfirms that KAIST is indeed a research-oriented graduate university that specializes in S&T. There are more than 31 000 KAIST alumni scientists and engineers in the Korean economy. Of the alumni, about three-quarters have graduate-level degrees. KAIST has also striven to recruit highly talented faculty members. In 2007 there were 442 faculty members and 323 technical and administrative staff members. Most faculty members earned their doctoral degrees from technologically advanced countries, especially the USA.

#### 5.3.2.3 Publications

There are some characteristics of education and research mechanisms that distinguish KAIST from other Korean universities, such as a flexible academic system, the laboratory as a center of learning and proactive collaboration with industry. KAIST has also utilized unique attributes for its education and research systems. For the undergraduate programs, KAIST has an autonomous and flexible academic system independent from the government. Most Korean universities are required to abide by government-directed admissions and curriculum requirements directed by the MOE which is responsible for higher education programs. The 'open-major system' for KAIST undergraduate programs allows a student to take classes for three terms before choosing a major depending on the student's own interest. In addition undergraduate students are free to change

*Table 5.4 Research publications of KAIST*

| | 1971–95 | 1996–2000 | 2001–05 | 2006 |
|---|---|---|---|---|
| International journals | 8780 | 8161 | 8757 | 1885 |
| (average publications per year) | (246) | (1632) | (1751) | (1885) |
| Domestic journals | 5896 | 3440 | 2678 | 461 |
| (average publications per year) | (366) | (688) | (536) | (461) |
| Total | 14676 | 11601 | 11435 | 2346 |

*Source:* KAIST internal data.

their majors at any time. Most KAIST undergraduate students come from science high schools, elite high schools specializing in educating talented young students. Out of 700 incoming freshmen, more than 75 percent come from these highly competitive high schools. Generally, Korean high school students can take a national college entrance examination in their third and final year. However KAIST provides second-year high school students with an opportunity to take an entrance examination, independent from a nationwide university examination adopted by all Korean universities.

For graduate-level programs, every student has to join a laboratory of an advising faculty where students learn state-of-the-art technologies in a specialized field, experience and participate in diverse research projects and cooperate with other colleagues under the same supervisor. KAIST students take research-oriented classes in which teaching and research activities are simultaneously conducted. KAIST requires doctoral candidates to publish their papers in globally renowned academic journals to ensure high quality of their doctoral dissertations. As shown in Table 5.4, KAIST has more actively published research papers in international journals than in domestic journals.

#### 5.3.2.4 Research expenditure

From its inception KAIST has emphasized close relationships with industry sectors. Most research areas and topics are related to the recent needs of industries. There are many ways to collaborate with the industrial sector, such as contract research, specialized training and education programs, and joint research. KAIST's research grants have gradually increased from 85 billion KRW in 2001 to 117 billion KRW in 2006. More than 85 percent of the research budget is from government sources and the remainder is from industry (Table 5.5). On average, each faculty member participates in about two to three projects per year. Major industry partners include leading global Korean firms in diverse sectors including Samsung Electronics,

*Table 5.5 Research grants received by KAIST*

| | | 2001 | 2002 | 2003 | 2004 | 2005 | 2006 |
|---|---|---|---|---|---|---|---|
| Government | Number of projects | 535 | 601 | 710 | 690 | 728 | 765 |
| | Value of grants (KRW billion) | 69.6 | 76.7 | 79.6 | 86.5 | 89.3 | 101.6 |
| Industry | Number of projects | 398 | 359 | 262 | 262 | 327 | 350 |
| | Value of grants (KRW billion) | 15.9 | 13.7 | 10.5 | 10.5 | 16.9 | 16.0 |
| Total | Number of projects | 933 | 960 | 972 | 952 | 1055 | 1115 |
| | Value of grants (KRW billion) | 85.5 | 90.4 | 90.1 | 97.0 | 106.2 | 117.6 |

*Source:* KAIST internal data.

Hyundai Motors and SK Telecom. Apart from contract research projects, KAIST offers specialized education and training programs for scientists and engineers in industry. Through these non-degree and open programs, researchers have opportunities to interact with S&T individuals from industry and understand the key problems and issues they face.

KAIST has successfully collaborated with various industries and explored research areas to meet social demands. In past years, a number of research projects from industry focused on the development of applied technologies. Applied technology-centered research has stemmed from the mission and objectives of KAIST, aiming to supply advanced technologies needed in Korean industries. In the early days, technologies developed at KAIST were not globally innovative technologies but relatively advanced technologies to solve critical problems faced by industry. Now as a group of leading firms become global players, there is an increasing need for more innovative knowledge to create new industries or more value and to compete with global innovators. In accordance with these changing demands, KAIST is also changing its research directions and topics to explore new areas with a very high level of uncertainty. Research at KAIST focuses on basic science and technological innovation that will have the greatest impact on industrial society. To meet this new social need, eight KAIST Institutes (KIs) have been set up to concentrate the university's research capabilities in specific advanced fields. Each KI is operated as an independent research center at the college level, receiving financial and facility support.

KAIST's education and research achievements have been favorably assessed by external institutions. In domestic assessments of universities, KAIST is the top-ranked university in terms of academic research and education in S&T. International evaluations are also impressive. KAIST was ranked 37th among the world's top 100 technology universities selected by *The Times of London* in 2006. *Asia Week* magazine named KAIST the 'Best Asian Science and Engineering College' for 1999 and 2000. The Accreditation Board for Engineering and Technology (ABET) conducted the first assessment of Korean universities in 1992, and evaluated KAIST's graduate programs as being equivalent to the top 10th percentile of colleges in the USA.

There has recently been intense competition between universities around the world to recruit talented students and faculty and to create new knowledge and technological innovations. Despite remarkable achievements, KAIST is still a long way from becoming a world-class university. It has to change its institutional systems to meet emerging industrial needs for innovations and to recruit prospective researchers from around the world.

#### 5.3.2.5 Institutions and organizations for technology commercialization

Technology commercialization at KAIST can be traced back to the 1980s when there was no formal legal registration or government support. Before 2000, under the Office of Research Affairs, there was only one staff member who dealt with intellectual property and technology licensing activities. An independent team was established in 2000. Many types of technology commercialization activities such as patenting and licensing were placed under the charge of this Technology Commercialization Team, similar to a TLO office in the USA. There are several individuals that handle administrative activities rather than strategic patenting and licensing management. Through trial and error processes in managing intellectual property, the management team gradually learned and accumulated experience and knowledge of technology commercialization. Still, there are currently only a few members who handle all processes in technology commercialization. There are no formal patent attorneys, few experts in business development or individuals experienced in business, and there are no technical experts to assess the value of technologies and negotiate with industry for licensing.

The Technology Business Incubation Team is another organization under the Office of University-Industry Cooperation reporting to the Office of Research. It handles the management of several business incubators on the KAIST campus. KAIST established the first Korean business incubator on its campus in 1994 after a feasibility study on Technology Business Incubator (TBI) establishment. In 1997, under the government's support for business incubators, KAIST's TBI expanded its office space to incubate more entrepreneurs and new ventures and was renamed the High-Tech Venture Center (HTVC), which incubated new start-ups and provided business services for tenants. With a US$10 million donation from KAIST alumni a new Alumni Venture Building was opened, hosting 35 offices for entrepreneurs. In 2002, the HTVC opened another new building with 40 offices and co-located all commercialization teams and activities within the same building.

Licensing processes at KAIST have been formalized and its key processes and activities linked to technology licensing have been institutionalized. To handle the increasing number of inventions and patents, a managing team documented general procedures which benchmarked the accumulated experiences and practices of American universities. As invention disclosures and patent applications are formalized, the managing team's members regularly offer education programs and workshops on how to write invention disclosures and patent applications. Also, there are formal incentive formulas to distribute royalties from technology licensing. 10 percent of all royalties goes to the commercialization team for administrative costs, another 63 percent is rewarded to the inventors directly and the remaining 27 percent is for university headquarters. Up to now, there are still no direct rewards for departments and colleges to which the rewarded researchers are affiliated. All of these intellectual property rights (IPR) management and licensing activities are dealt with by eight staff members in the Technology Commercialization Team, under the Office of University-Industry Cooperation. The staff members think they still lack expertise and experience in negotiating technology transfer and in business development of emerging technologies. Experts in negotiation and technologies could increase the value of inventions disclosed by university researchers and develop more effective claims on their patents. Although external patent attorneys help university researchers in writing patent filings and setting up patent claims, it takes much time and financial resources to file and manage these patents.

Institutional policies to manage intellectual property were established not long after the establishment of KAIST. The general processes and policies for technology commercialization followed these standard procedures. Most of the technological inventions have resulted in patent applications. Among the domestic patent applications, more than 96 percent of applications are successfully granted, compared with about 50 percent of grants for other Korean universities. For international patents, three out of four applications are successfully granted as a patent. The ownership of patents depends on research contracts. For instance, patents funded from government research belong to KAIST when a part of the license revenues are paid back to the funding agency. A group of patents funded from the private sector are filed under co-holders or under the funding firms, depending on the terms of the research contract.

General technology licensing agreements have exclusive and non-exclusive conditions and last ten years from the date of the licensing agreement. Most of the licensing fees consist of an initial payment – a minimum of 5 million KRW (US$5000) – and running royalties, depending on sales volume. Although the total number of licenses and the amount of contracted revenues has gradually increased, realized licensing revenues are still primitive. Of licensed technologies, there have been no realized licensing revenues exceeding more than US$1 million. One reason is the ongoing business practices. Korean industrial companies have traditionally regarded contract research funds as a part of licensing fees. They are reluctant to pay extra royalties for technology licensing. Another explanation for the low revenues from technology licensing is the types of licensed technologies. The applied technologies from KAIST contribute only limited value and revenues to existing businesses. Successful universities in the USA earn most of their licensing revenues from a small number of transformational inventions that created totally new markets and contributed to the development of highly profitable products, mostly in the biotechnology and pharmaceutical sectors. However, applied and 'follower' technological development can only earn limited value in these industries. This is one of the reasons why KAIST emphasizes development of exceptionally new innovations with a high risk and with high impact on industry and society.

To accelerate technology licensing KAIST changed its patent management systems and technology licensing mechanism. It developed new information systems to reduce administrative procedures and managerial costs that resulted from complex processes which also delayed patent filing. Another change is related to the technology licensing mechanism. Although only 4 percent of granted patents are transferred to industry, 70 percent of technologies are licensed to industry before patents have been filed for examination. This experience of technology licensing gives KAIST another opportunity to increase its licensing revenues from inventions.

## 5.4 UNIVERSITY COMMERCIALIZATION AT KAIST

### 5.4.1 General Process of Commercialization

Technology commercialization has been one of the most important issues at KAIST. To support the industrialization of Korea's economy, which

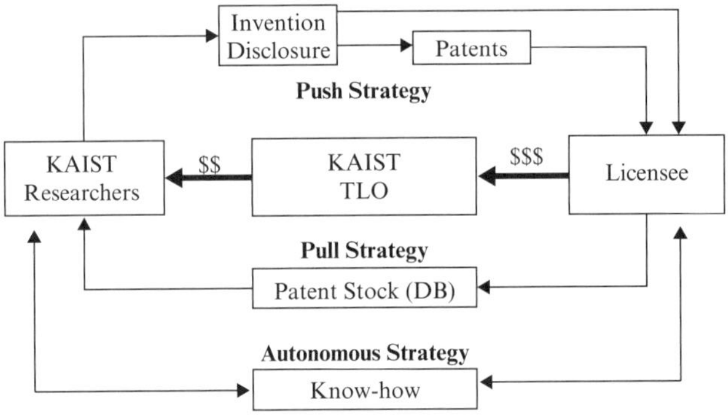


*Figure 5.2 Different mechanisms of technology transfer*

had a low level of technological capabilities, KAIST focused on the development of applied technologies for transfer to industry and commercialization. The performance of technology commercialization can be assessed by intellectual property and technology licenses. Among Korean universities, KAIST has shown the best performance record in technological commercialization.

When researchers invent new technology or knowledge, they must document their results and submit a formal invention disclosure with a view to issuing a patent and protecting this knowledge. With the inventor and other external experts, staff in the TLO office review the invention disclosure and assess whether this new technology is valuable enough to warrant applying for a new patent. For valuable inventions, inventors interact with external patent attorneys to write and claim patents and fill out relevant documentation. Some inventions are transferred to licensees before patenting. Other types of new knowledge can be diffused through diverse channels, including invited promotion and conferences, and conducting negotiations with potential licensees. After arranging a licensing agreement and receiving royalties, inventors can be rewarded with royalties, depending on their contribution to the invention.

There are three different ways to transfer technologies and knowledge developed at KAIST as shown in Figure 5.2. The first is through a normal technology transfer, called the 'Push Strategy'. When researchers at KAIST disclose their inventions and apply for patents, the TLO formally circulates this information and patent lists to potential licensees interested in emerging technologies. Potential licensees can then negotiate with the TLO and researchers regarding the details and conditions of the technology transfer. Another strategy is the 'Pull Strategy' in which potential licensees first look for their needed technologies and patents in KAIST's

*Table 5.6 Patents of KAIST*

| | | Until 2001 | 2002 | 2003 | 2004 | 2005 | 2006 | Total |
|---|---|---|---|---|---|---|---|---|
| Patent applications | Domestic | 1432 | 201 | 235 | 274 | 307 | 372 | 2815 |
| | Foreign | 741 | 130 | 86 | 149 | 115 | 160 | 1399 |
| | Sum | 2173 | 331 | 321 | 423 | 422 | 532 | 4214 |
| Patent grants | Domestic | 785 | 205 | 187 | 178 | 231 | 360 | 1945 |
| | Foreign | 206 | 56 | 101 | 74 | 68 | 73 | 578 |
| | Sum | 991 | 261 | 288 | 252 | 299 | 433 | 2533 |
| Number of invention disclosures | | NA | NA | NA | 414 | 436 | 576 | NA |

*Source:* KAIST internal data.

patent database and technology lists, and then negotiate with the TLO. In this case the licensees initiate the technology transfer processes. The final and most frequent method of technology transfer is the 'Autonomous Strategy'. Before documenting the detailed conditions of patents, both researchers and licensees interact with each other without the involvement of the TLO. In this case, the researcher and licensee know each other on a personal level and communicate their individual needs as well as the potential of the developed technology. Depending on the situation, there are many channels for informing potential licensees of new technologies, such as recruiting graduates, close relationships with alumni, academic conferences and public information on new inventions.

### 5.4.2 Patents

KAIST has emphasized the owning of regal rights and a technology portfolio of intellectual property and patents. Every year KAIST applies for domestic and international patents for its technologies. As of 2006 it held a total of 4214 patent applications and 2533 granted patents. It is the largest patent holder among Korean universities. As a result of continuous efforts in legalization on technologies, the number of patents of KAIST has grown. The patent applications and grants are shown in Table 5.6. The number of patent applications has gradually increased from a total of 331 patents in 2002 to 532 patents in 2006. Also, the total number of granted patents almost doubled from 261 patents in 2002 to 433 in 2006.

Of KAIST's international patents, about 60 percent were issued in the USA. There are 14 other patent-issuing countries including Japan, Germany and China. As the USA has the largest and most competitive market, KAIST researchers target US patents for their primary key technologies and then extend their applications into other countries. Technologically advanced countries such as the USA, Japan and Germany account for over 80 percent of all patents granted to KAIST. This shows that the patents of KAIST are valuable and technologically advanced. Recently the share of patents in China has been gradually increasing, indicating the emerging importance of China.

Patenting activity varies by departments and academic fields. Electronics and related departments are the most active in international and domestic patenting. Other top patenting departments are the mechanical engineering, biochemical engineering and material engineering departments. The electric and electronics department and mechanical engineering department have generated almost half of KAIST's patents. Relatively few patents have been granted in the natural sciences such as physics, chemistry and biology. This is correlative to the main knowledge and research topics at KAIST, which focuses on applied research rather than on basic sciences. This research orientation is strongly related to the general technology development trajectory of the whole Korean innovation system, from applied technology to basic research. It also explains why the amount of royalties from technology licenses is relatively low despite the large number of technology licenses.

### 5.4.3 Licensing

Technology licensing has been regarded as a key indicator in evaluating the performance of technology transfer. Of the total 191 technology licenses up to 2006, more than 100 licenses were contracted within just the last five years. The total contracted and expected technology license revenue amounts to more than 10 billion KRW, or about US$10 million. The performance of technology licensing has gradually improved in terms of the number of licenses and total revenues from licenses as shown in Table 5.7.

With regards to technology licensing at KAIST, the top-performing departments and academic fields include electric and electronics engineering, biotechnology engineering and mechanical engineering. In contrast to patents, licensing agreements come from many departments. For instance, biotechnology and computer science are generating a certain amount of licensing income although they only produce a small proportion of KAIST patents. One characteristic of technology licensing is the manner in which licensing arrangements are made and the appropriation of technologies. In the early 2000s most licensing was concerned with arrangements to utilize formal intellectual properties like patents. However, new technologies have recently been transferred to industry before inventors write formal documentation and appropriate the new knowledge as patents and legal

*Table 5.7 Performance in technology licensing*

| | Until 2001 | 2002 | 2003 | 2004 | 2005 | 2006 | Total |
|---|---|---|---|---|---|---|---|
| Total number of technology licenses | 82 | 23 | 10 | 15 | 22 | 39 | 191 |
| Total contracted amount for licenses (KRW million) | 5750 | 565 | 396 | 541 | 1232 | 2429 | 10913 |
| Total revenue from licenses (KRW million) | 2595 | 542 | 363 | 418 | 722 | 856 | 5496 |

*Source:* KAIST internal data.

property. Furthermore this transfer of unprotected knowledge, better known as 'know-how', is first arranged by the inventor – usually a faculty member – and the licensee; only then is formal documentation made through TLO offices. This represents the 'Autonomous Strategy' in Figure 5.2. In these cases the faculty member and industrial licensee have a close relationship, and interact with each other regarding technologies of interest and research progress before technology licensing and commercialization takes place. These informal mechanisms sometimes limit the opportunity to maximize value from emerging technologies by diversifying potential licensees and negotiating with them. There are big challenges concerning how to appropriate and claim new knowledge, and how to negotiate over emerging technologies to maximize the potential value of new technologies.

Of the faculty members, less than 20 percent have experience in technology licensing. The number of faculty members involved in licensing recently increased from 46 to 62 in four consecutive years. Despite this increase most faculty members have no experience in technology licensing and commercialization. This means that KAIST faculty members are still far from industry standards and have weak interest in technology commercialization.

## 5.5 ACADEMIC ENTREPRENEURSHIP AND NEW VENTURE CREATION AT KAIST

### 5.5.1 The Effect of KAIST on New Venture Creation in Korea

KAIST has been a major source of high-tech start-ups. Since the early 1980s more than 360 new ventures were spun-off from KAIST. It is the leading university in terms of university spin-offs, and this phenomenon corresponds to KAIST's culture which emphasizes practical application of knowledge.

The impact of start-ups from KAIST was crucial in fostering entrepreneurship and promoting new venture creation in Korea. KAIST alumni founders and chief executive officers (CEOs) organized KAISTCO in 1995 as a social community of CEOs who had founded start-up companies. KAISTCO became the cornerstone for establishing the Korea Venture Business Association (KOVA), which has played a pivotal role in policy-making and networking. It prepared many policy measures on issues such as an R&D support system, venture capital, capital market and taxes, and submitted them to the government. To overcome the Asian Financial Crisis in 1998, the Korean government accepted the majority of KOVA's recommendations to promote entrepreneurship and to create new jobs throughout the country.

The entrepreneurial culture prevails over the KAIST campus through diverse informal activities and some classes on entrepreneurship. On KAIST's campus, undergraduate students who were interested in new venture creation formed an student association called the KAIST Business (KB) Club. The KB Club became a cornerstone in its own right and created the Korea Venture Business Center (KVC), a nationwide student association composed of undergraduate students. The organization contributed to the creation of the venture boom in Korea. In terms of formal programs, KAIST opens some classes on entrepreneurship for invited successful alumni founders and CEOs to share their experience in seizing business opportunities and creating new ventures.

### 5.5.2 Entrepreneurs and Start-ups from KAIST

KAIST's entrepreneurial culture has been inherited from the early days of its history and many researchers have been interested in commercializing technologies and knowledge. The entrepreneurial history and entrepreneurship development of Korea is strongly linked to entrepreneurship and spin-offs of KAIST faculty members and researchers. Many entrepreneurs come from faculty and students, including undergraduates and graduates. A group of alumni established their ventures after accumulating several years of experience in their industries of interest.

#### 5.5.2.1 Faculty founders

Spin-offs by faculty were formally institutionalized in 1999 when a formal institutional policy was set and allowed faculty three years of leave. At that time there was a favorable business environment, including IT bubble periods

*Table 5.8 Faculty spin-offs from KAIST*

| | 1999 | 2000 | 2001 | 2002 | 2003 | 2004 | 2005 | 2006 | Total |
|---|---|---|---|---|---|---|---|---|---|
| Number of spin-offs | 3 (7.5%) | 23 (57.5%) | 5 (12.5%) | 3 (7.5%) | 0 (0%) | 1 (2.5%) | 3 (7.5%) | 2 (5.0%) | 40 (100%) |

*Source:* KAIST internal data.

and huge venture capital investments. This environment allowed new ventures to attract talented workers, angel and venture capital investment, and government research grants. Most faculty spin-offs were established during IT bubble periods just before 2001. After the burst of dotcom companies and the collapse of stock markets, there were few new ventures from faculty members. Faculty spin-offs are branched off from many departments, from engineering schools to business schools. The top three areas of faculty spin-offs are electric and electronics (seven spin-offs), computer science and software (six spin-offs) and mechanical engineering (four spin-offs). Other departments include chemical engineering and business (three spin-offs) and material and physics (two spin-offs). Most KAIST faculty founders prefer to maintain their faculty position and remain in the university when their permitted years of leave end. Instead of leaving universities, faculty members can both participate in their new ventures and conduct research, serving the dual purposes of academia and commercialization (Table 5.8).

#### 5.5.2.2 Start-ups by students and alumni

There are over 700 start-ups that were founded by KAIST students and alumni. Start-ups from KAIST are rooted in diverse departments, ranging from electrical and electronics through to the business school to natural sciences, like mathematics. This distribution of start-ups indicates that an entrepreneurial culture prevails over the departments. Of the student and alumni founders, 50 percent hold masters degrees and 36 percent hold doctoral degrees from KAIST. More than 20 start-ups from KAIST went public (Initial Public Offering, IPO) and became very successful. KAIST's new ventures have continuously created new industries and business models (Park and Bae 2004). Early new ventures were founded by students and alumni to commercialize the results of their research projects, such as substitution of imported goods. In the late 1990s and early 2000s a group of new ventures planned internet-related business opportunities such as internal portals, online games and social networking services. They pioneered emerging businesses including digital video recording and biotechnologies.

The most important effect of new ventures from KAIST is their social impact on entrepreneurship in Korea. For example, a group of leading entrepreneurs who were KAIST graduates formed an informal community to share information and create social networking. This group played a pivotal role in setting up new venture policies during the Asian Financial Crisis – the group evolved into the core of a new venture association to represent understanding and opinions of new ventures in Korea, and contributed to the enactment of a new venture promotion law. An entrepreneurial history and environment like that of KAIST can scarcely be found at other leading Korean universities. This entrepreneurial culture and heritage distinguish KAIST from other universities and help to build KAIST's prestige in technology commercialization.

### 5.5.3 Technology Business Incubator at KAIST High-tech Center

The Korean government is providing several policy measures to support high-tech venture firms. Among the policies on promoting entrepreneurship, business incubation has been one of the most important policy measures to facilitate widespread entrepreneurship across regions and other universities. MOST has been supporting KAIST in operating the TBI, providing technological and financial support. KAIST TBI was the first university-based business incubator in Korea. The KAIST HTVC, which was started in 1994 with the support of MOST, is the first and largest business incubator in Korea. HTVC hosts many high-tech start-up companies. Benefits for BI tenant companies are as follows: (i) cheaper rent for the office; (ii) business facilities, telecommunications and internet access; (iii) collaboration program with KAIST professors and researchers; (iv) special programs for financing, marketing and IPR management; and (v) networking among tenants. Since the first opening of KAIST's BIs, a total of 311 tenant companies have been incubated in the KAIST BI and 224 have graduated from their incubators. Seven of the graduated companies went public in the KOSDAQ markets.

## 5.6 CONCLUSIONS

### 5.6.1 Key Factors and Challenges in Technology Commercialization at KAIST

KAIST has contributed to providing highly qualified S&T personnel and developing applied technologies needed in the industrialization processes of the Korean economy. There are some key factors in technology commercialization at KAIST. First, strong and consistent support from the Korean government has played a critical role in developing a research-oriented university specializing in S&T. To meet societal needs for more skillful employees, and following careful recommendations of international experts, Korea's government established a new university having unique institutional structures independent from existing institutions and authority. The strategic choice to build a new university beyond the existing university system made it possible to establish a competent university with a limited financial budget and time constraints. This new system of university operation can operate more flexibly, enabling the university to recruit prospective students and focus on technological development in applied areas, which in turn leads to development of applied technologies.

Second, the research focus on applied technology and close interaction with industry is important. From its inception KAIST has focused on the needs of industry and extended its research areas into emerging fields. In the early 1970s and 1980s, when the Korean economy developed rapidly, the most demanding fields were applied engineering sectors such as electronics, mechanical and chemical engineering. Now the Korean economy demands more innovation in these areas and new emerging fields such as biotechnology. Technological development at KAIST was quite different from universities in the USA, where the focus was on basic science and inventions to create new industries. Close interaction with industry has helped KAIST understand the technological needs of industrial firms and develop applied technologies suitable for them. The interaction was tightened through an increasing number of research projects and diverse training programs. These relationships rapidly stabilized the position of KAIST within society, providing more opportunities for designing the academic curriculum and conducting research in areas of demand.

Third, there is a balanced approach between academic studies and industrial projects for university researchers to integrate two other types of innovation. Applied technology-centered research and interactions with private companies have provided faculty and students opportunities to understand what technologies are beneficial for industry and how to commercialize technologies within the private sector. University researchers have intensively experienced and learned how to write research proposals to attract industry sponsors, manage research projects in a timely manner and transform developed technologies into commercial products and services. From a number of research projects, university researchers can understand specific technological demands of private companies. This approach has also provided university researchers business opportunities to commercialize their own technologies. This entrepreneurial environment within KAIST has resulted in the creation of more than 400 spin-offs during the last 37 years. The experiences gained while studying provide important knowledge and insight for alumni and students to start their own ventures and develop more attractive technologies.

Fourth, faculty members play an important role in commercializing technologies. Each faculty member runs their research laboratory with a group of students. This laboratory is the center of technological development and academic research, like the technological project team or new product development team within private companies. Depending on the interests of faculty, research and operational directions of the laboratories can be very different from each other. For example, a laboratory under a faculty member interested in basic science conducts more research in basic sciences. Others directed by faculty members interested in commercialization give more attention to filing patents and interacting frequently with industry. The role of a faculty member at KAIST is very similar to that of a product champion in private companies. University faculty members impact the entrepreneurial environment of the laboratory, research topics and technological development undertaken with students, and obtaining research funds from external sponsors. These different interests of faculty members have resulted in disproportional commercialization performance among faculty members and research laboratories.

Fifth, research laboratories over the campus are central powerhouses of technology commercialization. All research and technological development is conducted within these laboratories. Under a faculty leader, members of a laboratory gain advanced knowledge in their fields and participate in diverse research projects. The learning mechanisms within a laboratory resemble a kind of apprenticeship, transferring knowledge from faculty through senior members down to junior members. Laboratory members have access to up-to-date knowledge in their specific fields and accumulate experience within the laboratory through diverse learning processes such as frequent seminars, collaboration on research projects and academic publications. They also build life-long human networks that will enable them to learn advanced knowledge and collaborate with others beyond KAIST. The organization environment and characteristics of research laboratories have a profound impact on technology commercialization. Teams of students and alumni from the same laboratory have founded spin-offs or new ventures to commercialize their technologies.

Sixth, KAIST has invested in building an entrepreneurial infrastructure over the campus. KAIST is the most successful Korean university to develop an interest in technology commercialization and an entrepreneurial environment. Although it has commercialized university technologies since its beginning, the level of commercialization was very primitive compared to the leading universities in the USA. Beginning in

1994 KAIST built technological BIs to nurture promising new ventures, with a goal of boosting entrepreneurship and technology commercialization. The consecutive legislative policies to promote technology commercialization in public research institutions changed shortly after the Asian Financial Crisis in 1998. Before these institutional changes KAIST was a dominant role model of how universities can proactively commit to entrepreneurship and commercialization. Also, business plan competitions and regular classes on entrepreneurship and technology commercialization have helped university researchers to understand the importance of business beyond technology development itself. KAIST offers many education programs for researchers interested in founding new ventures and executives of existing new ventures.

Despite its substantial contribution to commercialization, there are many challenges that KAIST is encountering in its efforts to become a world-class university with a large impact on global society. The growth and development of KAIST has been successfully aligned with social and industrial needs for a stable and qualified S&T workforce. Currently, Korea is struggling to change its economy from being a fast-growing catching-up country to become one of the most advanced countries. This change requires fundamentally different capabilities needed to generate more creative and innovative technology and knowledge. To meet these new social demands, KAIST has to build its research capabilities to the level of world-class universities. It should invest more resources into creativity and innovation rather than applied technologies. The new leadership of KAIST understands the current environmental changes and is pursuing new directions that will move KAIST towards being a world-class university.

Another challenge is the acceleration of technology commercialization. KAIST's previous performance in technology commercialization is remarkable when compared with other Korean universities. However there are still limited realized revenues from licensing and commercialization. In addition to the technology itself, the experience and expertise of the TLO is very important in increasing the value of invented technologies. More expertise and specialized skill sets in the TLO are required to maximize the value of technologies. KAIST should invest more resources and hire talented individuals to manage its intellectual property competently and negotiate with prospective licensees around the world.

The other challenge is the diffusion of an entrepreneurial environment beyond KAIST's boundaries. KAIST is located in a geographically concentrated region with numerous government-supported research institutes. Even though each research institute tries to promote entrepreneurship and technological commercialization within its organization, there are few efforts to build an entrepreneurial habitat to nurture technology commercialization and entrepreneurship beyond the walls of these institutes. KAIST should contribute to other research institutes and the regional economy by actively diffusing its entrepreneurial spirit.

### 5.6.2 Comparison Between Korea and the USA in University Technology Commercialization

There are some similarities and differences between the technology commercialization models of KAIST and those of US universities. Major differences are rooted in several factors such as socio-economic differences, institutional factors (including legislation and policies on commercialization), and organizational structure and characteristics of universities. In this section we briefly compare key differences between the KAIST system and US models of commercialization. First, there are different social expectations and economic roles of universities in the process of technological innovation. Traditionally universities have been regarded as a source of scientific knowledge and academic research. This tradition resulted in universities focusing on basic science to find natural laws, conducting theoretical research without any consideration of commercialization. It takes a long time to diffuse scientific findings to private companies and to transform theoretical findings into applications for real life. It is only in very recent years that universities have given their attention to commercializing their scientific knowledge and transferring innovative inventions to the industrial sector. Some leading universities such as MIT, Stanford, the University of California at San Diego and Columbia University are very successful in commercializing their technologies through licensing, spin-offs and joint research. They have played a major role in generating new innovations and theoretical developments which have had a large impact on global society. However KAIST was established to meet the social demand for a continuous supply of S&T personnel and for applied technologies. These were needed in the Korean economy, as a catching-up country with a relatively low level of technological capabilities. Throughout the development of the Korean economy, KAIST has actively interacted with industry by supporting a part of the R&D function of private firms, along with other government-supported research institutes. Having different expectations of universities within their respective social and economic environments, KAIST and US universities have different development histories and policies regarding research and technology commercialization management.

Institutional differences also affect the commercialization of university technology. Resources and government research grants for universities in the USA are plentiful. Competition among universities to obtain these research funds has resulted in an increase of scientific inventions and patent filing. Legislative and policy changes on the ownership of intellectual property (for example, the Bayh-Dole Act) ignited the commercialization of technology. Despite the debate on the Bayh-Dole Act's effect on technology commercialization (Mowery et al. 2004), it is clear that the institutional characteristics of universities and their research policies helped US universities to position themselves as global engines of technological innovation. In Korea there are policies and legislation to promote technology transfer from public research institutions. However these have not satisfactorily met the expectations of governments or industry. Even before these institutional changes KAIST had been successful in technology commercialization, and it has enhanced its commercialization mechanisms through entrepreneurial spin-offs. Another institutional difference between Korean and US universities is in IPR. In the US system, the actual date of the first invention (regardless of date of patent application) is considered more important for granting a new patent. This leads to inventors developing new and innovative items without spending their time and effort on filing documents for these patents. In the Korean system a patent is granted to the first applicant of the invention, so Korean inventors need to file their inventions as soon as possible. This difference has a significant influence on patenting strategy and protection of developed technologies. The quality of a patent is determined not only by the novelty of an innovation but also by effective claims against infringement.

### 5.6.3 Recommendations for Promoting University Technology Commercialization

To promote university technology commercialization, universities should make efforts in the following ways and prepare action plans. First, universities should clarify their mission of university-industry collaboration. Some universities can focus on applied technology and close interaction with industries. Second, a balanced approach to education and industrial R&D projects is needed. Universities should have a balance between their three basic functions of education, research and service/outreach. Third, championing roles of faculty members are critical in promoting university R&D commercialization. Entrepreneurial professors and students can lead commercialization efforts and create an entrepreneurial culture. Fourth, an entrepreneurial laboratory culture is a reservoir of R&D commercialization. Students can learn the system for commercialization and gain first-hand experience through hands-on research laboratories. Some laboratories produce many start-ups. Fifth, the government should provide strong and consistent support to promote commercialization. Legal infrastructure is very important. For example, the Korean government enacted the Korean version of the Bayh-Dole Act and adopted a new system permitting the establishment of R&D institution-based companies and technology holding companies. Sixth, the entrepreneurial environment and institutionalization as manifested in science parks, business incubators and technology licensing offices can accelerate R&D commercialization and technology transfer (Son and Kenney 2007). Finally, a significant role for the private sector and government R&D institutes is also needed for successful commercialization. In particular universities should evaluate the absorptive capacity of the private sector and select appropriate partners for technology commercialization.

## REFERENCES

Bae, Z.-T. (2008), 'Technological entrepreneurship and roles of universities in Korea', Paper presented at the International Symposium on Strategy for Future Development of Venture Ecosystem in Northeast Asian Countries. Seoul, Korea, [October.]

Etzkowitz, H., A. Webster, C. Gebhardt and B.R.C Terra (2000), 'The future of the university and the university of the future: evolution of ivory tower to entrepreneurial paradigm', *Research Policy*, **29** (2), 313–30.

Kim, L. (1997), *From Imitation to Innovation*, Cambridge, MA: Harvard Business School Press.

KRF (2007), *A White Paper on University Industry Cooperation in 2006*, Seoul, Korea: Korea Research Foundation.

Lee, K. (2005), 'Making a technological catch-up: barriers and opportunities', *Asian Journal of Technology Innovation*, **23** (2), 97–131.

Lee, J., Z.-T. Bae and D.-K. Choi (1988), 'Technology development processes: a model for a developing country with a global perspective', *R&D Management*, **18** (3), 235–50.

Ministry of Science and Technology (2007), *Statistics on R&D Activities in Korea,* Gwacheon, Korea: Ministry of Science and Technology.

Mowery, D., R. Nelson, B. Sampat and A. Ziedonis (2004), *Ivory Tower and Industrial Innovation: University-Industry Technology Transfer Before and After the Bayh-Dole Act*, Stanford, CA: Stanford University Press.

Park, S. and Z.-T. Bae (2004), 'New venture strategies in a developing country: identifying a typology and examining growth patterns through case studies', *Journal of Business Venturing*, **19** (1), 81–106.

Siegel, D., M. Wright and A. Lockett (2007), 'The rise of entrepreneurial activity at universities: organizational and societal implication', *Industrial and Corporate Change*, **16** (4), 489–504.

Son, D. and M. Kenney (2007), 'Universities, clusters, and innovation systems: the case of Seoul, Korea', *World Development*, **35** (6), 991–1004.

# 6. The Hong Kong University of Science and Technology: a case study in entrepreneurial university-led knowledge-based economic development

**Naubahar Sharif and Erik Baark**

## 6.1 INTRODUCTION

In this chapter we explore a new role for the university, one in which it contributes directly to the economic well being of the society in which it operates, contributing social and economic value not only indirectly, through training and knowledge generation, but also more directly by creating new avenues into the private sector. Here, in partnership with business interests, the university supports the application and exploitation of knowledge by lending its capabilities (including technology) to the creation of new market-ready products and services.

Hong Kong, where the higher education sector adapted in the 1990s to the former colony's new status as a Special Administrative Region of China, has witnessed its own version of this transformation. Hong Kong's higher education sector accounted for 80 percent of its total research and development (R&D) expenditure in 2000, indicating the central importance of universities in Hong Kong's innovation system (Hong Kong Census and Statistics Department 2007b). Even the 2003 figure of 56 percent is high compared with R&D spending in other advanced economies, where the figure is roughly 17 percent.

In common with other newly industrializing economies (NIEs) in Asia – notably Singapore – Hong Kong shows signs of developing a knowledge-based strategy for economic growth. Ever since the former Chief Executive, Tung Chee-Hwa, spelled out this strategy in 1998 and 1999 policy addresses, policy makers have been trying to effect a transition in Hong Kong's economy, replacing a one-dimensional service-based economy with one that relies on innovation as a critical economic driver.

As the economy depends more and more on the innovation system to fuel growth, the commercialization of intellectual capital to create value and jobs for Hong Kong industries must play a key role.

In this chapter we offer a case study that assesses how the Hong Kong University of Science and Technology (HKUST), the youngest research-based university in Hong Kong, has adapted to this new role against the backdrop of the rapidly globalizing knowledge economy.

## 6.2 CONCEPTUAL FRAMEWORK

The trends in university orientation that we have observed in Hong Kong have assumed global dimensions. Etzkowitz et al. (2000) argue that there is a worldwide movement among universities to adopt a more complex entrepreneurial model, one that stresses the commercialization of knowledge and the fueling of private enterprise in local and regional economies. Referring to the widely held belief that university research feeds economic growth like the magic seed from 'Jack's beanstalk', Miner et al. (2000) argue that entrepreneurial firms also drive economic growth like magic seeds. If universities are to play the role we have described as incubators of entrepreneurial resources in a vigorous and thriving innovation system, they will need to operate increasingly within the Triple Helix model of innovation-driven economic growth, in which they work closely in complex, interactive relationships with government agencies and private enterprise.

Wong et al. (2007) offer three reasons that the imperative requiring universities to shift from their traditional model to the new entrepreneurial model is expected to be even more urgent in a global market that includes such Asian NIEs as Singapore. We note, however, that only one of the reasons they offer applies to the Hong Kong situation, where they argue that

> the shift toward a knowledge-based economy, rather than one based on low-wage and low factor cost advantages, requires a significant increase in the indigenous capabilities of local enterprises to create and commercialize new knowledge, not just using knowledge imported from advanced countries. However, many of the local private enterprises that had developed in the earlier industrialization phases still tend to be laggards, rather than leaders, in engaging in technology development and innovation activities. Consequently, compared to their more technologically mature counterparts in the advanced economies, local industries in NIEs often have less experience, and lower capability, to commercialize knowledge generated from local universities. (Wong et al. 2007, p. 942)

As a result of the Asian Financial Crisis of 1997, economic reforms in mainland China and China's admission into the World Trade

Organization, Hong Kong's heavy reliance on its traditional capacity as a trading center can no longer sustain its economy. Long accustomed to relatively stable levels of robust growth, Hong Kong found itself in the late 1990s facing quarter upon quarter of negative growth, deflation and mushrooming unemployment. These developments spurred the Hong Kong government to participate actively, even urgently, in the transformation of Hong Kong's economic and innovation system so that it could create a new role for itself as well as a new engine for economic growth. In his inaugural policy address to the public on 8 October 1997 Tung said, 'My aim is to make Hong Kong an innovation centre. We . . . need to do more to stimulate the exchange of ideas between our university researchers, our businessmen and industrialists, and our customers, so as to drive forward innovation and turn technological development into commercial products' (Tung 1997).

Yet Hong Kong has been poorly positioned to leverage Triple Helix-style interactions in modernizing its innovation system. Private enterprises in Hong Kong lack the capacity for and consequently have hardly demanded rapid or extensive commercialization of university knowledge. The changing economic landscape in Hong Kong, which has been and continues to be reshaped by rapid growth in mainland China, has pressured industry and the government on several levels, making it difficult to harness Triple Helix interactions. A recent paper by Mok (2005) utilized the Triple Helix concept to examine policy changes related to entrepreneurship and university spin-off firms in Hong Kong. Mok observes increasing movement on the part of government in facilitating entrepreneurial linkages between universities and industry. This is encouraging because if Hong Kong's universities can play their part in a Triple Helix scenario they can serve as nodes in regional, national and even global systems. Global links with other major centers of learning can only enhance their capacity to generate and distribute innovative ideas and products.

These possibilities add to the urgency with which universities in Hong Kong must take on an entrepreneurial role, perhaps more so than universities in advanced economies, if they are to overcome the less favorable preconditions with which they contend and also optimize the potential advantage of increasing integration with mainland China. Universities in Hong Kong may, for instance, need to be more proactive in commercializing their inventions through spin-offs rather than relying on private enterprises to license them because of the latter's weakness in this respect.

We also believe that Hong Kong's universities should expand the focus on entrepreneurial culture beyond the role of technology commercialization emphasized by Etzkowitz et al. (2000) and Wong et al. (2007). The 'entrepreneurial university' model must take seed in sweeping curriculum reform, injecting university teaching with a stronger entrepreneurial dimension. In other words, the university's core function of teaching must be enlisted across multiple subject areas in the effort to reorient academic culture.

A 2004 study of entrepreneurship in Hong Kong reflects the reluctance of Hong Kong's relatively well-educated population to embrace entrepreneurism. New firms rely primarily on existing technology to avoid the risk of technology innovation (GEM 2004). Defined as the percentage of the population that is engaged in starting new companies, entrepreneurship in Hong Kong was found to be a meager 3 percent compared with a more 'normal' level of about 8 percent in emerging and developed economies. That figure applies to all new companies, not just entrepreneurs engaged in high technology (GEM 2004, p. 1). The latest report on entrepreneurship in Hong Kong (GEM 2007) indicates some increase in the propensity to engage in early-stage entrepreneurial ventures, with levels of engagement in new start-ups that are similar to those found in the USA. The GEM report nevertheless emphasizes that the weakness in R&D commitment and a relative lack of creativity in Hong Kong's educational system disadvantages local entrepreneurs. Greater China's increasingly competitive environment, in which there are fewer traditional jobs or opportunities to profit from speculative activities, requires universities in Hong Kong to urgently foster a more entrepreneurial mindset in students. Students must learn not only how to use new technologies but more importantly how to generate them for immediate commercialization.

## 6.3 OVERVIEW OF THE TERTIARY EDUCATION SECTOR IN HONG KONG, THE HKUST AND THE SHIFT TOWARD AN ENTREPRENEURIAL UNIVERSITY MODEL

### 6.3.1 Background to the Tertiary Education Sector in Hong Kong

The Education Bureau (EDB) is responsible for government policies pertaining to tertiary education, employment and human resource development. In this capacity it also oversees academic research policy, technical education and industrial training. The EDB determines and applies the government's funding for university education and research with the assistance of the University Grants Committee (UGC) and the Research Grants Council (RGC). The UGC is an advisory committee responsible to the government for the development and funding needs of higher education institutions. The RGC advises the government through the UGC on the research needs of such institutions.

Universities provide the education and research that lays the foundation for the science and technology base of Hong Kong's economy. Presently, by feeding the human capital base with technologically savvy personnel and conducting academic and contract research, universities constitute the single most important element in Hong Kong's innovation system. As Table 6.1 shows, over the period spanning 1995–2001 higher education R&D (HERD) constituted well over half of total R&D expenditures, focusing mainly on basic research while also conducting some applied research. Most of the R&D expenditure in the higher education sector from 1998 to 2002 was devoted to the physical sciences (between 23 percent and 25 percent), engineering and technology (between 22 percent and 23 percent), and medicine, dentistry and health (between 18 percent and 19 percent), with the arts and humanities and social sciences each showing a 10 percent share.

By 2007, however, business expenditure on R&D almost equalled that of higher education, so the proportion of R&D expenditure constituted by the higher education sector had fallen to 49 percent. Government R&D had in addition doubled over the 1995–2007 period and several public technology support organizations, such as the Science Park, Cyberport and the Applied Science and Technology Research Institute (ASTRI) were established.

Until recently over 65 percent of the total income of the eight UGC-funded universities came from government subventions (Figure 6.1). This figure has dropped steadily from 2003/2004 onwards, suggesting that the government is shrinking its role in subsidizing the higher education sector.

With 11 degree-granting organizations – eight of which are funded by the UGC – Hong Kong is now in a position to provide 14 500 first-year, first-degree places to approximately 18 percent of the 17- to 20-year olds in the population. One-third of these students eventually graduate with degrees in science and technology, contributing directly to the development of higher value-added activities in Hong Kong. Many of the remaining two-thirds, who graduate with degrees outside of science and technology, also contribute to the local innovation system insofar as they provide the knowledge and expertise needed to apply or use scientific ideas and technological products. These UGC-funded organizations offer a total of more than 45 000 degree places. Tertiary education accounted for approximately one-third of the government expenditure on education in 2004–05, which in turn was around 4 percent of GDP.[1]

The oldest of the eight UGC-funded universities, Hong Kong University (HKU) and the Chinese University of Hong Kong (CUHK), have the

*Table 6.1 R&D expenditure by performing sector, as a percentage of GDP 1995–2007*

| Year/ Sector | 1995 | 1996 | 1997 | 1998 | 1999 | 2000 | 2001 | 2002 | 2003 | 2004 | 2005 | 2006 | 2007 |
|---|---|---|---|---|---|---|---|---|---|---|---|---|---|
| BERD | 0.11[a] | 0.13[a] | 0.10[a] | 0.12[a] | 0.11[a] | 0.08[a] | 0.16[a] | 0.20[a] | 0.29[a] | 0.36[c] | 0.41[c] | 0.43[c] | 0.37[c] |
| HERD | 0.24[a] | 0.25[a] | 0.27[a] | 0.30[a] | 0.34[a] | 0.38[a] | 0.37[a] | 0.38[a] | 0.39[a] | 0.36[c] | 0.37[c] | 0.37[c] | 0.38[c] |
| GOVERD | 0.01[b] | 0.01[b] | 0.01[b] | 0.01[a] | 0.01[a] | 0.01[a] | 0.01[a] | 0.02[a] | 0.02[a] | 0.02[c] | 0.02[c] | 0.02[c] | 0.02[c] |
| Total (GERD) | 0.36 | 0.39 | 0.38 | 0.43[a] | 0.46[a] | 0.47[a] | 0.55[a] | 0.59[a] | 0.69[a] | 0.74[c] | 0.80[c] | 0.82[c] | 0.77[c] |

*Notes:*
BERD: Business Expenditure on Research and Development; HERD: Higher Education Expenditure on Research and Development; GOVERD: Government Expenditure on Research and Development; GERD: Gross Domestic Expenditure on Research and Development.

a. Figures in respect of BERD (from 1995 to 2003), HERD (from 1995 to 2003) and GOVERD (from 1998 to 2003) are extracted from the publication *Hong Kong as a Knowledge-based Economy: A Statistical Perspective, 2005 Edition*, C&SD, HKSAR.
b. Figures in respect of GOVERD from 1995 to 1997 are estimates based on 1998 and onward figures.
c. Figures in respect of BERD, HERD and GOVERD for 2004 to 2007 are extracted from Table 2 of the feature article, "Statistics on Research and Development of Hong Kong, 2003–2007," in the May 2009 issue of the *Hong Kong Monthly Digest of Statistics* published by C&SD, HKSAR. The figures are subject to revision.

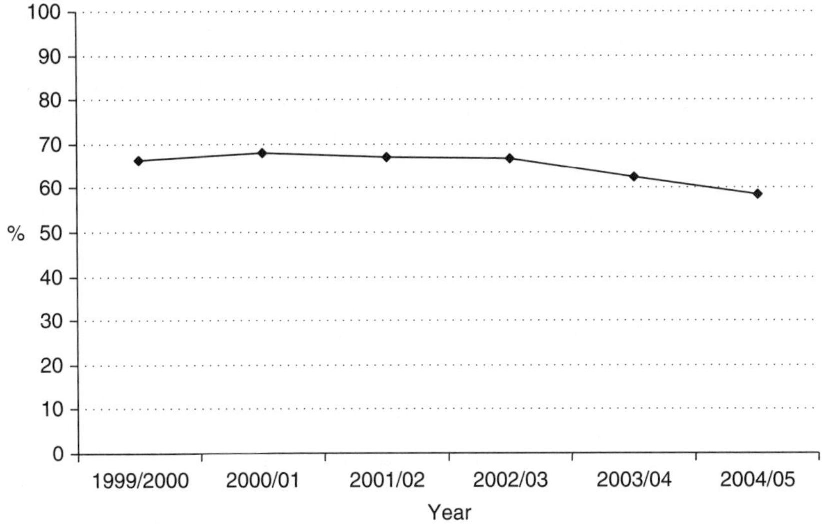


*Source:* Annual Reports of the University of Hong Kong, Chinese University of Hong Kong, Hong Kong University of Science and Technology, Hong Kong Polytechnic University, City University, Hong Kong Baptist University, Lingnan University and Hong Kong Institute of Education, from 2000 to 2006 (for the 1999–2000 academic year, income-related figures for the Hong Kong Institute of Education and Lingnan University were not available; for the 2000–01 academic year, income-related figures for Lingnan University were not available).

*Figure 6.1 Total government subventions as a percentage of total income (aggregated for all eight UGC-funded institutions)*

longest traditions of serving Hong Kong. They are pioneers in medicinal research, for example, being closely affiliated with Queen Mary and Prince of Wales Hospitals, respectively. In spite of its comparatively young age, HKUST has developed a healthy reputation, albeit within a much narrower range of specializations than HKU and CUHK: HKUST offers degrees and conducts research only in science, engineering, business, and humanities and social science. Considered together, these three universities have developed strong engineering schools and are the leading actors in engineering research in Hong Kong. During the latest (2006–07) allocation of RGC Competitive Earmarked Research Grants (CERG) for engineering projects, HKU was awarded 47 projects amounting to around HK$22 million, CUHK was awarded 48 projects amounting to HK$25 million and HKUST was awarded 78 projects amounting to more than HK$42 million.[2]

Of the remaining universities in Hong Kong, only Hong Kong Polytechnic University (HKPU) and City University receive substantial

*Table 6.2 Distribution of approved projects under the UICP Program as of 30 September 2009*

| Technology Area | Number of projects | HK$ million |
|---|---|---|
| Information technology | 45 | 49.7 |
| Electrical and electronics | 37 | 31.6 |
| Manufacturing technology | 37 | 22.3 |
| Biotechnology | 24 | 38.4 |
| Chinese medicine | 15 | 26.9 |
| Materials science | 14 | 7.5 |
| Environmental technology | 9 | 20.7 |
| Nanotechnology | 1 | 1.5 |
| Total | 182 | 198.6 |

*Source:* ITF website: http://www.itf.gov.hk (accessed on 9 November 2009).

engineering and technology research funding, with HKPU having won 93 CERG projects in the engineering field amounting to HK$42 million, and City University winning 53 CERG projects amounting to almost HK$30 million.

Citing cooperation between universities and industry as an explicitly defined objective, the government-administered University-Industry Collaboration Program (UICP) – one of the four Innovation and Technology Fund (ITF) programs – aims to stimulate private sector interest in R&D by leveraging the knowledge and resources of universities. As of 30 September 2009, there were 182 projects approved under the UICP with a total funding amount of HK$198.6 million (Table 6.2). UICP support is given as a grant, subject to a cash contribution by the company amounting to no less than 50 percent of the project cost. While no independent evaluation has yet been carried out to assess the effectiveness of this program, Table 6.2 shows that the majority of projects under the UICP fall into the information technology area (45 projects with a total funding amount of HK$49.7 million), biotechnology (24 projects with a total funding amount of HK$38.4 million) and electrical and electronics (37 projects with a total funding amount of HK$31.6 million).

### 6.3.2 Overview of HKUST Within Hong Kong's Tertiary Education Sector

The formation of HKUST marked a historical milestone in the development of Hong Kong's innovation system. When the HKUST project

*Table 6.3 Profile of HKUST*

| Indicator | 2000–01 | 2006–07 |
|---|---|---|
| Undergraduate students enrolled | 5578 | 5772 |
| Graduate students enrolled | 1754 | 3228 |
| Graduate students as % of total student's enrollment | 23.92% | 35.87% |
| Percentage of foreign students studying at HKUST[a] | 7.00% | 18.63% |
| Total research funding amount (HK$ million)[b] | 309.8 | 306.5 |
| Percentage of research sponsored by industry | 19.34% | 17.13% |
| Total number of research projects funded[c] | 484 | 551 |
| Percentage of research projects sponsored by industry partners[d] | 15% | 15% |
| Number of research publications[e] | NA | 764 |
| Percentage of research publications in refereed journals[e] | NA | 46.8% |
| Cumulative patents filed[f] | 146 | 509 |
| Cumulative patents granted[f] | 36 | 138 |
| Cumulative patents granted by USPTO | 34 | 94 |

*Notes:*

a. University Grants Committee (2002), 'Facts and figures', University Grants Committee, Hong Kong.

b. Office of Contract and Grant Administration (2008), 'Funding statistics: research funding', Hong Kong University of Science and Technology, http://www.ogca.ust.hk.

c. HKUST facts and figures, http://www.ust.hk/eng/images/about/pdf/hkust_facts_and_figures.pdf (p. 2).

d. There are no formal statistics available to ascertain this figure, but rather it is an educated estimate from HKUST's (Acting) Vice President of R&D, which has also been corroborated by HKUST's Director of Technology Transfer, Winter 2007/08.

e. Research Output by Institution and Research Output Category, 2004/05 to 2005/06 (2005–06 figures).

f. Information as of 31 January 2008 provided by the Technology Transfer Center, HKUST.

was first conceived in September 1986, it was envisioned as a world-class research university in certain key areas (science, engineering, business and management, and social science) that would serve the swiftly changing human resource needs of Hong Kong as it transitioned into an increasingly service-oriented economy. Founded in 1991, HKUST is the youngest research-based university in Hong Kong, with a total student enrollment of 9000, two-thirds of whom are undergraduates (see Table 6.3 for some relevant statistics).

The table shows that more than 17 percent of the research carried out at HKUST during the 2006–07 academic year was funded by industry. This is a relatively high proportion of total research funding, almost as high as the share of industry-funded research at MIT. Moreover, in terms of

*Table 6.4 HKUST's share of Hong Kong's R&D expenditures 1999–2005*

| Year | Total HK R&D expenditure[a] (HK$ million) | Higher education R&D expenditure[a] (HK$ million) | Total HKUST R&D expenditure (HK$ million)[b] | HKUST R&D expenditure as % of higher education R&D expenditure |
|---|---|---|---|---|
| 1999 | 5885.40 | 4316.90 | 567.73 | 13.15 |
| 2000 | 6218.40 | 4989.20 | 574.97 | 11.52 |
| 2001 | 7087.30 | 4846.60 | 656.87 | 13.55 |
| 2002 | 7543.60 | 4800.70 | 626.97 | 13.06 |
| 2003 | 8548.80 | 4796.20 | 677.15 | 14.12 |
| 2004 | 9505.20 | 4707.30 | 654.42 | 13.90 |
| 2005 | 10908.40 | 5085.00 | 661.79 | 13.01 |

*Notes:*
a. Hong Kong Census and Statistics Department (2007a).
b. Estimates derived from: figures provided by the R&D Branch, HKUST; figures provided by the Finance Office, HKUST; and figures from 'Facts and figures (1995–2005)', University Grants Committee, Hong Kong and Hong Kong Census and Statistics Department (2007c).

patenting activity HKUST enjoyed a lead over other Hong Kong universities during the 1990s and is still among the most prolific in terms of US Patent and Trademark Office (USPTO) patents granted per year. Despite being the newest and most poorly funded of this group of three universities, HKUST compares equally to its two counterparts in terms of spin-offs and patents granted. As of 31 January 2008, HKUST held a portfolio of 382 patents (including pending patent applications) of which 43 percent have been commercialized by various means such as contract research and licensing. This compares fairly well, for example, with the approximately 400 patents applications at HKU, of which approximately 10 percent have been licensed (data supplied during an interview with the Director of HKU's Technology Transfer Office (TTO), Dr Paul Cheung, 10 January 2008). Indeed no other local university contributes value to Hong Kong's science and technology base as singularly as does HKUST – the only tertiary education institution devoted to and heavily focused on science and technology training.

Table 6.4 provides a summary profile of HKUST up to the 2005–06 financial year. With an annual R&D budget of about HK$661 million HKUST alone constitutes about 6 percent of total R&D spending in Hong Kong.

In the latest Research Assessment Exercise (RAE) conducted in 2006,

HKUST achieved the highest total 'research index' among all eight UGC-funded universities, 87.12 percent. This is the percentage of eligible staff who submitted research output items for assessment that were judged by the relevant RAE panels to have generally reached or surpassed the quality standard set for the exercise. The results showed that HKUST had made remarkable improvements to achieve international excellence in research across all its disciplines.[3]

The results of the 2006 Competitive Earmarked Research Grant (CERG) validate HKUST's leading performance in the RAE. In June 2006 HKUST researchers were awarded HK$103.4 million – 21 percent of the total awarded to Hong Kong's tertiary institutions (for all disciplines). In this internationally peer-reviewed competition – based on the quality of proposals as perceived by international peers – HKUST achieved the highest success rate (61 percent of proposals were funded) for the 14th consecutive year and the highest funding per eligible faculty (HK$234 000).

As further evidence of HKUST's growing importance in Hong Kong's education sector, particularly in the areas of science and technology, Table 6.5 shows that HKUST has witnessed significant growth in its research human resource commitment. In 1996 there were a mere 54 people employed by HKUST to conduct research, including faculty and support staff, but this number had quadrupled by 2005 to well over 200. While this 200+ figure accounts for a relatively low 4.6 percent of total research personnel in UGC-funded institutions, this percentage had nevertheless increased by some 250 percent over the 1996–2005 period. Furthermore, if we consider HKUST research personnel employed in selected areas of research and teaching strength (with none in medicine), we find that HKUST's research personnel comprise a respectable 7 percent of all research personnel in Hong Kong's UGC-funded institutions. Considering only faculty in science and engineering, HKUST accounts for roughly 13 percent of research-related personnel among UGC-funded institutions in Hong Kong (according to 1999 figures).

HKUST's research capabilities and educational standards have also begun to attract international recognition. This is reflected in the university's commendable placing in international rankings. In 2000, a mere nine years after admitting its first batch of students, HKUST was ranked seventh in *Asiaweek*'s list of Asia's Best Universities (ahead of the Australian National University, the University of New South Wales and the City University of Hong Kong), underscoring its rapid rise toward excellence. More recently, HKUST achieved a surprisingly high ranking in the 2009 *Times Higher Education Supplement*'s ranking of top 200 universities in the world (Table 6.6), both overall as well as for individual faculties.

*Table 6.5 HKUST's share of research personnel in Hong Kong, 1996–2005*

| Year | Total R&D personnel in Hong Kong[a] | Research-related personnel in all areas[b] | Research-related personnel in selected areas[b,f] | HKUST all research personnel[c,d] | HKUST share of total research-related personnel in UGC institutions (%) | HKUST research personnel share of selected areas (%) | HKUST faculty in science and engineering[e] | UST research personnel and science and engineering faculty share of UGC research-related staff (%) |
|---|---|---|---|---|---|---|---|---|
| 1996 | NA | 2886 | 2172 | 54 | 1.90 | 2.50 | 317 | 12.90 |
| 1999 | 10118 | 3327 | 2430 | 150 | 4.50 | 6.20 | 276.5 | 12.80 |
| 2005 | 22054 | 4701 | 3111 | 217 | 4.60 | 7.00 | NA | NA |

*Notes:*

a. http://www.censtatd.gov.hk/hong_kong_statistics/statistics_by_subject/index.jsp?subjectID=7&charsetID=2&displayMode=T, Table No.132 (accessed 14 December 2007).
b. "Research-related staff (full-time equivalent) of UGC-funded Institutions by Departmental Cost Center Group and by staff grade: 2005/06 to 2006/07", http://www.ugc.edu.hk/eng/ugc/stat/staff_full.htm (accessed on 14 December 2007). For data before 2005, email reply from UGC secretariat.
c. Includes all staff in Research & Development Departments and Central Research Facilities.
d. 2005 data from 'Departmental staff number (full-time equivalent) of UGC-funded institutions by source of salary funding, staff grade, mode of employment and institution,' http://www.ugc.edu.hk/eng/ugc/stat/staff_full.htm (accessed 14 December 2007).
e. HKUST Faculty in Science and Engineering? 'HKUST Fact book' (various issues). For 1996 and 1999, staff include only senior staff.
f. Selected areas include: sciences; engineering and technology; business and management; social sciences; arts and humanities.

*Table 6.6 Ranking of HKUST in the World University Rankings by the* Times Higher Education Supplement *2005–09*

| | 2005 ranking | 2006 ranking | 2007 ranking | 2008 ranking | 2009 ranking |
|---|---|---|---|---|---|
| Overall | 43 | 58 | 53 | 39 | 35 |
| Science | 68 | 69 | 52 | 59 | 62 |
| Technology | 23 | 17 | 23 | 24 | 26 |
| Social sciences | 58 | 30 | 63 | 69 | 82 |
| Arts and humanities | 89 | >50 | n.a. | n.a. | 216 |

*Source:* *Times Higher Education Supplement* (various years).

### 6.3.3 Recent Shift Toward an Entrepreneurial University Model

Most public universities developed under the British Commonwealth tradition follow the orthodox model, emphasizing teaching over research. This has, for example, certainly been true of HKU. CUHK can also be said to have been following this traditional model. The 1980s and 1990s saw, however, an increasing emphasis on research, the British Commonwealth tradition notwithstanding. This change in emphasis was best reflected at HKUST where the shift toward research catalyzed similar shifts in other UGC-funded institutions in Hong Kong. HKUST signaled this shift by creating its TTO, which it called the Technology Transfer Center (TTC), upon its inception in 1991. Contrast this action with what has happened, for example, at HKU, where the TTO was set up only as recently as 1 September 2006.

The major impetus for change was due in no small measure to HKUST's founding vice-chancellor, Chia-Wei Woo, a Chinese-American from San Francisco State University (SFSU) who brought a distinctly North American ethos to HKUST in 1988. Professor Woo enjoyed the strong support of several senior executives in the then-colonial Hong Kong government (notably executive councilor Sir Sze-Yeng Cheung). Professor Woo not only introduced several initiatives to the tertiary education sector (for example, requiring all full-time faculty – without exception – to have earned at least a doctorate degree upon assuming their duties at HKUST), but also initiated a shift toward what Etzkowitz et al. (2000) described as the 'entrepreneurial university' model.

With the core HKUST departments comfortably established Professor Woo led the creation of the Entrepreneurship Program in July 1999. This program was introduced to assist faculty, staff and students in establishing technology start-up companies. HKUST promotes such activities to benefit Hong Kong economically and socially. HKUST features other entrepreneurial initiatives as well, including an entrepreneurship seminar series, a high-technology entrepreneurship course for senior engineering students, and also a dual-degree technology and management program.

The high-technology entrepreneurship training course is a three-credit course for entrepreneurially minded engineering and business school students who are interested in starting businesses in the Greater China region. The course features lectures and seminars on high-tech entrepreneurship and practical start-up exercises in a workshop environment under the guidance of departmental faculty and industry consultants. The dual-degree technology management program was first offered in 2003–04, and 30 to 40 students have enrolled in the program annually. Students who successfully complete the program receive a joint Bachelor of Engineering (BEng) and Bachelor of Business Administration (BBA) degree.

The HKUST Entrepreneurship Program incubates start-up companies in the privately funded Annex Building at HKUST. These new companies utilize serviced and furnished space at modest cost, with access to university facilities and resources including an Advisory Committee that assists them with business development. In 2007 HKUST opened its new Enterprise Center, constructed with a donation from the Hong Kong Jockey Club, which serves as the primary interface between the HKUST academic and research communities and the business and industrial environment of Hong Kong and the region. HKUST also introduces Entrepreneurship Program companies to or matches them up with potential investors in order to sell their technologies or to secure funding.

If a faculty member starts or joins a company that requires identifiable or patented intellectual property (IP), the company must obtain a license from HKUST, negotiated on a case-by-case basis, with the university paid by some combination of an up-front fee, deferred royalties or equity ownership. If, on the other hand, a faculty member starts or joins a company using expert knowledge, ideas, concepts and market insight but does not require HKUST to file patent material or otherwise document any IP, the university requires no royalty payment or equity ownership. In this latter type of case HKUST assumes a base equity share of 3 percent in companies accepted into the Entrepreneurship Program. As such a company increases its capital share over the years, HKUST's ownership position remains at 3 percent until the fully paid value of its shareholding reaches HK$300 000. The transfer of IP to the company usually increases equity. Such equity is held by HKUST's R and D Corporation Limited (RDC) on behalf of HKUST.

The Entrepreneurship Program focuses on several management issues. Most notably, companies under the auspices of the program maintain

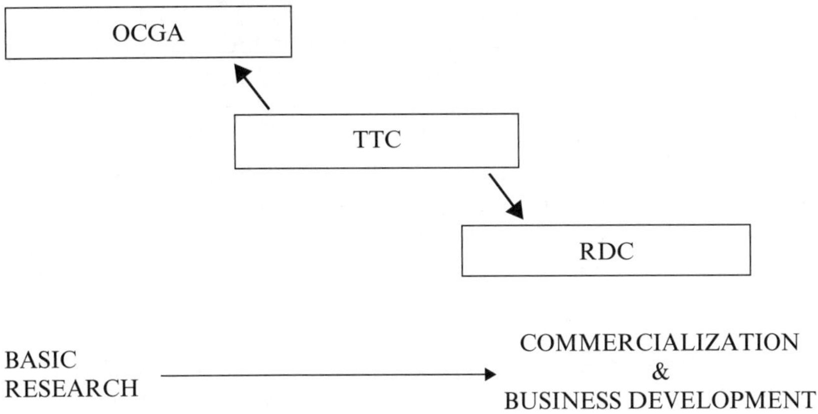


*Source:* https://www.ust.hk/vpaao/ADP/Annex%201-6%20Pdf/Annex%203.pdf, p. 40 (accessed 8 February 2008).

*Figure 6.2 HKUST's technology transfer support structure*

clear separation between HKUST's academic, educational and research environments and its own commercial and business activities. This feature is of course not unique to HKUST as eight of Hong Kong's leading universities are majority funded by the Hong Kong government. But this separation of responsibilities means that the Entrepreneurship Program must be sensitive to issues of accountability, conflict of interest, conflict of commitment and use of public funds for private gain. Thus the Entrepreneurship Program requires unambiguous policies defining the areas that fall within the governance purview of its companies.

Since February 2000 HKUST's RDC has also managed a modest Venture Capital Fund of HK$10 million. Start-up companies may seek such funding (as equity capital) to advance their business development plans. The RDC provides seed money and seeks financial partners to co-invest in its start-ups. While the Entrepreneurship Program entry process gives companies the opportunity to prove their potential, an application to the Venture Capital Fund is subject to a more rigorous review process as this involves investing the assets of the RDC.

HKUST's technology transfer support structure can be summarized as per Figure 6.2. The Office of Contract and Grant Administration (OCGA), the TTC and the RDC work closely to facilitate R&D and to transfer the benefits to industry and society. The TTC handles 'upstream' activities, namely, partnership/project development and IP evaluation, whereas the RDC handles 'downstream' activities, namely, negotiation, signing of contracts, billing and monitoring of R&D contracts and licensing agreements (that is, anything of a legal nature). OCGA acts as a one-stop center for information on funding agencies and research opportunities and assists in developing proposals (for funding agencies) and preparing publications and reports (for reporting agencies).

Notwithstanding the success of activities related to IP development and licensing, contract research and promotion of entrepreneurship, the TTC has sought to intensify the university's efforts to develop sustainable routes for commercialization of the technology developed at HKUST. Taking a more proactive approach, the center has been developing a new framework for the adoption of technology by industry that moves beyond the simple model of creating spin-offs. This approach emphasizes the mobilization of a social network of business managers who are keen to expedite commercialization and willing to cooperate closely with the university's faculty members and post-graduate students to bring a given technology to the stage from which it can be launched on the market. The managers bring some investment to the table while the RDC provides a substantial matching investment (earning a larger equity share). These new forms of joint venture establish a much more professional environment for the commercialization of the technologies.

HKUST's experience shows that by modestly expanding the role of TTOs it is possible to proactively bridge the gap between university and industry. While HKU's and City University's TTOs concentrate on the traditional 'early-stage' role of TTOs – making academic research accessible to local industry – HKUST's Entrepreneurship Program demonstrates a viable alternative means of operation whereby university research can be made practically useful. Furthermore, HKUST's experience shows that the TTO can lead in this process by acquiring an equity stake in companies (and increasing the equity stake with the transfer of IP). The TTO's modest Venture Capital Fund further evinces HKUST's creativity, illustrating the key finding that there is much room for expanding the roles of other Hong Kong tertiary institutions.

## 6.4. CHANGES IN HKUST'S PERFORMANCE AS A RESULT OF THE SHIFT TOWARD THE ENTREPRENEURIAL UNIVERSITY MODEL

### 6.4.1 Overview of Changes

The shift toward the entrepreneurial university model has only recently begun; it may be premature to assess its impact on the university's performance in terms of the broader set of objectives that the new model implies. It is also not clear what mix of performance benchmarks may be appropriate for such an assessment. Finally, any change in performance in recent years may have been due to factors other than those we have associated with this shift.

Notwithstanding these caveats, Table 6.3 provides a summary overview of the key changes that HKUST has undergone between the 2000–01 and 2006–07 academic years. In essence Table 6.3 shows only a moderate expansion of the university in terms of conventional educational and research output but a more dramatic change can be observed in entrepreneurship promotion, technology commercialization and collaborative R&D and contract research. Progress in these latter areas is further elaborated below following a brief discussion of HKUST's human resources development role.

### 6.4.2 Human Resources Development

In 2005 in Hong Kong there were 10 580 graduates from science and technology-related fields of study – representing 37 percent of all graduates. This figure has remained fairly steady in both absolute numbers and percentages compared with the period of 2001 (with 10 476 students representing 35 percent of total graduates) to 2005 (with 10 580 students, representing 37 percent of the total). In 2005 14 percent of all science and technology-related graduates in Hong Kong earned degrees from HKUST (University Grants Committee 2008). These graduates represent an important human resource component that contributes directly to the development of higher value-added activities.

While the figures for HKUST may seem rather low at first glance it is worth remembering that it has no medical school, unlike HKU and CUHK. The figure for science and technology-related graduates includes the following fields: medicine, dentistry, health sciences, biological sciences, physical sciences, mathematical sciences, computing and information technology, and engineering and technology. HKUST can claim competence in only the latter five categories. HKUST is moreover the youngest of all universities in Hong Kong and until plans for post-2010 expansion are realized its absolute contribution to the science and technology-related landscape is likely to remain small but vital.

While the number of graduates coming out of HKUST appears to have stabilized in the near term at around 9000, HKUST has even more importantly shifted the emphasis of its contribution to human resource development toward the training of graduate students, in line with its increasing emphasis on knowledge creation. HKUST has seen a steady

*Table 6.7 HKUST research publications 1990–2006*

| A<br>Year | B<br>Journal publications covered by SSCI, AHI and SCI-Extended [a] | C<br>Refereed articles [b,c] | (B/C)<br>(%) |
|---|---|---|---|
| 1990 | 9 | NA | NA |
| 1991 | 20 | NA | NA |
| 1992 | 45 | NA | NA |
| 1993 | 112 | NA | NA |
| 1994 | 186 | 879 | 21.1 |
| 1995 | 386 | 1001 | 38.6 |
| 1996 | 532 | 1285 | 41.4 |
| 1997 | 658 | 1173 | 56.1 |
| 1998 | 783 | 1148 | 68.2 |
| 1999 | 864 | 1542 | 56.0 |
| 2000 | 1012 | 1437 | 70.4 |
| 2001 | 1046 | 1101 | 95.0 |
| 2002 | 962 | 1429 | 67.3 |
| 2003 | 1127 | 1394 | 80.8 |
| 2004 | 1227 | 1613 | 76.1 |
| 2005 | 1155 | 1519 | 76.0 |
| 2006 | 1302 | 1556 | 83.7 |

*Notes:*
a. ISI Thompson Web of Science.
b. Research Grants Council (RGC) of Hong Kong, *Annual Reports*.
c. Does not include books, non-refereed journal articles, conference papers, book chapters or reports.

annual increase in the proportion of graduate students comprising its total student population. As shown in Table 6.3 graduate student enrollment increased from 1754 (comprising 24 percent of the total student population) in 2000–01 to 3228 in 2006–07 (comprising 36 percent of the total student population).[4]

### 6.4.3 Knowledge Creation and Commercialization

The output of research publications (as measured by the number of research articles) authored by HKUST faculty has grown since the mid-1990s, standing at 1556 in 2006 as compared with 879 in 1994 (Table 6.7), with even more rapid growth in internationally refereed journal publications, as its share has risen from about one-fifth in 1994 to over four-fifths in 2006. This has no doubt resulted from HKUST's intensive efforts to develop into one of the leading universities in science, technology and business management.

Concomitant with the increase in the rate of new knowledge creation during 1994–2006, the pace of knowledge commercialization also changed just as significantly. We examine the key areas of invention disclosures, patenting and licensing in detail below.

#### 6.4.3.1 Invention disclosures

The availability of data reflecting invention disclosures are patchy but they show increasing numbers. As with other universities invention disclosure documentation fluctuates because not all faculty members disclose all their inventions to their universities. The absolute number of invention disclosures is small if compared with what occurs, for example, at the National University of Singapore, and generally there are three reasons for the small number of disclosures: first, HKUST is small, with a mere 7000 students and approximately 400 faculty members compared with HKU's 18 000 students and roughly 1000 faculty members; second, HKUST is still a relatively young university and as Table 6.8 shows, invention disclosures are increasing annually as it acquires the necessary expertise; third, some invention disclosures result in multiple patent applications (this is truer for the more recent data – 2005–06 onwards).

#### 6.4.3.2 Patenting

Like other NIEs such as Singapore, Hong Kong undertook innovation activities rather late as compared with advanced Organization for Economic Cooperation and Development (OECD) countries (Hong Kong entered the domain of science and technology even later than Singapore, Korea or Taiwan). This accounts for the relatively low number of Hong Kong-based inventions receiving patent grants. Nevertheless Hong Kong's patenting performance improved dramatically between 1970–1974 and 1995–1999. Table 6.9 shows that the average number of US patents granted to Hong Kong's inventors jumped from an average of 12 per year to an average of 114 during this period. In addition to filing US patents Hong Kong's inventors increasingly file patents in China and the European Union, leaving only a small minority to apply locally.

HKUST has played a role in this increased patenting activity primarily as a result of its encouragement of entrepreneurial activity among its faculty members (for example, through the establishment of the TTC). Since its founding in 1991 HKUST has implemented an IP policy whereby all IP created by HKUST staff are assigned to HKUST, with the TTC tasked to license the IP and distribute any return from commercialization

*Table 6.8 Number of invention disclosures, patents filed by and granted to HKUST 1992/93–2006/07*

| Financial Year | Number of invention disclosures | Patent applications | Patents granted |
|---|---|---|---|
| 1992/93 | 3 | 1 | 0 |
| 1993/94 | 9 | 3 | 0 |
| 1994/95 | 11 | 3 | 1 |
| 1995/96 | 14 | 5 | 2 |
| 1996/97 | 29 | 19 | 1 |
| 1997/98 | 20 | 19 | 3 |
| 1998/99 | 22 | 25 | 7 |
| 1999/00 | 34 | 38 | 12 |
| 2000/01 | 29 | 33 | 10 |
| 2001/02 | 17 | 26 | 8 |
| 2002/03 | 13 | 29 | 8 |
| 2003/04 | 28 | 37 | 16 |
| 2004/05 | 31 | 58 | 14 |
| 2005/06 | 48 | 89 | 22 |
| 2006/07 | 69 | 124 | 34 |
| Total | 377 | 509 | 138 |

*Note:* Information as of 31 January 2008 provided by the Technology Transfer Center, HKUST. Fiscal year runs from 1 July to 30 June. Patent applications include provisional applications for which formal invention disclosure forms have not been submitted yet.

*Table 6.9 US patents granted to Hong Kong's inventors 1970–99*

| Period | 1970–74 | 1975–79 | 1980–84 | 1985–89 | 1990–94 | 1995–99 |
|---|---|---|---|---|---|---|
| US patents granted | 59 | 75 | 113 | 177 | 279 | 570 |

*Source:* Mahmood and Singh (2003, p. 1034).

among the inventor, their department and HKUST central administration. For patents commercialized with a value lower than HK$1 000 000 the inventor receives 50 percent of the returns from commercialization, the home department receives 25 percent, and the university the remaining 25 percent. For patents commercialized with a value above HK$1 000 000, commercialization returns are distributed equally among the inventor, the department and the university.[5]

The total number of HKUST patent applications has for the most part grown steadily over the 1995–2006 period, with over 124 patent applica-

*Table 6.10 HKUST licensing agreements 1995–2006*

| Year of license | Number of licenses |
|---|---|
| 1995 | 2 |
| 1996 | 5 |
| 1997 | 2 |
| 1998 | 5 |
| 1999 | 6 |
| 2000 | 8 |
| 2001 | 9 |
| 2002 | 9 |
| 2003 | 20 |
| 2004 | 15 |
| 2005 | 11 |
| 2006 | 29 |
| Total | 121 |

*Source:* HKUST R and D Corporation.

tions in 2006–07 (see Table 6.8). The number of patents granted also registered a similar increase with a high of 19 patents granted in 2005–06.[6] Over three-quarters of the patents granted to HKUST have been filed in the USA while 10 percent have been filed in the People's Republic of China.

### 6.4.3.3 Licensing

Table 6.10 shows an increase in the intensity of technology commercialization from 2000 onwards. As of 30 June 2006 HKUST had secured 121 licensing agreements. The majority of these agreements were signed between 2003 and 2006.

In spite of this increase in technology licensing, the actual income generated from licensing remains modest. Figure 6.3 shows the expenditures/revenues from the protection and licensing of IP over the past 13 years, indicating a deficit situation for most years. The year 2006 represented a rare exception as a portfolio of 22 electronic patents was licensed for over HK$4 million. It is not uncommon in university technology licensing to run an annual deficit. The money that universities make from technology transfer comes mostly from a small number of 'blockbuster' patents or growth companies. While the gestation period between licensing and royalty generation may diminish such income, interviews with the current and ex-director of HKUST's TTC as well as representatives of the RDC suggest that encouraging technology commercialization rather

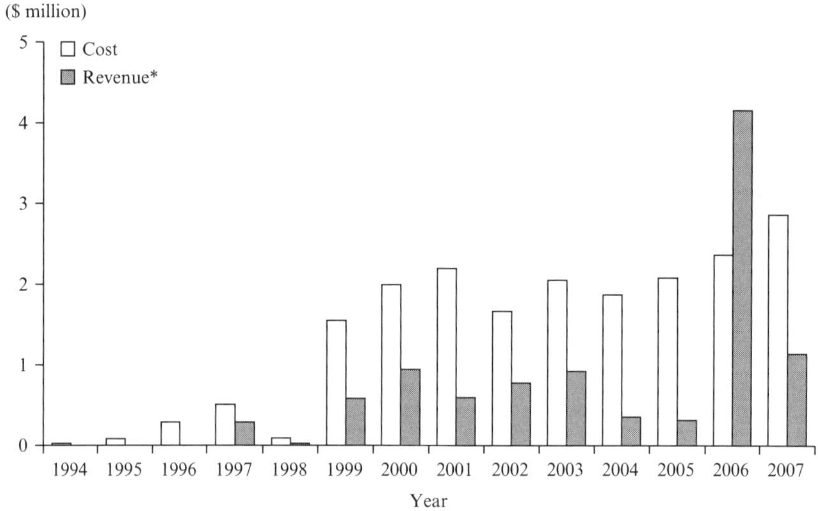


*Notes:* * HKUST's R and D Corporation Limited has received equity in a number of companies to which technology has been transferred.

*Source:* https://www.ust.hk/vpaao/ADP/Annex%201-6%20Pdf/Annex%203.pdf, p. 42 (accessed on 8 February 2008).

*Figure 6.3 Direct costs of IP protection and maintenance and licensing/ technology transfer revenues*

than maximizing royalties is a matter of policy. Nevertheless, the potential income from anti-malarial drugs, the use of nanotechnology for wrinkle-resistant fabrics and health foods based on cultivated cordyceps (a fungus) derived from traditional Chinese medicine is promising.

#### 6.4.3.4 Entrepreneurial start-ups and spin-offs

Table 6.11 shows the results of HKUST's embracing of technology commercialization through spin-offs and start-up formation following the establishment of the Entrepreneurship Program in July 1999. The number of start-ups (HKUST-based companies) incubated at HKUST has been relatively stable at an average of five or six start-ups in each of the four most recent years for which figures are available. There were also 14 spin-offs (companies that use HKUST technology but operate independently). The HKUST spin-off rate of around two per year is significantly lower than at such American universities as MIT, Stanford or Harvard, but the total of 61 spin-offs and start-ups represents a creditable return.

In spite of its successes, the HKUST Entrepreneurship Program has

*Table 6.11 Number of HKUST start-ups and spin-offs 1999–2006*

| Year | Start-ups* | Spin-offs |
|---|---|---|
| 1999–2000 | 25 | |
| 2000–01 | | |
| 2002–03 | NA | 14 |
| 2003–04 | 5 | |
| 2004–05 | 6 | |
| 2005–06 | 5 | |
| 2006–07 | 6 | |
| Total | 47 | 14 |

*Note:* * Estimates from HKUST R and D Corporation.

*Source:* HKUST R and D Corporation.

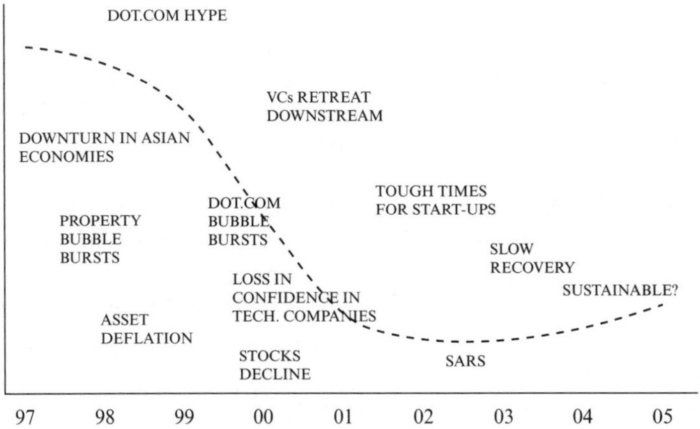


*Source:* Eastham et al. (2007).

*Figure 6.4 Pattern of start-up formation and spin-offs from HKUST Entrepreneurship Program*

been a victim of poor timing, as it was established just prior to the bursting of the 'dot.com' bubble in 2000. Figure 6.4 from Eastham et al. (2007) shows how the number of start-ups and spin-offs has fluctuated over the years in large part in response to locally felt global macroeconomic changes such as the SARS health crisis.

HKUST has an opportunity nevertheless to learn from its experience with start-up and spin-off companies that have not achieved commercial sustainability. Although activity in the HKUST Entrepreneurship Program has recently picked up, it remains to learn how to sustain these companies, as HKUST must maintain operations and achieve growth beyond the inception phase.[7] Although a few highly successful companies such as Perception Digital Limited have moved off campus, several companies remain headquartered in the incubator facilities of the university's Enterprise Center.

#### 6.4.3.5 Collaborative R&D and contract research

HKUST encourages faculty members to engage in collaborative R&D, often funded by the Innovation and Technology Commission (ITC) or the newly formed ITC R&D Centers. Since the inception of the ITC, HKUST has received 86 projects to which the ITC has contributed HK$372 million – more than to any other university in Hong Kong. In addition two out of five Innovation and Technology Commission (ITC) R&D Centers are hosted or co-hosted by HKUST: the R&D Center for Nanotechnology and Advanced Materials Limited and the R&D Center for Logistics and Supply Chain Management Enabling Technologies Limited. The number of R&D contracts continues to grow, as shown in Figure 6.5, reaching 149 projects worth HK$45.6 million in 2006–07.[8] Table 6.12 lists current R&D project clients.

All HKUST R&D contracts include overhead costs, but there is flexibility in terms of the percentage of overhead applied. The norm is 30 percent on direct costs (excluding equipment and subcontracts), a figure that is sometimes reduced under special circumstances – if the contract is with a small or medium-sized enterprise, unusually large (multimillion HK dollars) or involves strategic partners.

In these situations HKUST seeks ownership of IP but when clients resist the university defines what HKUST-owned IP is being brought into the project. R&D contracts commonly restrict publications, requiring the submission of draft publications to clients for review, say 30 days ahead of submission, to protect confidential information.

## 6.5 UNIVERSITY-INDUSTRY LINKAGE CASES: PERCEPTION DIGITAL AND RAYMOND INDUSTRIAL LIMITED

In this section we illustrate the pattern of university-industry linkages and entrepreneurship at HKUST by focusing on two actual cases: Perception Digital Limited. and Raymond Industrial Limited.

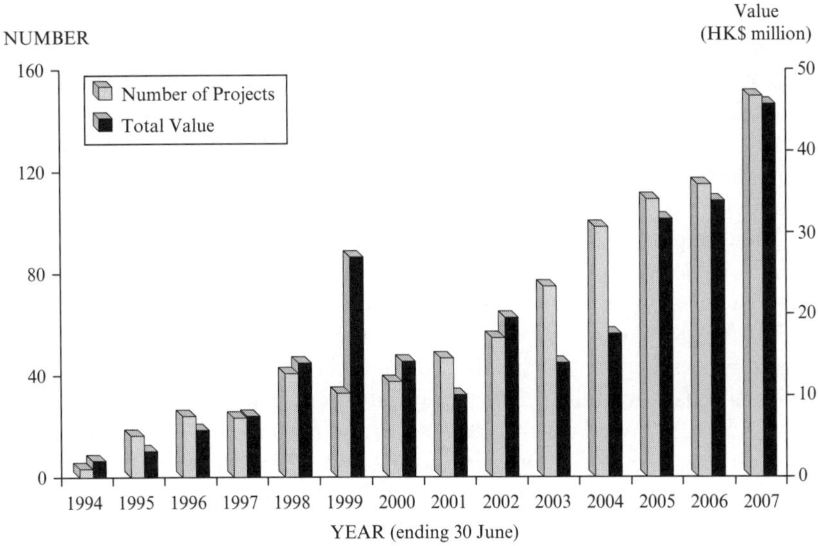


*Source:* https://www.ust.hk/vpaao/ADP/Annex%201-6%20Pdf/Annex%203.pdf, p.43, (accessed 8 February 2008).

*Figure 6.5 New HKUST R&D contracts by year*

*Table 6.12 Clients of HKUST R&D contracts as of 30 June 2008*

| R&D contract client | By number of projects (%) | By contract amount (%) |
|---|---|---|
| Hong Kong business and industry | 50 | 23 |
| Hong Kong public sector | 20 | 5 |
| Mainland organizations | 15 | 10 |
| Overseas companies | 15 | 62 |

*Source:* Technology Transfer Center & HKUST R and D Corporation Limited. *Annual Report* 2007–08.

### 6.5.1 Perception Digital: A HKUST Start-up

Perception Digital Limited was established in 1999 by three faculty members from the Department of Electrical and Electronic Engineering at HKUST. The Chief Executive Officer (CEO), Dr Jack Lau, has retired from his position at the university to devote all his time to managing

Perception Digital. His co-founders remain faculty members but also contribute to the company as consultants. Perception Digital, with an annual turnover of around US$75–80 million and almost 300 people employed in Hong Kong, the Chinese mainland and the USA, provides turnkey and firmware solutions in the areas of portable MP3 players with flash or hard disk memory, portable hard disks and various wireless audiovisual products.

The company was founded on the basis of research at the Consumer Media Laboratory of HKUST, eventually moving into the university's incubator facilities to design MP3 products to be manufactured in Guangdong Province. The unpatented technologies for displays, digital signal processing and other software that were developed at HKUST provided the background for market penetration. The company's engineers have since developed other technologies and designs that are implemented in products sold under Original Equipment Manufacture (OEM) agreements to major brand-name firms or sold as consumer products under its own brand. Increasingly, the firm operates as a design supply chain integrator/manager capitalizing on its core competencies including algorithms design, wireless communication, digital signal processing, power management, precise electro-mechanics design and expertise in integrating various technologies.

According to Lau (interviewed June 2008) HKUST offers fairly liberal technology transfer and entrepreneurship policies. Faculty are encouraged to involve themselves in start-ups and are granted leave to pursue technology commercialization, although they must avoid heavy management responsibilities and remain on the university's teaching staff. HKUST assumed an initial equity stake in Perception Digital but has not interfered in firm management and it helped Lau procure venture capital and business services.

Lau underscores the importance of maintaining strong linkages with HKUST, and Perception Digital remains a major employer of HKUST engineering graduates. Perception Digital's scholarships for engineering students facilitate an important channel for developing the firm's knowledge base.

### 6.5.2 Raymond Industrial: Long-term Collaboration

The collaboration between Professor I-Ming Hsing in the HKUST Department of Chemical Engineering and Raymond Industrial Limited (RIL) was initiated in 2001 as a result of a previous successful R&D project with the Chung Nam Corporation related to fuel cell technology. Professor Hsing had been introduced to Chung Nam by the HKUST TTC in 1999. Chung Nam is the second-largest OEM watch manufacturer in Hong Kong and the project was sponsored jointly by the corporation and the ITF. The director of the firm then introduced Professor Hsing to Raymond Wong, the Managing Director of RIL, another important OEM manufacturer from Hong Kong.

The contact with RIL resulted in a still thriving, long-term mutual commitment. RIL was also interested in developing nanotechnology applications for fuel cells in a collaborative project with ITC funding under the 90-10 rule (with HKUST keeping the intellectual property rights) at around HK$1.3 million. Although the project did not succeed in commercializing a new fuel cell technology (and RIL is no longer involved in fuel cell development), some of the technologies developed could be used for high precision coating and machining of components for electric shaver heads (a key RIL product).

RIL has also established a joint laboratory with HKUST on its industrial estate in Nansha to assist in projects carried out at the Center for Bioengineering and Biomedical Devices located at the Nansha campus of the HKUST Fok Ying Tung Graduate School, which is directed by Professor Hsing. At this wet laboratory local researchers conduct research in an environment that is not yet available at the Nansha campus.

When Professor Hsing, an MIT graduate who worked with DuPont and other firms, came to Hong Kong he thought that projects funded by the RGC were not very exciting so he applied his experience and enthusiasm to his new setting. Professor Hsing believes that HKUST should emphasize its comparative advantage in advanced applied R&D over the long term rather than in solving short-term technical problems.

## 6.6 CONCLUSIONS

HKUST began making significant strides toward fulfilling its envisioned promise as a progressive entrepreneurial university in the late 1990s. Since finding its identity it has steadily increased its entrepreneurial profile, effecting a qualitative change in its contributions to economic development in Hong Kong. Today we are able to observe that it is no longer content merely to provide traditional human resources and new knowledge, assuming a more visible role in knowledge commercialization through increased patenting, licensing to private industry, start-up incubation and spin-offs.

The empirical evidence of HKUST's changing role supports the findings of Wong et al. (2007) that universities can more effectively contribute to economic growth by adopting the entrepreneurial university model. HKUST's embrace of this model may not fully explain its own increasing share in this entrepreneurial engagement but the programs it has employed, particularly the Entrepreneurship Program, may prove instructive to other East Asian universities.

We have been limited to some extent in this study by the relative lack of official statistics. This is, however, a common problem related to science and technology statistics across all sectors in Hong Kong. These developments pertaining to innovation and commercialization performance are so recent that they may not represent sustainable, long-term changes. We have yet to see substantial licensing revenue or significant commercial spin-off successes. Thus we cannot claim with utter confidence that HKUST's commitment to the entrepreneurial university model will eventually bring significant economic payoffs.

In any event we believe that a sustained focus on developing the capacity for research collaboration and personnel exchange with both new and established firms can yield significant benefits for the region and for the universities. In other words, we should not pursue technology transfer as an end in itself but rather as a means of creating or strengthening research collaboration between universities and the private sector. HKUST must nevertheless be careful not to oversell its catalytic powers in the face of heightened industry expectations. As Sharif and Baark (2008) show, the gap between universities and industry is substantial and universities must avoid widening that chasm.

To these conclusions we add several lessons, which are serving to reorient the Entrepreneurship Program's work. First, establishing successful spin-offs and start-ups is difficult. Macroeconomic factors related to the program and even to the university can potentially affect any such program's success, as illustrated by the 'dot.com' bust of 2000. Second, the road from identifying a discovery or an innovation to creating a viable commercial opportunity for that technology is long and arduous. Human capital is as vital as the technology itself to creating a successful technology-based start-up. Start-ups need people with the relevant business knowledge, experience and acumen, without which anyone can fail even if their idea makes sense. Finally, a start-up venture must bring into the process a strongly market-oriented focus. A technologically brilliant concept will fail if there is no market for it.

In response to these lessons the Entrepreneurship Program will institute three concrete changes. First, the program is working to create a better network of early-stage investors to improve funding availability. Second, it is more actively engaging in joint ventures between its start-ups and well-established companies. Finally, it is concentrating on identifying cutting-edge technologies with real market potential rather than enlisting existing technologies.

HKUST's transition to the entrepreneurial university model is a work in progress that represents a concerted effort to contribute more effectively to the process through which a regional economy is being transformed into a knowledge-based economic hub. The university is, however, setting itself up for public scrutiny of the pace and reach of its success in commercializing technologies, not least because of the considerable extent of public funding of Hong Kong's universities. HKUST faces stubborn resistance in transforming technology commercialization into a university-wide focus, and it is hard to predict success in light of heightened public expectations. Still, considering Hong Kong's slow transition into the information age, it is equally difficult to see how the transition to a knowledge-based economy can be completed without the contributions of its university system. In that respect all eyes will remain on HKUST.

## NOTES

1. See statistics posted on the website of the EDB, available at http://www.emb.gov.hk/index.aspx?langno=1&nodeid=1032 (accessed 14 June 2007).
2. For more detailed figures, see http://www.ugc.edu.hk/eng/rgc/result/cerg/cerg.htm (accessed 23 September 2007).
3. As it did for RAE 1999, the UGC again adopted a very broad definition of research in RAE 2006 by using a definition of 'scholarship' developed and refined by the Carnegie Foundation, covering 'discovery, integration, application and teaching'. This definition was adopted to accommodate a wide spectrum of research disciplines, from sciences and technology to business to arts to education. Available at the UGC website, http://www.ugc.edu.hk/eng/ugc/publication/press/2007/pr02032007.htm (accessed 28 December 2007).
4. In line with this increase in graduate student numbers has been the increase in foreign students, from 7 percent in 2000–01 to 19 percent in 2006–07. It is not clear from the statistics, however, whether this increase in foreign students studying at HKUST can be attributed to the undergraduate level or graduate level, although anecdotal evidence suggests it is the latter.
5. As regards to joint patenting with industry, almost no patents have creators from both HKUST and industry – the vast majority have inventors from HKUST alone, whereas a few have inventors from HKUST and one or more other local tertiary institutions.
6. Citations are not a widely used measure of patent quality. Rather, the utility of a patent is often measured in terms of whether the patent is licensed or sold.
7. There are some emerging successes such as Perception Digital Limited, Googol Technology (HK) Limited, Advanced Packaging Technology Limited, Radica Systems Limited and TIM EDPlatform Limited. All of these start-ups were founded by faculty members and their students as partners.
8. R&D contracts are typically of the same format as R&D contracts in the USA, except for some local customizations made in some cases. Overhead costs are included at 15 percent.

## REFERENCES

Eastham, T., M. Yuen, C. Xu, R. Law and S. Woo (2007), 'From start-ups to joint ventures', paper presented at the Triple Helix Conference 2007, Singapore.

Etzkowitz, H., A. Webster, C. Gebhart and B.R.C. Terra (2000), 'The future of the university and the university of the future: evolution of ivory tower to entrepreneurial paradigm', *Research Policy* **29** (2), 313–30.
GEM (2004), *Global Entrepreneurship Monitor: Hong Kong & Shenzhen 2004*, Hong Kong: Center for Entrepreneurship, The Chinese University of Hong Kong.
GEM (2007), *Global Entrepreneurship Monitor: Hong Kong 2007*, Hong Kong: Center for Entrepreneurship, The Chinese University of Hong Kong.
Hong Kong Census and Statistics Department (2007a), *Hong Kong Monthly Digest of Statistics: Feature Article – Statistics on R&D of Hong Kong, 1998–2002 & 2001–2005*, C&SD, HKSAR.
Hong Kong Census and Statistics Department (2007b), *Hong Kong Monthly Digest of Statistics: Feature Article – Statistics on Research and Development of Hong Kong, 2000–2004*, C&SD, HKSAR.
Hong Kong Census and Statistics Department (2007c), *Hong Kong Statistics Digest – Feature Article – Government Funding of Research and Development, Innovation and Technology Upgrading, 1995/96–2001/02*, C&SD, HKSAR.
ISI Thompson Web of Science, ITF (2007), available at http://www.itf.gov.hk (accessed 24 April 2007).
Mahmood, I.P. and J. Singh (2003), 'Technological dynamism in Asia', *Research Policy*, **32**, 1031–54.
Miner, A.S., M. De Vaughn, D. Easley and T. Rura (2000), 'The Magic Beanstalk vision of university venture formation', in K. Schoonhoven and E. Romanelli (eds), *The Entrepreneurship Dynamic*, Stanford, CA: Stanford University Press, pp.109–46.
Mok, K.H (2005), 'Fostering entrepreneurship: changing role of government and higher education governance in Hong Kong', *Research Policy*, **34**, 537–54.
Office of Contract and Grant Administration (2008), *Funding Statistics: Research Funding*, Hong Kong University of Science and Technology, http://www.ogca.ust.hk (accessed January 2008).
Research Grants Council (RGC) of Hong Kong (1994–2006), *Annual Reports*, Hong Kong: RGC.
Sharif, N. and E. Baark (2008), 'Mobilizing technology transfer from university to industry: the experience of Hong Kong universities', *Journal of Technology Management in China*, **3** (1), 47–65.
*Times Higher Education Supplement* (various years).
Tung, Chee-Hwa (1997), Annual Policy Address, 'Building Hong Kong for a New Era', HKSAR government, Hong Kong.
University Grants Committee (2002), 'Facts and figures', University Grants Committee, available at http://www.ugc.edu.hk/english/documents/figures/index.html (accessed January 2008).
University Grants Committee (2008), 'Graduates of UGC-funded programmes by institution, level of study, mode of study and academic programme, 2004/05 & 2005/06', available at http://www.ugc.edu.hk/eng/ugc/publication/report/figure2006/figures/09.pdf, and http://www.ugc.edu.hk/eng/ugc/publication/report/figure2006/figures/08.pdf (accessed 15 January 2008).
Wong, P.K., Y.P. Ho and A. Singh (2007), 'Towards an "entrepreneurial university" model to support knowledge-based economic development: the case of the National University of Singapore', *World Development*, **35** (6), 941–58.

# 7. Towards a 'Global Knowledge Enterprise': the entrepreneurial university model of the National University of Singapore

**Poh-Kam Wong, Yuen-Ping Ho and Annette Singh**

## 7.1 INTRODUCTION

As highlighted by Wong (2006), Singapore has achieved among the highest economic growth performance among newly industrializing economies (NIEs) in the 40-year period 1960–2000, with average GDP growth of 8 percent per annum. This rapid economic growth has been primarily driven by a direct foreign investment (DFI)-leveraging strategy (Wong 2003) that relied on attracting foreign investment of increasing technological sophistication. However in more recent years this heavy reliance on foreign investment and technology transfer has become insufficient to sustain economic growth, which has moderated to 4.6 percent per annum over 2000–09.

To cope with increasing global competition Singapore has shifted towards a knowledge-based strategy for economic development since the late 1990s (Wong and Singh 2008). Singapore's policy makers have charted a course for Singapore's transition from an investment-driven economy to an innovation-driven economy, emphasizing the building of intellectual capital and its commercialization to create value and jobs. While the role of Singapore's universities in nurturing talent has always been recognized, increasing prominence has been given in recent years to their role in stimulating economic growth through industrially relevant research, technology commercialization, high-tech spin-offs, attraction of foreign talent and injecting an entrepreneurial mindset among their graduates. This chapter examines how the National University of Singapore (NUS), the leading university in Singapore, is changing its role in the national innovation system (NIS) of Singapore.

Singapore's case is of particular interest to other small NIEs because of her unique status as a relatively small city-state, where the pressure for globalization and the pace for change towards a knowledge-based economy to sustain economic survival are particularly intense. As such, the challenges that her university system faces are likely to be symptomatic of those that other small NIEs will face in the near future. In particular, the experience of NUS is of relevant interest to study how the role of local universities in the NIS of NIEs can be transformed by adopting an 'entrepreneurial university' model.

## 7.2 NATIONAL INNOVATION SYSTEM POLICY CONTEXT

### 7.2.1 Innovation System Challenges and Policy Changes

From an NIS capability development perspective, Singapore's development strategy over the last 45 years since political independence can be summarized as shifting from an emphasis on developing operating capability in the 1960s–1970s to focusing on process adaptive capability development in the 1970s–1980s, followed by an emerging shift towards innovative capability development since the early 1990s. This shift towards innovation accelerated after the 1997 Asian financial crisis in tandem with the new emphasis on developing a knowledge-based economy (Wong and Singh 2008).

The primary focus of Singapore's NIS in the current emerging knowledge economy development phase since the late 1990s is the creation and commercialization of knowledge protected by intellectual property (patented high-tech innovations and trademarked designs, proprietary specialized knowledge assets and processes, and copyrighted creative content). However in this shift towards a knowledge-based economy Singapore needs to overcome four key transformational challenges. Firstly, it requires a shift of Singapore's capabilities in terms of using technology towards capabilities in creating technology (Wong and Singh 2008). While Singapore's success in leveraging foreign multinational corporations (MNCs) in earlier development phases cannot be denied, one consequence is that her reliance on imported technology has to some extent crowded out the development of indigenous innovative ability. Going forward Singapore needs to achieve a more balanced NIS that emphasizes both the attraction of foreign MNCs to bring more innovative activities to Singapore, as well as the development of local innovation capabilities through greater public investment in research and development (R&D) and technology commercialization.

Secondly, Singapore still lacks a sufficient number of highly skilled knowledge professionals, especially research scientists and engineers (RSEs). In some ways this is to be expected given Singapore's small size and thus limited manpower pool. However it is a major constraint to the future development of its NIS, given the critical mass of science and technology manpower needed for its high-tech industrial drive, especially in the life sciences.

Thirdly, there is still a lack of an entrepreneurial culture and mindset among Singaporeans. Again this is somewhat to be expected in a stable, rapidly growing economy with secure, high-paying jobs available in MNCs. However it was exacerbated by a government policy that provided incentives for the best and brightest graduates to work in the public sector and in government-linked companies (GLCs). Consequently a culture that eschews risk and failure has developed over the years, and entrepreneurship has not been generally considered as a viable career option (Wong et al. 2005).

The final challenge is that the innovation system in Singapore is characterized by a relatively underdeveloped technology commercialization ecosystem. Firstly, there is still a lack of private enterprises capable of developing product or process innovations from the R&D coming out of public research institutes and universities. This lack is the result of the underdevelopment of local high-tech enterprises due to the high reliance on foreign MNCs in the past. At the same time even though foreign firms are increasing their innovation activities in Singapore, they still tend to have weak linkages with the local innovation actors and their propensity to license locally produced technologies is still in a nascent stage (Wong and Ho 2007). Secondly, the venture ecosystem for supporting new start-ups is also weak. Unlike advanced high-tech hubs like Silicon Valley (Lee et al. 2000), Singapore's venture capital (VC) industry is still in its early development stage and the established VC funds tend to focus on less risky later-stage financing. A critical mass of sophisticated business angel investors is also lacking and the dense social networks of venture professionals, mentors and deal-making institutions (start-up law firms, serial entrepreneurs and so on) that characterize places such as Silicon Valley are only starting to emerge. Consequently even if the public sector were able to significantly increase its R&D output, the underdeveloped venture ecosystem would make it difficult for such innovations to be commercialized via spin-offs.

The Singapore government has made a concerted effort to shift its policy focus to overcome these four transformational challenges since the late 1990s. For example, the public policy focus on intensifying investment in R&D received a big boost from a decision announced in 2000 to make

Singapore a life science hub in Asia, in the hope of diversifying Singapore from its high dependence on electronics/interaction and communications technology (ICT) (Finegold et al. 2004). A US$1 billion fund was initially allocated to boost public investment in several new life science research institutes, to co-fund new R&D projects by global pharmaceutical firms, as well as to initiate the building of a new life science complex called Biopolis. Besides increasing R&D funding in existing public research institutes under the Agency for Science, Technology and Research (A*STAR), the government also established a new S$500 million National Research Fund (NRF) to fund new R&D fields and their commercialization, including biomedical translational research, water technology, alternative energy and clean technologies, and interactive digital media (IDM).

Public programs to promote high-tech entrepreneurship were also launched in 1999 with the announcement of the Technopreneurship 21 (T21) Initiative that included a new US$1 billion fund to boost the growth of VC industry, as well as a number of liberalization policies such as allowing entrepreneurs to start up their ventures in their home, relaxation of listing requirements on the local stock exchange and making bankruptcy laws less onerous (Wong et al. 2005). However, in the wake of the dotcom crash in early 2000s, the T21 Initiative was scaled back and modified to take on a broader focus on promoting entrepreneurship in general, not just technology start-ups. For example, in a report of the Subcommittee on Entrepreneurship and Internationalization (part of the larger Economic Restructuring Committee (ERC) Report released in September 2002), six broad areas of policy changes were recommended to make the Singapore economy more conducive for entrepreneurial development:

1. Culture: to influence the cultural values of Singaporeans towards entrepreneurship by providing students and working professionals more opportunities to learn about entrepreneurship.
2. Capability building: to attract more entrepreneurial talent from overseas and encourage greater mobility of talents between the public and private sector.
3. Conditions: to reduce government regulatory red-tape and review the role of GLCs in the domestic economy.
4. Connectivity: to enhance the global connectivity of Singaporeans to the world.
5. Capital: to improve start-ups and small and medium-sized enterprises (SMEs) access to capital.
6. Catalyst role of government: to extend investment and tax incentives currently available for large MNCs to smaller enterprises as well (ERC 2002).

In the context of such major public policy changes to shift the Singaporean economy towards a knowledge economy, it is inevitable that increasing policy attention has also been paid to changing the role and function of the country's higher education system, given its pivotal role in the NIS. In particular in early 2009 the NRF set aside a pool of funding (the University Innovation Fund (UIF)) to support university research commercialization and academic entrepreneurship activities, over and above its research funding.

### 7.2.2 Changing Role of University in the National Innovation System: The Need for an 'Entrepreneurial University' Model

As argued by Etzkowitz et al. (2000) and Etzkowitz (2003), universities around the world are increasingly shifting from their traditional primary role as educational providers and scientific knowledge creators to a more complex 'entrepreneurial' university model that incorporates the additional role of commercialization of knowledge and active contribution to the development of private enterprises in the local and regional economy. As a result universities become an increasingly important component of the NIS and need to operate increasingly within a Triple-Helix nexus involving close interaction with government and private industry.

Wong et al. (2007) argued that the preconditions for Triple-Helix dynamic interactions in NIEs are much weaker than in the advanced economies due to three factors: a greater involvement of the state in economic activities; a lower base of research and inventive output coming from the university; and lower demand and ability of private enterprises to commercialize university knowledge. Thus universities in NIEs arguably face even greater urgency to take on an 'entrepreneurial' role than universities in the advanced economies in order to compensate for these less favorable preconditions.

In the context of a small NIE such as Singapore, Wong et al. (2007) further suggest that the local university system should take on an additional economic role: attracting foreign talent. Given the small local population, Singapore needs to be able to tap first-class foreign talent to help staff the top echelons of specialized knowledge workers required in a knowledge economy. The experience of regions like Silicon Valley, Boston and London strongly suggests that the competition for global talent not only works through the increasingly globalized market for qualified technical professional specialists, but actually starts before the talent reaches the labor market – through the attraction of top students from overseas. In addition many of those who return to their home countries often continue to build economic links with the regions where they went to school and indirectly contribute to the vibrancy of those regions. This type of brain circulation dynamics has been well documented by Saxenian (2000, 2002) in the context of Silicon Valley and its growing links with Asia. In contrast to this foreign talent attraction role of Western universities (particularly universities in the Anglo-Saxon world), East Asian universities have traditionally been much less globalized, primarily focusing on the education of the local elites. Although the use of the local language as the medium of instruction represents an obstacle to attracting foreign students, the more important reason is the lack of a political commitment to allocate scarce university places to foreign students.

Wong et al. (2007) also emphasize the need for universities in NIEs to inject a stronger dose of entrepreneurship in the content of university education itself. Besides teaching entrepreneurial skills, they argued that the university should play an active role to foster a more entrepreneurial mindset among students. This is especially important in the context of Singapore, where entrepreneurial propensity among the highly educated population has been low (Wong et al. 2005) because consistently high economic growth in the past had generated relatively full employment and bright career prospect in the corporate (mainly MNC subsidiaries) and the public sector. With stable job opportunities and steady corporate careers no longer guaranteed in the increasingly competitive global economy of the new millennium, the university sector in Singapore urgently needs to reorientate student expectations of the job market and to prepare them for a more uncertain world.

How has the university educational system in Singapore responded to the transformational challenges of Singapore's NIS since the late 1990s? To what extent has it moved towards the entrepreneurial university model as advocated by Etzkowitz, and in what ways has it incorporated the additional roles of attracting foreign talents and nurturing entrepreneurship into its core educational and research missions as advocated by Wong et al. (2007)? In the rest of this chapter we will examine the strategic changes made by one of the universities in Singapore, the NUS, in pursuit of an entrepreneurial university model.

## 7.3 OVERVIEW OF NUS AND ITS TRANSITION TOWARDS AN ENTREPRENEURIAL UNIVERSITY MODEL

### 7.3.1 Overview of NUS

Established in 1905, NUS is the oldest and largest among the three public universities in Singapore, and the only one that can be regarded

*Table 7.1 Profile of National University of Singapore (NUS) (AY 2008–09)*

| Indicators | AY 2008–09 |
|---|---|
| Faculty members | 2207 |
| Research staff | 2098 |
| Undergraduate students enrolled | 23 822 |
| Graduate students enrolled | 7163 |
| Total research funding | S$469 million[a] |
| Journal publications in SCI/SSCI/A&HCI | 3565[b] |
| Patents filed | 56 |
| Patents granted | 25 |
| Cumulative number of US patents granted (calendar years 1990–2009) | 266 |
| Cumulative journal articles in Thomson-Scientific-indexed journals (1999–2009) | 30 780 |

*Notes:*
a. FY 2008–09.
b. Calendar year 2008.

*Source:* NUS *Research Report 2007–2008*; *NUS Annual Report 2009/2010*; USPTO Database of Issued Patents; ISI Essential Science Indicators.

as having the status of a 'Doctoral/Research University-Extensive' under the Carnegie Classification of Institutes of Higher Learning in terms of the breadth of its faculties and the depth of its research output. Table 7.1 provides a summary profile of NUS for the academic year 2008–09. With a total teaching faculty of over 2200 and a total student population of 31 000, NUS is among the largest universities in Asia. With an annual R&D budget of about S$469 million, NUS alone constitutes about 8 percent of total R&D spending in Singapore; with 266 US patents granted as of 2009, NUS is also the fourth-largest US patent holder in Singapore.

Despite its early establishment more than a century ago, NUS was for many years primarily a local teaching institution, and it was not until the 1980s that it began to take on some research functions. NUS' intake of overseas students also only began to expand significantly since the late 1980s. Despite being a relatively latecomer in the international higher education scene, NUS' research capabilities and educational standards began to attract international recognition since the 1990s, as evidenced by the university's commendable placing in a number of international rankings of IHLs (Institutes of Higher Learning). For example, since 2004 NUS has generally been ranked among the top 30 universities in the world, both

*Table 7.2 Ranking of NUS in the World University Rankings by the Times Higher Education Supplement, 2004–08*

| | 2004 | 2005 | 2006 | 2007 | 2008 | 2009 |
|---|---|---|---|---|---|---|
| Overall | 18 | 22 | 19 | 33 | 30 | 30 |
| Biomedicine | 25 | 15 | 9 | 12 | 17 | 20 |
| Science | 35 | 34 | 22 | 25 | 31 | 27 |
| Technology | 9 | 9 | 8 | 10 | 11 | 14 |
| Social Sciences | 10 | 13 | 11 | 20 | 18 | 20 |
| Arts and Humanities | 17 | 56 | 22 | 21 | 30 | 23 |

*Source:* Knowledge Enterprise Online, various issues, available at http://newshub.nus.edu.sg/; *Times Higher Education Supplement* (various years); THE-QS World University Rankings, http://www.topuniversities.com/worlduniversityrankings/.

overall and by key faculties, according to the *Times Higher Educational Supplement* (THES) World University Rankings (WUR) (Table 7.2). In another ranking, the Academic Ranking of World Universities (ARWU) by Shanghai Jiaotung University's Institute of Higher Education, NUS' overall rank was lower (within the 101–151 band in 2008) due to the methodology's heavy emphasis on Nobel Prizes; in the engineering and information technology (IT) fields where no Nobel Prizes are awarded, NUS was ranked 29th by ARWU in 2008.

### 7.3.2 NUS' Shift Towards an Entrepreneurial University Model: The 'Global Knowledge Enterprise' (GKE) Strategy

#### 7.3.2.1 Impetus for change

Like most other public universities developed under the British Commonwealth tradition, the primary mission of NUS has been teaching, with research as a secondary mission (Etzkowitz et al. 2000). While the 1980s and 1990s saw increasing emphasis on research, mirroring the steady shift in the Singapore economy towards R&D (Wong 2006), it was not until the mid-1990s that NUS began to establish a technology licensing office. Even then patenting activities were relatively low (only 21 US patents granted cumulatively up to 1997) and involvement of professors and students in spinning off companies was relatively rare before 2000.

The major impetus for more rapid growth in technology commercialization activities came only in the late 1990s, partly precipitated by the Asian financial crisis in 1997 that caused a recession in the Singaporean economy in the following year. A major economic restructuring strategy was formulated by the government in response that highlighted the need to accelerate the shift towards an innovation-driven, knowledge-based economy and recognized the impetus for the local university to play a greater role in technology commercialization (ERC 2002).

The impetus for change from the external environmental coincided with the imminent retirement of the incumbent vice-chancellor of NUS, thus providing an opportunity for the government to effect change at NUS at a much faster pace by bringing in an external candidate for the top leadership role at NUS (now retitled 'president' to reflect the shift towards the US university model as the global benchmark). Singapore-born and Harvard-trained, the new president Professor C.F. Shih came to NUS with strong credentials, having prior work experience in a leading US corporation (General Electric) and a research leadership role at the helm of a major research lab in an Ivy-League university (Brown University) in the USA. With strong government support, Shih provided the crucial leadership to accelerate significantly the pace of change of several initiatives that were already in motion, but more importantly he initiated a strategic shift towards embracing an 'entrepreneurial university' model. As is typical of visionary leaders, he did this by establishing a new strategic vision for the university that encompassed an explicit recognition of entrepreneurial development, and by creating a new organizational vehicle to spearhead the change to the new vision.

#### 7.3.2.2 Setting strategic direction for change: the 'Global Knowledge Enterprise' vision

As part of his strategy to set NUS on a new direction for development, President Shih articulated a vision for the university to become a 'Global Knowledge Enterprise'. This vision statement signaled two strategic dimensions for change: firstly, the need for NUS to compete in the global arena rather than within the national economy; and secondly, the need to inject an entrepreneurial dimension to NUS' research and educational activities, particularly in terms of accelerating the commercialization of the university's knowledge and producing more entrepreneurially minded graduates. The two dimensions in the new vision statement – global and enterprise – are closely intertwined, giving rise to both synergy and tensions. On the one hand, since the domestic market of Singapore is small, success in knowledge commercialization requires mindsets and competencies to compete globally and hence entrepreneurial development activities must have a global orientation. On the other hand, the overall drive towards academic excellence at a global level also exerts significant constraints on what can be done on the enterprise dimension – the pursuit of enterprise activities cannot be at the expense of achieving global excellence in research and education. In effect, with this vision statement NUS aimed to level up substantially in its global competitive standing in terms of academic research and educational excellence, while at the same time differentiating itself from other academically excellent universities in the world by a special emphasis on being entrepreneurial. Politically, the vision statement reassured the academic faculty and government of the priority that NUS would place on pursuing global academic excellence, while allowing room for new initiatives towards entrepreneurship to be explored.

A major development in the drive to become globally competitive is the shift of emphasis away from local manpower development towards making the university a global educational hub that attracts top foreign students and faculty in increasing competition with other leading universities in the world. In line with this globalization drive NUS also began revising faculty compensation and policy, making it more flexible so that the university could pay more to attract top faculty and also reduce pay for the underperforming faculty. Tenure and promotion policy was made much more stringent and performance-based in line with the benchmarks of leading universities in the USA. Intake of foreign students also increased, while more local students were encouraged to go on exchange programs abroad for at least a semester to gain international exposure.

To give NUS the needed institutional flexibility to become globally competitive, the Singapore government allowed NUS to become corporatized in 2006. Although public funding support remains significant, with clear strategic goals to be achieved every three to five years, NUS was given much greater autonomy in how it operated to achieve the targeted outcomes. Moreover after corporatization NUS was encouraged to aim high, with the government indicating that it was willing to commit higher funding if chances of achievement were realistic.

### 7.3.3 Organizational Strategy: The Creation of NUS Enterprise

Complementing the new vision statement, President Shih created a new division in the university to spearhead the pursuit of activities to make NUS entrepreneurial. To serve as the first chief executive officer (CEO) of the new organization, President Shih personally hand-picked a professor from the engineering school who had been among the first to start a venture to commercialize his inventions. Under the broad mission to inject a more entrepreneurial dimension to NUS education and research, the CEO, Professor Jacob Phang, was tasked to make NUS Enterprise the primary organizational vehicle for coordinating and managing all major university activities related to technology commercialization and entrepreneurship promotion within NUS. As there were already several existing units within the university that were related to that broad mission, the new CEO was asked to take these under his wing and restructure them as he

*Table 7.3 Functional divisions of NUS Enterprise, 2009*

| Units of NUS Enterprise as of 2009 | Core functions |
|---|---|
| NUS Entrepreneurship Center (NEC) | • Experiential entrepreneurship education<br>• Outreach<br>• Innovation and entrepreneurship research<br>• Entrepreneurship support services, including seed-funding of NUS-related start-ups, enterprise incubation and mentorship programs |
| Industry Liaison Office (ILO) | • Technology licensing and intellectual property(IP) management<br>• Industrial liaison |
| NUS Overseas College (NOC) | • Overseas high-tech start-up internship cum education program |
| NUS Extension (NEX) | • Continuing education |
| NUS Publishing (NPU) | • University press |

saw fit, but he was otherwise given great latitude to define and implement new initiatives to make the university 'more enterprising'. An annual operating budget equivalent to less than 1 percent of the university's overall budget was set aside to kick-start the new division.

The fact that the CEO of NUS Enterprise would report directly to the university president and have a broad mission was significant because it gave him a mandate to adopt an integrated, coordinated approach to perform the technology commercialization roles of an entrepreneurial university. Such an integrated approach was indeed taken up by the CEO, who after some early experimentation, assembled and incorporated into the new NUS Enterprise cluster seven major operating units as of 2004. In early 2006 the founding CEO decided to relinquish his position so that he could devote full time to growing the high-tech venture that he founded before he took on the NUS Enterprise CEO position. A new CEO, Dr Lily Chan, was recruited from outside NUS. She had significant prior technology commercialization experience: having been a biomedical researcher and having run a biomedical start-up, she later joined the Economic Development Board to promote biomedical investment in Singapore before becoming a key member of the team running BioOne, a $1 billion VC Fund set up by the government to invest in life science start-ups. Building upon what was established earlier, Dr Chan streamlined the various operating units into five as shown in Table 7.3.

The five units can be divided into two groups. The first group comprises the NUS Entrepreneurship Center (NEC), the Industry Liaison Office (ILO) and the NUS Overseas College (NOC). The second group comprises NUS Extension (NEX) and NUS Publishing (NPU) which were incorporated as independent corporate entities and operate as relatively autonomous businesses. We will focus our discussion on the first group of operating units as they are integral to NUS Enterprise's core mission of facilitating university technology commercialization, as opposed to the second group which is somewhat peripheral. While many of the functions performed by these three core operating units are conventional and could be found in many universities, what distinguish the NUS Enterprise model are (a) its effort in coordinating and creating synergies across functions and (b) its entrepreneurial approach in spawning a number of new initiatives that represented significant innovations for NUS. We highlight below a number of these new initiatives.

### 7.3.4 Key Innovative Programs of NUS Enterprise

#### 7.3.4.1 Entrepreneurship education initiatives of NUS Entrepreneurship Center (NEC)

NEC was established in NUS Enterprise with three aims: (i) promoting entrepreneurial learning among the NUS community through experiential educational programs and various outreach activities; (ii) providing entrepreneurship support services to NUS professors, students and alumni wanting to start up their own ventures; and (iii) advancing knowledge in the policy and practice of technology venturing in Singapore and the region through research. The center was originally established in 1988 as a university-level center called the Center for Management of Innovation and Technopreneurship (CMIT). In 2001 it became a division of NUS Enterprise and was renamed as the NUS Entrepreneurship Center (NEC).

Like entrepreneurship centers established in many universities, NEC offered a wide range of educational programs, both within the classroom and extra-curricula. What distinguished NEC's non-classroom educational activities was an explicit attempt to help evangelize and catalyze the development of the external venture ecosystem, and to make NUS a magnet for entrepreneurial networking activities that link the NUS community with the external venture ecosystem. As highlighted earlier, this venture ecosystem catalysis role was important because the external venture ecosystem was still relatively underdeveloped in Singapore. For example, NEC pioneered the annual national business plan competition, Startup@Singapore, in Singapore in 1999. Although the competition drew its initial inspiration from the MIT $50K Competition, NEC opted to open the competition to the entire Singapore community rather than NUS students only. The director of the center also helped found and chaired a business angel networking group (Business Angel Network Southeast Asia (BANSEA)) in Singapore in 2002 that operated like the Band of Angels in Silicon Valley, thereby helping to plug NUS into the nascent angel investment community in Singapore.

Another distinctive approach adopted by NEC was to launch a Technopreneurship Minor Program that was open to all NUS undergraduate students, albeit with a strong focus on students in science and engineering. The goal was to make students in technical disciplines more business-savvy and entrepreneurially minded, and helped address a weakness in the existing educational programs of NUS where students in technical disciplines had few opportunities to learn entrepreneurial skills. Interestingly, NUS Enterprise initially approached the Business School in NUS to offer such courses to technical students but the school declined as it was uncertain of the sustainability of demand. Although NEC was not an academic department, thanks to the strong championing of NUS Enterprise, it was granted a special academic teaching unit status by the provost to mount the courses. Through an active marketing program that leveraged on the center's various non-classroom educational activities, NEC was able to increase enrollment rapidly in its technopreneurship courses from less than 200 in the first year to over 1300 by 2005–06. Having been shown to be viable, the minor program was transferred to the NUS Business School in 2007. NEC subsequently launched a new entrepreneurship learning program called Innovative Local Enterprise Achiever Development (iLEAD) that offers a seven-month internship opportunity in selected local high-tech start-ups for NUS undergraduates, followed by a two-week overseas study visit trip to Silicon Valley or China.

With new funding support from the NRF UIF in 2009, the center launched several new educational initiatives to broaden its entrepreneurship education offerings. These include a new experiential learning initiative for PhD students called the Extra Chapter Challenge, which offers a six-month fellowship extension to PhD students who have made discoveries or inventions in their thesis research that have commercial potential. The program gives these students the opportunity to develop a commercialization plan under the supervision of a mentor with significant business experience in the industry where the inventions could be exploited. Another innovative initiative is the Innovation and Entrepreneurship Practicum grant scheme that provides seed-funding of up to S$10000 each to students to develop their innovative ideas into viable product prototypes or business plans. A third initiative involves a collaboration with Grameen Creative Lab (GCL) to provide incubation and educational training for social innovators and entrepreneurs to launch social businesses with scalable, sustainable social impacts on low-income groups at the Bottom of the Pyramid (BOP).

Another function of NEC was to provide targeted assistance to NUS spin-offs. Because of the relative underdevelopment of the venture ecosystem, academic entrepreneurs who wanted to spin off their own ventures faced a severe 'valley of death'. Thus one of NUS Enterprise's first acts was to establish an NUS Venture Support (NVS) Division that offered incubator facilities on campus to give physical infrastructural support as well as selected advisory services to NUS-related start-ups. However it quickly recognized that this was not enough. Through the active championing of the then CEO, NUS Enterprise was able to convince NUS senior management to provide an allocation of S$5 million (in two tranches) for NVS to establish a seed-funding scheme, the NUS Venture Support Fund (NVSF). NVSF provided very early-stage seed-funding of up to S$300 000 each to promising NUS spin-off companies by NUS faculty, students and alumni. The aim was to incubate these ventures so that they could become viable candidates for external funding by venture capitalists, angel investors or corporate investors, or until they achieved significant organic growth. Between 2004 and 2006 NVSF funded a total of 11 NUS start-ups, out of over 40 that applied. Additionally a separate student start-up fund known as the Fund for University Student Entrepreneurs (FUSE) was also established in 2005 to provide developmental funding to new ventures started by students or recent graduates. A total of 21 student start-ups were funded under FUSE by the end of 2006. Co-funding for these two schemes was obtained from two public funding sources – SPRING SEEDS, a government co-funding scheme which provided 1:1 matching for NVSF (capped at S$300 000) and the SPRING Entrepreneurial Talent Development Fund (ETDF), another scheme targeted at student entrepreneurs that provided 3:1 matching for FUSE (capped at S$45 000 per company).

Although NVS was initially established as a separate unit from NEC, it received close advisory and management support from NEC, including the formation of the investment committee. After the new CEO came on board in 2006, NVS was merged into NEC as part of the organizational streamlining. Besides expanding the physical incubation services, NEC also initiated a mentorship program that brought experienced entrepreneurs, investors and senior executives to campus to provide regular coaching to NUS spin-off companies. A number of technology-specific seed-funding schemes were also introduced, including one for IDM and another for clean-tech. NEC has also developed a program to assist NUS professors and students to apply for various new government grant schemes that have been launched in recent years to support university technology commercialization, such as the SPRING Proof of Concept (POC) and Proof of Value (POV) grants. More recently NEC has initiated an overseas launch-pad program to help NUS spin-offs enter key overseas markets quickly, with the first such launch-pad located in Silicon Valley.

#### 7.3.4.2 The NUS Overseas College (NOC) program

The NOC program was introduced by NUS Enterprise in 2001 as a new initiative that integrated both dimensions of globalism and entrepreneurship. Conceived originally by the university president, the basic concept of the NOC program was to send the brightest and most entrepreneurially minded NUS undergraduate students to different entrepreneurial hubs in the world to work as interns in high-tech start-up companies for one year, during which time they would also take courses related to entrepreneurship at partner universities in each of the regions. In essence the NOC program represented an experiment in learning entrepreneurship by 'immersion', that is; by immersing students as 'apprentices' in high-tech start-ups or growth enterprises in foreign locations to expose them to the tacit aspects of entrepreneurial practice and foreign business culture. The program was conceived as a long-term investment in grooming future entrepreneurial leaders with global mindsets and network connections; as such it was not expected that the students would start their own ventures right after graduation, but rather that they would be infused with an entrepreneurial mindset that would orient their future research towards commercializable innovations and influence their future career choices towards more entrepreneurial pursuits. The program also aimed to help them establish valuable life-long social networks with the entrepreneurial communities of high-tech hotspots overseas so that they would be more inclined towards, and better equipped for, working in or starting new high-tech start-ups with global aspiration.

The first such NOC program was launched in Silicon Valley in 2002, followed by Philadelphia in 2003, Shanghai in 2004, Stockholm in 2005, Bangalore in 2006 and Beijing in 2009. In steady state, the program aimed to accommodate a total of about 200–250 students per year. NEC worked closely with NOC to provide the educational support to these overseas internship programs by helping to develop the academic collaborations with the selected partner universities in the respective overseas location (for example, Stanford in Silicon Valley, Fudan in Shanghai, the Royal Institute of Technology (KTH) in Stockholm and Tsinghua University in Beijing), and by conducting additional entrepreneurship courses not available from the partner universities at those overseas locations as well as providing academic supervision of internship-based student projects.

Through the organizational mechanism of NUS Enterprise, NUS

was able to launch the NOC program in a relatively short time, cutting through red-tape that would otherwise have ensnarled the program at various faculty-level committees. The strong sponsorship of the president, who emphasized that the NOC program was to be the flagship program that would give NUS high visibility in supporting entrepreneurship education, coupled with the entrepreneurial drive of the NUS Enterprise CEO, helped in speeding the approval process. The close cooperation between NEC and NOC also helped in ensuring smooth coordination and compatibility of academic course content between the NOC and Technopreneurship Minor programs.

Despite the high cost of the program (substantial subsidies are provided by NUS to each NOC participant), the NOC program appears to have achieved its objective of making the overall NUS Enterprise initiative highly visible, not only in the local media but also internationally. An increasing number of NOC participants who have completed the program and returned to NUS are full of praise and appreciation of the transformational impacts of the program. Their word-of-mouth viral marketing has spread through the Singaporean student communities and raised the profile of NUS as an entrepreneurial university, attracting the more entrepreneurially minded high school students to apply to NUS. As a significant proportion of the NOC participants are foreign students from other Asian countries, this visibility is also spreading to the student communities in these countries as well.

#### 7.3.4.3 Restructuring of technology licensing functions

NUS' technology transfer office was originally set up in 1992 to handle technology transfer matters as part of the Industrial and Technology Relations Office (INTRO), including the protection and licensing of NUS intellectual property (IP), and advising of industry, faculty and staff on matters relating to IP. In the initial years the office did not have a full complement of experienced IP professionals, and so it was not perceived as being a major source of assistance by either NUS inventors or industry seeking IP for licensing. A number of instances were also reported of the office making tough demands in terms of extracting licensing fees. For these reasons, one of the first acts of the CEO of NUS Enterprise upon taking over INTRO was to introduce a number of major initiatives to make the office more 'inventor friendly', with less emphasis on maximizing licensing revenue, and a stronger focus on getting greater and faster deployment of NUS technology to the marketplace whether through licensing to existing firms or spinning off new firms. NUS Enterprise also set a policy of taking equity in lieu of royalty payments, and of simplifying the process of negotiation on equity valuation, for start-ups by NUS professors and students. INTRO was renamed Industry Liaison Office (ILO) in 2006 and reorganized to facilitate greater synergy between university-industry research collaboration and technology transfer facilitation. The professional staffing of the office was strengthened, and a more proactive approach was implemented to map out areas of technological strengths of the university so as to guide future innovation directions and to facilitate patent portfolio aggregation to achieve critical mass in selected technology areas.

### 7.3.5 Evolution of Organization and Strategy

Reflecting the entrepreneurial orientation of the founding CEO, the first three to four years of NUS Enterprise were rather fluid and fast-paced, at times even chaotic, as the organizational leadership emphasized a bias for action and getting things done rather than putting in place well thought out standard operating procedures and management systems. By empowering operating unit directors with strong ideas to champion them, this entrepreneurial organizational culture facilitated the rapid introduction of a number of innovative programs that have been highlighted in Section 7.3.4 earlier. As with all entrepreneurial organizations, however, there was a tendency for too many initiatives to be started without enough systems and control procedures in place to ensure timely documentation and consistent application of management policy.

The departure of the founding CEO in early 2006 was an opportune time for the president to bring in a new leader, Dr Lily Chan, to help manage the transition of NUS Enterprise to a more mature and stable organization. With a strong track record in managing VC investment and a prior background in running a biotech start-up, the new CEO quickly put in place a more balanced organizational development approach with a stronger emphasis on management systems. Reaffirming the integrated NUS Enterprise approach, she emphasized the need for greater coordination and cooperation among the units of NUS Enterprise by consolidating some of the separate units. For example, the Venture Support function was absorbed into NEC, INTRO was reorganized into ILO and cooperative activities were institutionalized across the different operating units rather than leaving it up to the initiative of their directors. She also pushed ILO and NEC to take a more proactive approach to seek out inventive and entrepreneurially minded faculty members, rather than waiting for them to apply. Finally the strategic goals of NUS Enterprise have also been refined over time to focus on two core missions: embedding entrepreneurial learning as an integral part of an NUS education by providing experiential entrepreneurial education; and translating NUS research into innovation by providing industry partnership and entrepreneurship support.

Through these two strategic thrusts, the ultimate aim of NUS Enterprise

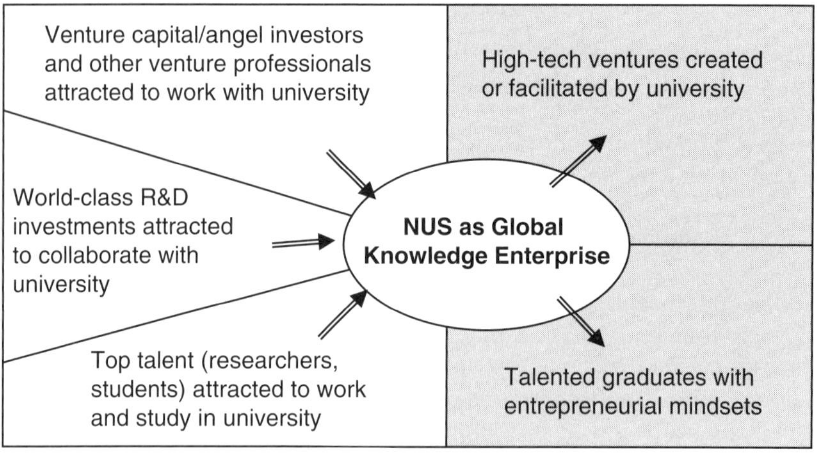


*Figure 7.1 Vision of NUS' pivotal role in Singapore's knowledge economy*

is to help NUS to achieve its vision of becoming a GKE and playing a pivotal role in Singapore's knowledge economy by attracting and transforming the relevant innovative resources into entrepreneurial output (Figure 7.1).

In summary, notwithstanding organizational changes, leadership transition and strategy evolution, it is fair to say that NUS has persevered in pursuing an integrated, coordinated approach to promoting technology commercialization and academic entrepreneurship through the organizational vehicle of NUS Enterprise, which is autonomous from the conventional administrative and academic departments and reporting to the university president through a high-level Entrepreneurship Committee that he co-chairs. Despite a change in the university presidency in 2008, NUS Enterprise has continued to receive sustained support from the senior management of NUS.

## 7.4 IMPACT OF NUS' SHIFT TOWARDS THE ENTREPRENEURIAL UNIVERSITY MODEL

### 7.4.1 Overview of Impact

The shift towards the 'entrepreneurial university' model has occurred in less than ten years, with much of the changes really gaining momentum only in the last five years. While it may thus be premature to assess the long-term impacts of these changes, some visible outcomes can be discerned especially in the last two or three years.

*Table 7.4 Profile of changes in NUS, before and after shift to entrepreneurial university model*

| Indicator | AY 1996–97 | AY 2008–09 |
|---|---|---|
| Teaching staff | 1414 | 2207 |
| percent foreign | 39.0% | 51.9%[a] |
| Research staff | 843 | 2098 |
| percent foreign | 70.1% | 78.6%[a] |
| Undergraduate students enrolled | 17960 | 23822 |
| Graduate students enrolled | 4478 | 7163 |
| Graduate students as percentage of total student enrollment | 20.0% | 22.7% |
| Percentage of foreign students studying at NUS | 13%[b] | 35.2% |
| Total research funding | S$102 million | S$469 million[c] |
| Journal publications in SCI-Expanded/SSCI | 1307[d] | 3545[e] |
| Patents filed | 13 | 56 |
| Patents granted | 4 | 25 |
| Total patents granted by USPTO | 21[f] | 266[g] |
| Total spin-offs using NUS IP | 6[f] | 47[g] |

*Note:*
a. Percentage for FY 2004–05.
b. Percentage of total student intake for 1997–98.
c. Figure for FY 2008–09.
d. CY 1996.
e. CY 2009.
f. CY 1990–97.
g. CY 1990–2009.

*Source:* NUS *Research Report* (various years); NUS *Annual Report* (various years); USPTO Database of Issued Patents; ISI Web of Science; NUS Office of Research.

Table 7.4 provides an overview of the key changes that NUS has undergone between Academic Years 1996/97 and 2008/09, showing that NUS has not only achieved considerable progress in the conventional performance dimension of education and research output, but has performed even better in the new dimensions of foreign talent attraction, entrepreneurship promotion and technology commercialization. These are further elaborated below.

### 7.4.2 Manpower Development and Attraction of Foreign Talent

Between 1996/97 and 2008/09 undergraduate enrollment in NUS increased by about 30 percent, but graduate enrollment grew much faster (60 percent) reflecting an increasing focus by NUS on the training of graduate research students, the bulk of whom were enrolled in the engineering and science faculties. The number of faculty staff likewise increased at a faster rate (56 percent), with research staff increasing even faster (149 percent). An important part of the growth in NUS students and faculty/researchers over the last ten years can be attributed to the ability of NUS to attract foreign talent. As can be seen from Table 7.4, between AY1997/98 and AY 2008/09 the proportion of foreign students in NUS' student population nearly tripled from 13 percent to 35 percent. While less dramatic, a significant increase is also evident in the rising share of foreigners among faculty and research staff recruitment since 1997, although the latest data is for 2005 only. The share of foreigners among faculty members increased from 39 percent in 1997 to over 50 percent by 2005, while the foreigners' share among researchers increased from 70 percent to almost 80 percent. The biggest source of foreign researchers came from mainland China, representing more than half of all foreign researchers by 2005.

### 7.4.3 Knowledge Creation through Research Publications

The output of NUS research publications, as measured by internationally referred publications covered by Thomson Reuter's SCI/SSCI/AHI, has grown fairly steadily over the years, from 1556 in 1997 to 3565 in 2009 (an increase of 129 percent). More importantly the increase in quantity of research publications has been achieved with a simultaneous improvement in quality as measured by the ratio of citations/publications. According to a study by the Higher Education Evaluation and Accreditation Council of Taiwan (HEEACT), NUS' global ranking in 2008 in terms of average citations per paper was 150th (out of 499 universities ranked) when analysed over the last 11 years but rose to 54th if analysed over only the last two years (HEEACT website). Indeed the slowdown in the rate of growth of publications in recent years partly reflects a changing emphasis of the university on quality over quantity of publications.

### 7.4.4 Knowledge and Technology Commercialization

In tandem with the growth in new knowledge creation between 1997 and 2009, the pace of knowledge commercialization has also increased. These changes are analysed in the next two sub-sections.

#### 7.4.4.1 Patenting

The average number of NUS patent applications per year was 77 for the period 1997–99. This has nearly doubled to 145 in the period 2003–09. The

*Table 7.5 Number of patents filed by and granted to NUS, FY 1997–2009*

| Financial Year | Patent applications | | Patents granted | |
|---|---|---|---|---|
| | Total no. | Average no. per year | Total no. | Average no. per year |
| 1997–99 | 230 | 76.7 | 39 | 13.0 |
| 2000–02 | 304 | 101.3 | 72 | 24.0 |
| 2003–09 | 1018 | 145.4 | 226 | 32.3 |

*Note:* Figures include patents filed in multiple countries.

*Source:* NUS *Research Report* (various years); NUS ILO; NUS *Annual Report*.

*Table 7.6 Composition of NUS patents by technology category, 1990–2009*

| Technology category | Number of patents | % |
|---|---|---|
| Electrical and electronic | 83 | 31.2 |
| Computers and communications | 65 | 24.4 |
| Chemical | 45 | 16.9 |
| Drugs and medical | 44 | 16.5 |
| Mechanical | 18 | 6.8 |
| Other | 11 | 4.1 |
| Total | 266 | 100.0 |

*Source:* Calculated from USPTO Database of Issued Patents.

number of patents granted per year has increased even more significantly, from 13 per year over 1997–99 to 32 per year over 2003–09 (Table 7.5). By the end of 2009 NUS had become the fourth-largest holder of Singapore-based inventions granted US patents, accounting for 4.4 percent of all US patents granted to Singapore-based inventors.

The engineering faculty's dominance in NUS patenting is apparent from the technological fields in which the patents fall. Of the 266 unique US patents granted to NUS by the end of 2009, almost one-third were in the electric and electronic technological category, with another 24 percent in computer technology. The other two major technology fields were chemicals and biomedical, both around 17 percent (Table 7.6).

The patenting data also reveals that there has been an increase in collaborative innovation activities between NUS and external organizations

*Table 7.7 NUS licensing agreements 1987–2009*

| | No. | % |
|---|---|---|
| 1987–96 | 29 | 10.8 |
| 1997–99 | 31 | 11.5 |
| 2000–05 | 178 | 66.2 |
| 2006–09 | 31 | 11.5 |
| Total | 269 | 100.0 |

*Source:* NUS ILO.

in Singapore from 2000. The proportion of NUS patents jointly owned with external organizations increased from 18 percent over 1995–1999 to 30 percent over 2005–09.

#### 7.4.4.2 Licensing

Although the proportion of NUS inventions that were licensed out remains relatively low, an increase in the volume of technology licensing activities since 2000 is evident from Table 7.7. As of the end of 2009 NUS had made 269 technology licensing agreements. Of these, less than one-quarter were issued before 2000; the bulk of licenses were signed in the ten years between 2000 to 2009. The number of licensing deals concluded in the more recent period (2006–09) is lower than the 2000–05 period, reflecting a changing policy of ILO; focus is now given to signing a smaller number of licenses with higher revenue potential rather than going for quantity regardless of revenue potential.

Indeed in terms of licensing income generated, total revenue has been on an upward trend since 1999, notwithstanding the reduction in the number of licensing deals in the more recent period. Although the amount appears to be somewhat erratic from year to year, average licensing royalties have increased from an average of S$83 800 per year before 2000 to over S$250 000 over 2000–05, and further to S$508 000 over 2006–09 (Table 7.8). As highlighted in Section 7.3.4.3 earlier, ILO has consciously been pursuing a 'balanced' approach to technology licensing in recent years, with priority placed on promoting diffusion of university technology for impacts rather than licensing income maximization. To support academic spin-offs ILO has also opted to take equity in lieu of royalties when licensing technology to spin-offs by the academic inventors. Both these measures have the effect of reducing licensing income, at least in the short and medium term.

*Table 7.8 NUS licensing royalties, 1996–2009*

| | Total licensing royalties $'000 | Average licensing royalties per year $'000 |
|---|---|---|
| 1996–99 | 335.0 | 83.8 |
| 2000–05 | 1515.0 | 252.5 |
| 2006–09 | 3557.5 | 508.2 |

*Source:* NUS ILO.

*Table 7.9 Research collaboration agreements (RCAs) in NUS, 1995–97 versus 2005–09*

| | No. of RCA | Average no. per year | No. of RCA with industry | % of RCA with industry |
|---|---|---|---|---|
| 1995–97 | 109 | 36.3 | 47 | 43.1 |
| 2005–09 | 704 | 140.8 | 196 | 27.8 |

*Source:* NUS ILO.

### 7.4.5 Industry Collaboration

Another avenue for translating university research into technologies with commercialization potential is through collaborative research projects with industry. As can be seen from Table 7.9 the number of research collaboration agreements (RCA) undertaken by NUS has grown substantially over the last decade, rising from an average of 36 a year over 1995–97 to 141 over 2005–09. The absolute number of RCAs with industry has also increased, from an annual average of 16 over 1995–97 to 39 over 2005–09. Although the share of RCAs with industry has fallen over this time, it is possible that this is due to the smaller numbers of RCAs in the initial period. Aside from industry, RCAs have been formed with governmental bodies, public research institutes and foreign universities. In addition to the RCAs, NUS also interacts with industry through consultancy contracts. Significant consultancy work has been undertaken by NUS faculty, averaging about 700 consultancies over 2003–04. However no reliable statistics are available in more recent years due to the decentralization of such data collection to individual faculties.

Our findings of visible improvement in NUS' technology commercialization and industrial collaboration performance in recent years are consistent with a number of prior research on university-industry collaboration in Singapore. As reported by Wong (1999), a survey of manufacturing firms in Singapore in 1997 found that the propensity for technological collaboration between universities and private industry was relatively low, and that universities were not highly regarded by industry as an important source of technology. In a more recent study, Ho et al. (2006) found a visible increase in the propensity of university-industry collaboration as reported by industry.

As an example of recent successes arising from RCA between NUS and industry, the Kikkoman-NUS Collaborative R&D Lab, first established in 2006, successfully commercialized its first product (herbal soy sauce) in 2008. Several major new collaborative labs with industry were also established on campus over the last two or three years, including the NUS-Airbus Collaboration and the NUS-GE Water Technology Center.

### 7.4.6 Academic Entrepreneurship

As mentioned earlier, the lower ability of local firms to engage in knowledge commercialization means that universities in Singapore need to take a more direct role in commercializing their inventions by forming spin-offs, leading to a somewhat different commercialization model from the USA, for example, where technology licensing is emphasized more heavily (Heher 2006).

The results of NUS' change in policy to encourage technology commercialization through spin-off and start-up formation after 2000 are evident from Table 7.10, which shows the number of NUS-related spin-offs that have received some form of support from NUS Enterprise including seed-funding, incubation and exclusive licensing of IP to the founders. While some spin-offs received only one form of support, many received multiple forms of support, resulting in considerable overlaps between the various categories. For example, 58 out of the 62 incubatee companies received some form of seed-funding, while 14 out of the 40 spin-offs that received exclusive IP licensing also received some form of seed-funding support. Excluding such overlaps, more than 100 start-up companies have received at least one form of support services from NUS Enterprise over the years up to the end of 2009. Most of the NUS spin-offs, especially those that received seed-funding from NUS Enterprise, have been assigned a specific mentor to provide coaching and advice on how to cope with various start-up challenges such as finding customers and closing sales, finding and making pitches to VCs and angel investors, and so on.

Given that there were only 11 NUS spin-offs with exclusive IP licensing from NUS before 2000, and given that the seed-funding schemes and

*Table 7.10 NUS-related spin-offs supported by NUS Enterprise as of end of 2009*

| Nature of NUS Enterprise support | No. |
|---|---|
| No. of spin-offs founded by NUS faculty or students/alumni that received one form or other of NUS Enterprise's seed-funding schemes | 58 |
| No. of spin-offs founded by NUS faculty or students/ alumni that license IP from NUS (with or without NUS seed-funding) | 40 |
| No. of companies incubated in NUS Incubator (including companies that have graduated) | 62 |
| No. of NUS spin-offs (in any of the above categories) that have received follow-on investments by VCs/angel investors/corporate investors | 13 |
| No. of NUS spin-offs (in any of the above categories) that have received Technology Enterprise Communication Scheme (TECS) or NRF POC/POV grants | 11 |

*Source:* NUS Entrepreneurship Center.

physical incubation program did not commence until after 2000, it is clear that the volume of NUS spin-off activities increased significantly after the formation of NUS Enterprise.

Because most of the start-ups supported by NUS Enterprise are quite recent, many are in the early stages of growth and there have yet been few cases of significant commercial success. It is encouraging to note, however, that 13 of the companies supported by NUS Enterprise have received follow-on investment by an external investor (VC fund or angel investor) or have been acquired at valuation generating positive returns to the early equity investors. The total amount of external follow-on investments received by these 13 companies exceeded S$18.3 million. In addition NUS Enterprise also assisted 11 of its portfolio of spin-offs to successfully bid for a number of government POC grant schemes as of the end of 2009, with total grant values exceeding S$4.2 million.

Table 7.11 lists some examples of NUS spin-offs that can be deemed to have successfully exited from the various NUS Entrepreneurship support programs as of late 2010, either because they have received substantial follow-on investment by external investors or because they have been acquired by other companies at a valuation that gave positive returns to the original investors, including NUS Enterprise. An early example was WizLearn, which was spun off from the NUS Computing Center in 2001 to commercialize the learning management system that the center has originally developed

*Table 7.11 Selected examples of NUS spin-off companies that have successfully exited from NUS Entrepreneurship support program*

| Name of spin-off | Founders | Notable developments |
|---|---|---|
| Semicaps | Professor in Engineering School | Received follow-on investment from a VC |
| WizLearn | Spin-out from NUS Computer Centre | Merged with a local private company and later acquired by a large Indian corporation |
| Cadi-Scientific | Alumni | Seed-funded by NUS Enterprise, subsequently received follow-on investment from a VC |
| Fatte Telecoms | Professor in Engineering School | Received angel investment, later merged with a local company that became the first company quoted on Singapore's new over-the counter (OTC) capital market |
| Mozat | Two PhD students | Seed-funded by NUS Enterprise, subsequently received follow-on investment from an angel investor group and then a VC |
| tenCube | Three alumni, two of whom are NOC returnees | Winner of StartUp@Singapore business plan competition, seed-funded by NUS Enterprise, subsequently received follow-on investment from a corporate group in India and later acquired by US firm McAfee |
| MXR | Professor in School of Computing and his former PhD students | Seed-funded by NUS Enterprise, subsequently received angel investment |
| PEM | Professor in School of Computing and his former graduate students | Seed-funded by NUS Enterprise, subsequently received investment by a VC |
| Jitcomm | Former PhD student | Seed-funded by NUS Enterprise, subsequently received POC grants and investment by a VC |
| ClearBridge Biomedics | Professor in Bioengineering | Received investment from an external Incubator Fund |

*Source:* NUS Entrepreneurship Center.

for use by NUS students and faculty. Recognizing the commercial potential of this in-house software platform, NUS Enterprise made the initial investment to set up the company to market the software to other universities and organizations. Since the staff of the Computing Center preferred to stay on in NUS, NUS Enterprise also seconded one of its managers to run the business. The company was later merged with a private Singaporean company in educational software (Ask&Learn) and was subsequently acquired by a much larger Indian educational software company.

A more interesting case is tenCube, which benefited from several forms of help from NUS Enterprise. Founded by three students, two of whom gained their entrepreneurial exposure from the NOC program in Silicon Valley, the start-up participated in and won first prize at the StartUp@Singapore business plan competition. It was provided with incubation space in NUS, and went on to secure seed-funding from NUS Enterprise and received mentoring from a number of NUS Enterprise mentors on business strategies and gaining access to the local telecom operators. The company subsequently received follow-on investment from an Indian corporate group in 2009 and was acquired by the US firm McAfee in 2010.

Another interesting case is Personal e-Motion (PEM), founded by a professor in the School of Computing and two of his former students. The start-up was initially seed-funded and incubated by NUS Enterprise, which also provided extensive mentoring and introduction to potential venture investors. This helped guide the start-up to change its business model substantially, and eventually resulted in the company securing a VC investment and a significant partnership deal with a large overseas publisher.

A more recent example is Clearbridge Biomedics, a spin-off to commercialize a biochip invention by a professor in bioengineering. By trapping circulating tumor cells from just 1 ml of blood from a cancer patient, the biomedical device has the potential to change the entire spectrum of cancer therapy from diagnostics to treatment and post-treatment care. With the help of – and introduction by – NUS Enterprise, in 2010 the inventor was able to secure an investment from Clearbridge Accelerator, one of several incubators in Singapore co-funded by a new government scheme called the Technology Incubation Scheme (TIS).

As these examples illustrate, NUS Enterprise took a flexible approach to each start-up case and sought to find ways to add value depending on the special needs of each case. In deciding whether to provide the initial seed-funding, a key consideration is whether this would help the start-up to reach a stage of development that would significantly increase its chance of getting follow-on investment. Being an educational institution, NUS Enterprise recognizes that its role is not to help the start-up to grow all the way to success, but to plug the initial gap before outside investors are willing to invest. Where there is already external investment interest, as in the case of Clearbridge Biomedics, NUS Enterprise was content to play the match-making role. In other cases like WizLearn and PEM, NUS Enterprise provided significantly more hands-on help, while in the case of tenCube the contribution of NUS Enterprise arguably started with the experiential learning opportunities provided to the founders through the NOC program in Silicon Valley and the StartUp@Singapore business plan competition.

Over and above NUS Enterprise's targeted start-up support programs such as seed-funding, incubation and mentoring, it is likely that the various entrepreneurship education and outreach programs of NUS Enterprise have also had a broader impact on increasing the interest and propensity of NUS professors and students to be involved in entrepreneurial activities, although such impact is harder to quantify. Nevertheless a number of indicators are instructive. Firstly, based on recent annual graduate employment surveys, graduates from the various NOC programs appear to have exhibited three times the propensity to start their own business or to be employed in small start-up firms, compared with their peers. Secondly, the number of participants in the annual StartUp@Singapore business plan competition has been steadily increasing over the years, with a growing number of the participating teams subsequently applying for seed-funding or incubation at NUS. Thirdly, while there are no reliable statistics available to track the number of NUS-related start-ups that did not receive any form of NUS Enterprise support, one gauge of the level of such activities is the number of enquires and applications received by NEC that did not subsequently translate into actual support provided. This figure is more than six times larger than the number that actually received some form of support. Many of these potential start-ups, which were judged by NUS Enterprise to be non-viable, did not materialize; however it is likely that some may have gone on to become actual start-ups. Anecdotal evidence also suggests that some actual start-ups by NUS professors, students and recent alumni never approached NUS Enterprise for help in the first place.

## 7.5 BENCHMARKING OF NUS WITH OTHER UNIVERSITIES

While the earlier sections have highlighted the changes in knowledge creation and commercialization output of NUS over the years, it is useful to provide a more thorough benchmarking of NUS' technology commercialization performance versus other leading universities in the world.

Tables 7.12a and 7.12b provides a comparison of NUS against selected American and British universities on a number of technology

*Table 7.12a Comparisons of NUS versus leading universities, 2006*

| | Total academic staff | R&D expenditure, US$ million | % industry sponsored | No. of patents issued | No. of disclosures | No. of licenses | License income (US$'000) | No. of spin-offs |
|---|---|---|---|---|---|---|---|---|
| NUS | 1944 | 117.1 | 7 | 45 | 73 | 6 | 107[a] | 2 |
| MIT | 1805 | 1212.8 | 6.0 | 121[b] | 523 | 121 | 43 500.0 | 23 |
| Stanford University | 1772 | 699.2 | 5.1 | 118[b] | 518 | 109 | 61 310.7 | 7 |
| Georgia Institute of Technology | >900[c,d] | 467.7 | 8.9 | 39[b] | 366 | 25 | 1 817.3 | 8 |
| University of Pennsylvania | 3491[e] | 640.3 | 8.1 | 49[b] | 306 | 37 | 8255.1 | 3 |
| Cambridge University | 1558 | 335.7 | 10.8[c] | 41[f] | 118 | 60 | 5881[c] | 4 |
| Imperial College | 1114 | 360.6 | 13.4[f] | NA | 284 | 20 | 3002[c,g] | 6 |
| Oxford University | 1627[h] | 535 | 8.3[c] | 34[f] | 158 | 50 | 3992[f] | 7 |

*Notes:*
a. Data is for FY 2005/6.
b. US-issued patents only.
c. Data is for 2005.
d. Teaching staff only.
e. Data is for 2008, FTE.
f. Data is for 2003.
g. License income and patent recovery.
h. Data is for 2009.

*Sources:* AUTM STATT database; University websites; NUS *Research Report* 2006.

*Table 7.12b Comparisons of NUS versus leading universities, 2006 (selected ratios)*

| | Patents/ disclosure (%) | Licenses per 100 patents | Spin-offs per 100 patents | Disclosures per $million R&D spending | Patents per $million R&D spending | Licenses per $million spending | License income/ R&D spending (%) | Disclosures per 100 acad. staff | Patents per 100 acad. staff | Licenses per 100 acad. staff | Spin-offs per 100 acad. staff |
|---|---|---|---|---|---|---|---|---|---|---|---|
| NUS | 61.6 | 13.3 | 4.4 | 0.6 | 0.4 | 0.1 | 0.1 | 3.8 | 2.3 | 0.3 | 0.1 |
| MIT | 23.1 | 100.0 | 19.0 | 0.4 | 0.1 | 0.1 | 3.6 | 29.0 | 6.7 | 6.7 | 1.3 |
| Stanford University | 22.8 | 92.4 | 5.9 | 0.7 | 0.2 | 0.2 | 8.8 | 29.2 | 6.7 | 6.2 | 0.4 |
| Georgia Institute of Technology | 10.7 | 64.1 | 20.5 | 0.8 | 0.1 | 0.1 | 0.4 | 40.7 | 4.3 | 2.8 | 0.9 |
| University of Pennsylvania | 16.0 | 75.5 | 6.1 | 0.5 | 0.1 | 0.1 | 1.3 | 8.8 | 1.4 | 1.1 | 0.1 |
| Cambridge University | 34.7 | 146.3 | 9.8 | 0.3 | 0.1 | 0.1 | 1.8 | 7.6 | 2.6 | 3.9 | 0.3 |
| Imperial College | NA | NA | NA | 0.4 | NA | 0.2 | 0.8 | 25.5 | NA | 1.8 | 0.5 |
| Oxford University | 21.5 | 147.1 | 20.6 | 0.8 | 0.1 | 0.1 | 0.7 | 9.7 | 2.1 | 3.1 | 0.4 |

commercialization indicators for the year 2006, the latest year for which we could get comparable data. As can be seen, taking into account NUS' smaller R&D expenditure, its performance in translating R&D dollars to invention disclosures and patents is actually comparable or even higher than some of these benchmark universities, although it still lags behind in terms of licensing to patent ratio and licensing income to R&D spending. As highlighted earlier, this is understandable in view of the lack of readiness of private industry in latecomer NIEs like Singapore.

## 7.6 CONCLUSION

In summary, our empirical analysis indicates that NUS' contribution to Singapore's NIS has changed qualitatively in the period before and after 2000, shifting from being primarily a manpower provider and knowledge creator to take on a more visible role in knowledge commercialization through increased patenting, licensing to private industry and spinning off new ventures. Although the shift towards an 'entrepreneurial university' model may not be the sole cause, and some elements of the model may actually have begun in the mid-1990s, the experience of NUS is consistent with the hypothesis that universities in NIEs need to adopt key elements of the entrepreneurial university model in order to contribute more effectively to their NIS. While the specific governance model (for example, the NUS Enterprise cluster approach) and initiatives/programs (for example, the NOC program and the seed-funding programs) adopted may be unique to the Singapore context, the reform experience of NUS may nonetheless be instructive for other East Asian universities seeking to develop their own entrepreneurial university model.

While there is already some visible impact in terms of output of technology commercialization and academic entrepreneurship since the formation of NUS Enterprise, it is still unclear whether they represent long-term sustainable changes. Indeed to date there has been no outstanding success achieved in terms of major technological breakthroughs leading to substantial licensing revenue, or spin-offs that have become significant commercial success. It is thus by no means certain that NUS' shift towards the entrepreneurial university model will eventually lead to significant economic payoffs. Nonetheless we can take heart from the fact that some of the leading US universities in terms of technology commercialization also took a long time to achieve commercial viability in terms of their technology licensing office (Shane 2004).

Besides knowledge commercialization, our findings of a high and increasing level of recruitment of foreign students, researchers and faculty members by NUS are also consistent with our argument that an entrepreneurial university model for universities in small NIEs needs to incorporate the additional role of attracting foreign talent. While the level of involvement of foreigners in NUS is probably exceptional by the standards of East Asian universities and perhaps even when compared to Anglo-Saxon universities, it does suggest that the ability to compete for talent on a global scale is likely to be an important feature of any entrepreneurial university model for NIEs.

Last but not least, NUS' experiment in injecting a more entrepreneurial dimension to the educational experience of its students, particularly those in technical fields, may be instructive for many universities facing a similar challenge of making their technical graduates more business-savvy and entrepreneurially minded. The conventional solution of concentrating on technical specialization and leaving the acquisition of business skills and entrepreneurial acumen to a later stage (for example, through an MBA education) may not be optimal for the increasingly dynamic labor market of a global, knowledge-based economy where creativity, entrepreneurial mindset, social skills and international networking are increasingly important.

In conclusion, the case of NUS in Singapore illustrates an attempt by a university from a small NIE to transform itself to contribute more effectively to the dynamic shift of the country towards a regional knowledge-based economic hub. The approach adopted is admittedly risky: making technology commercialization and academic entrepreneurship an explicit goal of the university invites inevitable public and international scrutiny of the actual performance of the university in commercializing technologies. Given that many leading universities in the world have not yet been successful in promoting technology commercialization, the risk that NUS will fail to meet raised public expectation is significant. Moreover the pursuit of technology commercialization and academic entrepreneurship as an objective of the university requires extensive organizational changes and will encounter considerable resistance. These risks are nonetheless worth taking, for the rewards of success will be substantial.

## REFERENCES

Association of University Technology Managers (AUTM) (2006), AUTM STATT Database, available at http://www.autm.net/source/STATT/index.cfm?section=STATT (accessed December 2010).

Economic Review Committee (ERC) (2002), *Report of the Entrepreneurship and Internationalization Subcommittee*, Singapore: Ministry of Trade and Industry.

Etzkowitz, H. (2003). 'Innovation in innovation: the Triple Helix of university-industry-government relations', *Social Science Information*, **42** (3), 293–337.

Etzkowitz, H., A. Webster, C. Gebhardt, B. Regina and C. Terra (2000), 'The future of the university and the university of the future: evolution of ivory tower to entrepreneurial paradigm', *Research Policy*, **29** (2), 313–30.

Finegold, D., P.K. Wong and T.C. Cheah (2004), 'Singapore's emerging biotech cluster: old strategy replication or new approach?', *European Planning Studies*, **12** (7), 921–41.

Heher, A.D. (2006), 'Return on investment in innovation: implications for institutions and national agencies', *Journal of Technology Transfer*, **31** (4), 403–14.

Higher Education Evaluation and Accreditation Council of Taiwan (HEEACT), '2008 performance ranking of scientific papers by world universities', available at http://ranking.heeact.edu.tw/.

Ho, Y.P., P.K. Wong and A. Singh (2006), *Survey of Innovation, Intellectual Property Creation and Usage Among Singaporean Firms*, Singapore: IP Academy.

ISI Essential Science Indicators, available at http://thomsonreuters.com/products_services/science/science_products/a-z/essential_science_indicators (accessed December 2010).

Lee, C.M., W.F. Miller, M.G. Hancock and H.S. Rowen (2000), *The Silicon Valley Edge: A Habitat for Innovation and Entrepreneurship*, Stanford, CA: Stanford University Press.

National University of Singapore (NUS) (various years), *Annual Report*, Singapore: NUS.

National University of Singapore (NUS) (various years), 'Knowledge enterprise Online', available at http://newshub.nus.edu.sg/ (accessed December 2010).

National University of Singapore (NUS) (various years), *NUS Research Report*, Singapore: NUS Office of Research.

Saxenian, A. (2000), *Silicon Valley's New Immigrant Entrepreneurs*, San Francisco, CA: Public Policy Institute of California.

Saxenian, A. (2002), *Local and Global Networks of Immigrant Professionals in Silicon Valley,* San Francisco, CA: Public Policy Institute of California.

Shane, S. (2004), *Academic Entrepreneurship: University Spin-offs and Wealth Creation*, Cheltenham, UK and Northampton, MA, USA: Edward Elgar Publishing.

*Times Higher Education Supplement* (various years), 'THE-QS world university rankings', available at http://www.topuniversities.com/worlduniversityrankings/ (accessed December 2010).

United States Patent and Trademark Office (USPTO) website, available at http://www.uspto.gov/patft/index.html (accessed December 2010).

Wong, P.K. (1999), 'University-industry technological collaboration in Singapore: emerging patterns and industry concerns', *International Journal of Technology Management*, **17**, 270–84.

Wong, P.K. (2003), 'From using to creating technology: the evolution of Singapore's national innovation system and the changing role of public policy', in S. Lall and S. Urata (eds), *Foreign Direct Investment, Technology Development and Competitiveness in East Asia*, Cheltenham, UK and Northampton, MA, USA: Edward Elgar Publishing, pp.191–238.

Wong, P.K. (2006), 'The re-making of Singapore's high-tech enterprise ecosystem', in H. Rowen, W. Miller and M. Hancock (eds), *Making IT: The Rise of Asia in High-tech*, Stanford, CA: Stanford University Press, pp. 123–74.

Wong, P.K. and Y.P. Ho (2007), 'Knowledge sources of innovation in a small open economy: the case of Singapore', *Scientometrics*, **70** (2), 223–49.

Wong, P.K. and A. Singh (2008), 'From technology adopter to innovator: the dynamics of change in the national system of innovation in Singapore', in C. Edquist and L. Hommen (eds), *Small Economy Innovation Systems: Comparing Globalization, Change and Policy in Asia and Europe*, Cheltenham, UK and Northampton, MA, USA: Edward Elgar Publishing, pp. 71–112.

Wong, P.K., L. Lee, Y.P. Ho and F. Wong (2005), *Global Entrepreneurship Monitor 2004: Singapore Report*, Singapore: NUS Entrepreneurship Center.

Wong, P.K., Y.P. Ho and A. Singh (2007), 'Towards an "entrepreneurial university" model to support knowledge-based economic development: the case of the National University of Singapore', *World Development*, **35** (6), 914–59.

# 8. University technology commercialization in Taiwan: National Taiwan University (NTU) and National Taiwan University of Science and Technology (NTUST)

**Paul C.B. Liu, Dar-Zen Chen and Jen-Diann Chiou**

## 8.1 INTRODUCTION

In order to promote national economic growth and improve the quality of life for its citizens, Taiwan is devoting great efforts to increase scientific and technological (S&T) resources, accelerate research and development (R&D) and innovation, cultivate S&T personnel and develop key S&T industries. Moreover Taiwan makes good use of its strong information and communications technology (ICT) infrastructure, excellent R&D capabilities in the academic and research sectors and a successful transformation from an efficiency-driven industry to innovation-driven industry.

In Taiwan, S&T and industry-academia collaboration related policies are mainly formulated through important conferences such as the Science and Technology Meetings of the Executive Yuan, National Science Council Meetings, National Science and Technology Conferences, Science and Technology Advisory Board Meetings of the Executive Yuan and the Industry Technology Strategy Review Board Meetings. The consensus reached at these meetings are used to guide the formation of the policies above. These policies, regulations and government agencies involved will be introduced in this chapter. In the context of this socio-economic environment, the infrastructures and case studies of National Taiwan University (NTU) and National Taiwan University of Science and Technology (NTUST) on technology commercialization and industry-academia collaboration will also be covered.

# 8.2 SOCIO-ECONOMIC ENVIRONMENT AND POLICY

## 8.2.1 Resources of Scientific and Technological Development

### 8.2.1.1 Expenditure on R&D

Over the past five years, Taiwan's higher education expenditure on R&D has maintained growth rates of more than 4.5 percent each year. The government was the main provider and contributed more than four-fifths of the funds for R&D. It is worth noting that enterprise contributions rose from 3.2 percent in 2001 to 5.8 percent in 2005, proving that industry-academia collaborations have become more important in the business community.

### 8.2.1.2 Expenditure on industry-academia collaboration

Focusing on the the funding for industry-academia collaboration, the Taiwanese government also contributed more funds than enterprises. The funds from the government are 1.5 times more than the non-government sources. The statistics in Tables 8.1 and 8.2 divide schools into four

*Table 8.1 Taiwan expenditure on industry-academia collaboration, 2006–07 (funding from government)*

| School | Year | Government | | |
|---|---|---|---|---|
| | | Industry-academia collaboration | Vocational training | Total |
| National higher education | 2007 | 4314859 | 863455 | 5178314 |
| | 2006 | 5772921 | 345830 | 6118751 |
| Private higher education | 2007 | 1100521 | 232601 | 1333122 |
| | 2006 | 1489204 | 318914 | 1808118 |
| National professional education | 2007 | 944272 | 95727 | 1039999 |
| | 2006 | 805499 | 180835 | 986334 |
| Private professional education | 2007 | 614488 | 226732 | 841220 |
| | 2006 | 998832 | 181898 | 1180730 |
| Total | 2007 | 6974139 | 1418515 | 8392654 |
| | 2006 | 9066457 | 1027477 | 10093934 |

*Note:* Unit: NT$ thousands.

*Source:* Higher Education Evaluation and Accreditation Council of Taiwan (2008).

*Table 8.2 Taiwan expenditure on industry-academia collaboration, 2006–07 (non-government funding)*

| School | Year | Non-government | | | | |
|---|---|---|---|---|---|---|
| | | Industry-academia collaboration from enterprises | Vocational training from enterprises | Industry-academia collaboration from others | Vocational training from others | Total |
| National higher education | 2007 | 1 119 492 | 25 140 | 1 324 595 | 51 740 | 2 520 967 |
| | | ▲ 4.30% | ▲ 60.61% | ▲ 17.12% | ▼ –61.95% | ▲ 7.00% |
| | 2006 | 1 073 381 | 15 653 | 1 130 974 | 135 992 | 2 356 000 |
| Private higher education | 2007 | 531 524 | 16 458 | 774 297 | 18 761 | 1 341 040 |
| | | ▲ 32.52% | ▼ –69.47% | ▲ 30.35% | ▼ –6.28% | ▲ 25.45% |
| | 2006 | 401 091 | 53 904 | 593 997 | 20 018 | 1 069 010 |
| National professional education | 2007 | 352 543 | 49 164 | 125 562 | 21 771 | 549 040 |
| | | ▼ –0.15% | ▼ –52.57% | ▼ –29.46% | ▼ –9.35% | ▼ –16.65% |
| | 2006 | 353 065 | 103 649 | 177 990 | 24 015 | 658 719 |
| Private professional education | 2007 | 657 455 | 6764 | 147 102 | 22 384 | 833 705 |
| | | ▲ 31.93% | ▲ 53.88% | ▲ 4.86% | ▲ 6.11% | ▲ 25.54% |
| | 2006 | 498 329 | 4396 | 140 280 | 21 095 | 664 100 |
| Total | 2007 | 2 661 014 | 97 525 | 2 371 556 | 114 655 | 5 244 750 |
| | | ▲ 14.41% | ▼ –45.09% | ▲ 16.07% | ▼ –42.99% | ▲ 10.47% |
| | 2006 | 2 325 866 | 177 602 | 2 043 241 | 201 120 | 4 747 829 |

*Note:* Unit: NT$ thousands.

*Source:* Higher Education Evaluation and Accreditation Council of Taiwan (2008).

systems and divide fund purposes into industry-academia collaboration and vocational training. The funds from enterprises for industry-academia collaboration rose, and this may show that the industry is increasing its interest in this issue.

### 8.2.2 Taiwanese Bayh-Dole Act: Fundamental Science and Technology Act

In 1999 the Taiwan government enacted the Fundamental Science and Technology Act to set guidelines and principles for the promotion of S&T development. The Act releases the ownership of R&D results from the government to universities, research institutes and private industries, in order to make it easier for industry to commercialize R&D results. Before this, there was no regulation over ownership of R&D results in Taiwan, and assignment of rights was limited to the provisions contained in employment contracts. Moreover S & T policies in Taiwan were mostly executed by administrative orders instead of legislation, and the R&D results from projects sponsored by the government usually belonged to the government. The formulation of this act was proposed by the fifth National Science and Technology Conference in September of 1996, and adopted most of the provisions of the Bayh-Dole Act of the USA.

### 8.2.3 Government Agencies Involved in Industry-Academia Collaboration

The government agencies responsible for promotion of scientific and technological development in Taiwan are Academia Sinica under the Office of the President, and various agencies under the Executive Yuan, including the Ministry of the Interior (MOI), Ministry of National Defense (MND), Ministry of Education (MOE), Ministry of Economic Affairs (MOEA), Ministry of Transportation and Communications (MOTC), Department of Health (DOH), Environmental Protection Administration (EPA), Atomic Energy Council (AEC), National Science Council (NSC), Council of Agriculture (COA), Council of Labor Affairs (CLA) and National Communications Commission (NCC). The Taiwanese government guides and implements policies on scientific and technological development from budget planning to execution at each agency.

#### 8.2.3.1 NSC, MOE, MOEA

The agencies that have specific focus on industry-academia collaboration are NSC, MOE and MOEA. The mission of NSC is to sponsor the Technology Licensing Offices (TLOs) in the academia, and related measures which include providing industry-academia collaboration research

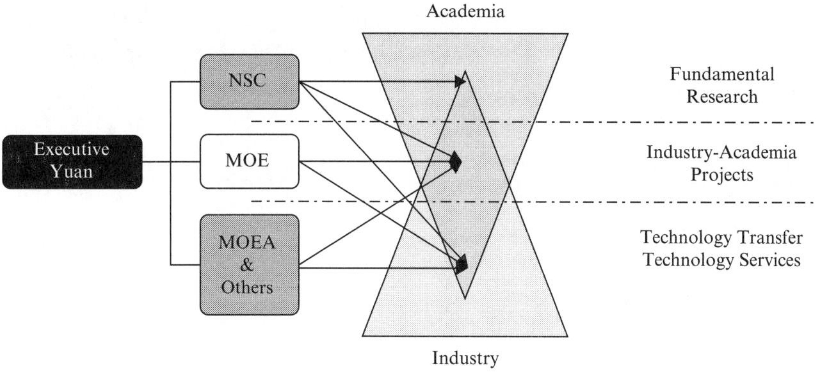


*Source:* Chen (2008).

*Figure 8.1 The government agencies framework*

projects, upgrading industrial technology and enhancing human resources. MOEA serves to sponsor the incubation centers and related measures which include small business innovation research (SBIR), traditional industry upgrade projects, technology development program for academics and leading product development. The mission of MOE is to sponsor the technology development centers and industry-academia collaboration. Related measures undertaken by MOE include establishling criteria for university assessments of faculty promotion, industry-academia collaboration law and regulations.

Apart from playing the key role in directing S&T policy, the government also serves an important function in conducting S&T development. Basic research is conducted by Academia Sinica and universities under the MOE. Applied research and experimental development are conducted by the research units of government agencies under the Executive Yuan, the R&D departments of state-run businesses and specially commissioned non-profit research institutes. Experimental development and commercialization are conducted by private enterprises. Responsibilities for actual implementation of S&T development are shared by the research units of universities, research organizations, non-profit organizations and public and private enterprises.

#### 8.2.3.2 Science and Technology Advisory Group of the Executive Yuan (STAG)

In 1979 the Science and Technology Advisory Group of the Executive Yuan (STAG) was established separately. The mission of STAG is to provide recommendations for national S&T development, sponsor and organize

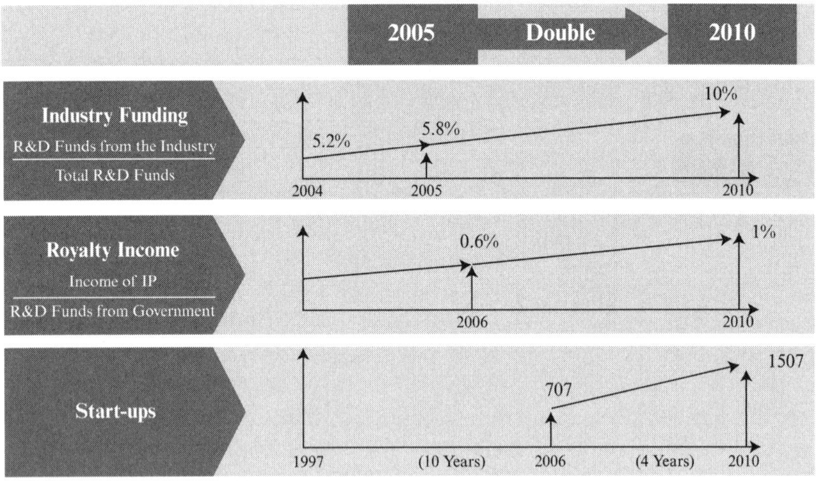


*Source:* Adding value to joint innovation from industry-academia (research sector) cooperation (2007), National Chengchi University.

*Figure 8.2 Double effects of industry-academia collaboration in 2010*

STAG Board Meetings and various Strategic Review Board (SRB) meetings, steer the promotion of S&T development programs or projects, and collect important S&T development information and provide consultations.

Following discussions at the (eighth) National Science and Technology Conference, the vision of STAG is to double the effects of industry-academia collaboration in 2010 for the higher education sector, including industry's funding share (the higher education sector's share of R&D funds from the industry), royalty income share (the incomes generated from IP) and start-ups in innovation incubators.

### 8.2.4 Adding Value to Joint Innovation from Industry-Academia (Research Sector) Cooperation

To enhance the impact of industry-academia cooperation, the Executive Yuan of Taiwan drafted the *Economic Development Vision for 2015 – First-Stage Three-year Sprint Program (2007–2009)*, in which the key plan includes 'Adding value to joint innovation from industry-academia (research sector) cooperation' in the Industrial Manpower Package (Figure 8.2). The main target of this plan is to analyse industry-academia policies and impacts of various countries for introspection to Taiwan. Furthermore the plan aims to integrate the performance of industry-academia cooperation and double the higher education sector's share of R&D spending provided by industry, income generated from intellectual property (IP) and number of entrepreneurs within three years of the launch of the plan.

'Adding value to joint innovation from industry-academia (research sector) collaboration' was directed by Dr Paul Liu, the lead author of this chapter. Dr Liu observed from his research that industry-academia collaboration in Taiwan is confronted with some challenges including the immature environment, institutional obstacles and lack of communication mechanisms. To solve these problems, there were guidelines related to five issues – policy, regulation, technology, organization and personnel – suggested in this plan.

On the issue of policy, the first job was integrating the industry-academia collaboration programs of NSC, MOE and MOEA, and then setting up a single unit. Also suggested in the plan are raising funds for technology investment, encouraging universities to develop R&D specialties and checking that superior R&D achievements will build up a mechanism for R&D input and output.

Regarding regulation, it is recommended to amend the University Act to add industry-academia collaborative effects as an important conference to the teacher review system to create positive incentives for professors to commercialize their inventions. To reinforce these incentives, untying accounting and feedback systems are also necessary reforms. Moreover in order to bring about more resources, regulatory obstacles to directing corporate capital to industry-academia collaboration need to be solved.

On the issue of technology, developing the techniques that enterprises want is the first step of technology commercialization. Meeting the needs and preferences of enterprises and organizing R&D teams with multi-discipline members are needed for technologies which have become increasingly more complicated. In addition, in order to generate interest and increase participation in R&D among professors from different schools, the R&D database for all schools should be expanded to create a platform for professors to share knowledge and experience.

On the issue of organization, the most important task is to integrate every industry-academia center, technology transfer center and incubator into a single unit. In addition, inviting professionals who are familiar with business models as full-time managers can truly bring to the forefront the thinking of industry and help bridge industry and academia.

On the issue of personnel, there are actually several differences between a professor and a businessman by nature, such as their personalities, philosophies of thinking, the culture of their surroundings and so on. Therefore eliminating the gap between the professors and industry will be the essential point for technology commercialization. On the other hand, connecting teaching contexts and industrial demands from the beginning can also link both sides more closely.

This plan represents the policy that the Taiwan government currently emphasizes, and the above suggestions are the exact measures the government and universities are now promoting to increase the effects and impacts of technology commercialization. The execution of this plan not only supervises if universities exercise the resources on industry-academia collaboration well, but also attempts to transform the whole S&T circumstances. Under this policy, it is expected that the resources will not be wasted, and the results will be beneficial to the economy.

## 8.3 FOCAL UNIVERSITY: NATIONAL TAIWAN UNIVERSITY (NTU)

### 8.3.1 Brief Profile of NTU

NTU comprises colleges of liberal arts, science, social science, medicine, engineering, bioresources and agriculture, management, public health, electrical engineering and computer science, law and life science. NTU is also famous for its performance in various areas, such as health care, information technology and social science. NTU also earns the same recognition in its research results from industry-academic cooperation and technology transfers in many industries, such as agriculture, manufacturing, information technology (IT) and biotech industries.

NTU has always been a leading comprehensive university and is famous for its academic achievements in various areas in Taiwan: 3566 essays in SCI, SSCI journals and five essays in *Nature* and *Science* were published from NTU in 2006. Total citations of essays from NTU were 9456 during 2002 to 2006, and the number of high citations (HiCi) accumulated over the past ten years is 135. NTU is also famous for its research projects sponsored from enterprises. The number of research projects in 2006 was 3143 and total funds of these projects amounted to NT$5.4 billion. Moreover NTU has filed 589 patents during the period 2001 to 2007, and 235 of them had been issued as of December 2007. At the same time, there have been 378 technology transfers with NT$156 million as total royalties and revenues generated.

### 8.3.2 Indicators of Education, Research Output and Technology Commercialization

For NTU, every performance index has linkages to the university governance system after receiving a five-year NT$50 billion funding from the Minister of Education. The governance covers every aspect, such as admission, internationalization, faculty, teaching, research and industry-academy cooperation (Table 8.3). NTU is required to set quantitative indices of each aspect in every year during the five-year funding period.

In industry-academia cooperation, NTU expects to grow 5 percent over a three-year average every year in the future. The same growth in patents and return from technology transfer is also expected. However the number of incubated companies in NTU every year cannot be more than 30 due to limited space at NTU. Thus instead of increasing the quantity of incubated firms, NTU is trying to improve the quality of the 30 companies incubated.

NTU has always strived to promote technology transfer and industry-academy cooperation to its professors. With the support of administrative staff, NTU had made strategic simplification to the patent application process of research results to increase inventors' motivation to file patents. Moreover NTU has introduced an incentive policy for patent filing. As a result, growth of patent filings has been much higher than 5 percent. The number of patent filings during 2006 was in fact three times that recorded during 2005. The incentive policy provides four alternatives for inventors through which they may choose to undertake different levels of risk and ratios of patent filing fees.

### 8.3.3 Technology Commercialization Institutions

NTU formed its R&D Board on 1 April 1996. Due to the need for scale, it was restructured as the Office of Research and Development on 1 August 2007. The Office of R&D was originally divided into four divisions: the Planning group; the Technology Transfers Group; the Sponsored Research Projects Affairs Group; and the Innovation and Incubation Group (also the Industry-Academy Cooperation Group). The Planning Group is responsible for management of research units; the Technology Transfers Group consists of six full-time staff and is responsible for management of patents and technology transfers; the Sponsored Research Projects Affairs Group is responsible for industry-academy cooperation including cooperative education; and the Innovation and Incubation Group works with the Innovation and Incubation Center, which is an independent company consisting of about eight full-time staff and is responsible for incubation of start-ups. Summing up the number of staff, there are about 14 full-time people responsible for technology transfer and incubation of start-ups in NTU.

In the near future, according to the development plan of NTU, there

*Table 8.3 NTU Indicators of education, research output and technology commercialization*

| Performance index | 2006 | 2007 goal | Achievement in September 2007 |
|---|---|---|---|
| **Admission** | | | |
| 1. Undergraduate students enrolled | 3671 | 3771 | 3771 |
| 2. Teacher.student ratio | 14:57 | 14:49 | 14:26 |
| **Internationalization** | | | |
| 1. Number of foreign students | 373 | 385 | 327 |
| 2. Number of international exchange students | 242 | 249 | 325 |
| 3. International cooperation with contracts and funds | 88 | 93 | 91 |
| 4. Programs in foreign language[a] | 247 | 350 | 350 |
| 5. Hosts to major international conference | 54 | 56 | 48 |
| 6. Visits of foreign scholars | 102 | 120 | 70 |
| 7. Number of people involved in operations of academic organizations | 292 | 300 | 344 |
| **Faculty and research results** | | | |
| 1. Growth[b] rate of international essays[c] | 2964 | 3112[d] | 2955 |
| 2. Growth in citation of essays[b,c] | 9456 | 9800 | 9684[e] |
| 3. Number of academic press books | 23 | 24 | 24 |
| 4. Editors of major international journals | 209 | 215 | 326 |
| 5. Members of major academic organizations | 80 | 83 | 86 |
| 6. Number of researchers | 51 | 51 | 55 |
| 7. Accumulated number of major awards | 81 | 90 | 174 |
| 8. Number of high citations (HiCi)[f] | 135 | 140 | 160 |
| **Teaching** | | | |
| 1. Number of undergraduate students taking multi-discipline programs[g] | 2261[h] | 2345[i] | 2345[i] |
| 2. Ratio of undergraduate students who passed CEF B1 | 42.22% | 50% | 40.22% |
| 3. Ratio of course rated below 3.0 in teaching survey feedbacks in past three years | 0.54% | 0.53% | 0.52% |
| 4. Ratio of course rated above 4.0 in teaching survey feedbacks in past three years | 64.87% | 65% | 64.87% |
| 5. Number of online course in Collaborative Enhanced Instruction By Asynchronous learning (Ceiba) | 1924 | 2224 | 1592 |

*Table 8.3* (continued)

| Performance index | 2006 | 2007 goal | Achievement in September 2007 |
|---|---|---|---|
| **Industry-Academia Cooperation** | | | |
| 1. Cash flow from industry-academia cooperation | NT$5.3 billion | NT$5.5 billion | NT$4.65 billion |
| 2. Number of patents | 59 | 63 | 32 |
| 3. Revenue from technology transfer | NT$16.43 million | NT$17 million | NT$13.63 million |
| 4. Number of incubated companies | 30 | 30[j] | 25 |

*Notes:*
a. The number involves the number programs which should essentially be taught in a foreign language.
b. The growth rate is calculated on the basis of the past five-year average.
c. International essays are SCI, SSCI and A&HCI essays.
d. The goal is to grow 5 percent on the basis of the past five-year average.
e. The number is an average from 1 August 2002 to 31 August 2007.
f. The number involves the accumulation of the past ten years.
g. Multi-discipline programs include double major and minor.
h. The ratio of this number to the number of undergraduate students is 13.5 percent.
i. The ratio of this number to the number of undergraduate students is 13.3 percent.
j. The number cannot be higher than 30 due to limited space on campus.

*Source:* Office of Research and Development, National Taiwan University, 2007.

will be a management team under the Tech Transfer Manager to promote and market the IP of NTU and represent NTU to its partners, thus optimizing the IP value of NTU. The Tech Transfer Manager is a full-time position that also carries the role of vice president of the Office of R&D. The recruitment of the Tech Transfer Manager is different from faculties in colleges. Thus it would be difficult for the faculty to volunteer as a Tech Transfer Manager. In order to attract the most outstanding to be the Tech Transfer Manager, NTU sets no limit to the salary of the job. The Tech Transfer Manager will lead an executive team constituting licensing associates, IP marketing managers and Incubator managers to cooperate with the existing divisions and play a unique role in the operation of the R&D Office (Figure 8.3).

NTU does not set any restriction on the education level or degree qualifications for the Tech Transfer Manager. The top priority of NTU for the recruitment of a potential Tech Transfer Manager is his or her experience. Due to the shortage of professional information, Tech Transfer Manager is not a popular profession in Taiwan. It is difficult for NTU to find a

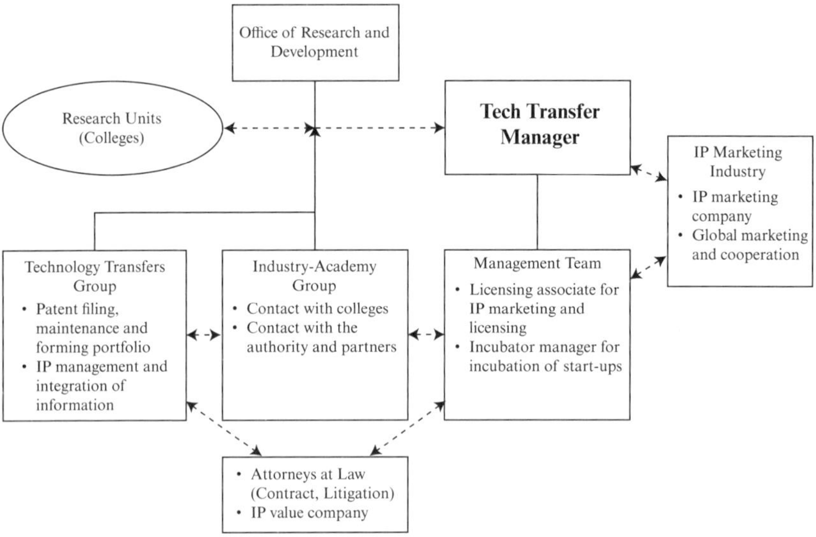


*Source:* Office of Research and Development, National Taiwan University, 2007.

*Figure 8.3 Organization framework for technology commercialization in NTU*

qualified Tech Transfer Manager and so the school is not in a position to set strict requirements of recruitment. Additionally it is impossible to ask university faculty in Taiwan to give up their research and become a full-time manager. Given the dilemma, NTU offers performance bonuses for encouragement. Moreover the more opportunities given to the Tech Transfer Managers to spin-off with new technology, the more appealing NTU may become to qualified professionals.

Outsourcing can be an option as well. Since the current instruments are insufficient for NTU to promote its IP, NTU cooperates with many organizations and companies. The Industrial Technology Research Institute of Taiwan (ITRI) is an example of a collaborative partner. NTU may commission out its marketing to these organizations and companies.

The Tech Transfer Manager will play the most important role in the development of NTU. Experience sharing and professional training is another important issue. Given the fact that Tech Transfer Managers are encouraged to spin-off with new technology, it will be crucial to find qualified successors and to have professional training for the smooth operation of the technology transfer function.

### 8.3.4 Technology Commercialization Cases

#### 8.3.4.1 Dr Liang-Gee Chen's MPEG-4 codec architecture

*Introduction* Dr Liang-Gee Chen, a member of NTU's Graduate Institute of Electronic Engineering, had been leading a research team for video codec and announced this technology. The research team has been putting all efforts into industry-academia cooperation and transferring the research results to local companies. As a result, the digital image and video industry in Taiwan has matured. The team developed a new MPEG-2 video decoder in 1998. In 1999 they announced a high-efficiency JPEG codec and transferred it to Global Unichip Corporation. The codec was commercialized to be chips applied in digital still cameras (DSC), and large volumes of shipments were made. During the period 2004 to 2006, after the MPEG-4 video codec was announced, the team announced the high speed Motion JPEG codec, high-efficiency H.264/AVC codec, low power-consumption MPEG-4 codec and Motion JPEG2000 codec requiring low memory in International Solid-State Circuits Conference (ISSCC). All of these great technologies have been transferred to enterprises in the ICT industry and are having an impact in the industry.

*About MPEG-4 codec* MPEG-4, officially called ISO/IEC 14496, is a video compression technology developed by the MPEG organization. In this case, the content of the technology is a platform-based MPEG-4 video codec silicon IP, and the transferred items include verilog code, MPEG-4 software, a software development kit (SDK) for controller in the architecture and documents. This technology was transferred to ITRI, WiseWare and Global Unichip for royalties of NT$91 million in 2002, and cumulative to the present day, this technology has brought in more than NT$100 million to NTU.

Dr Chen and his team have been continuously developing the technology and have filed 29 patents in many countries. The NSC licensed the technologies to industries and research institutes, such as Jeilin Technology, Global Unichip, Sunplus Technology, AverMedia Technology, Quanta Computer and ITRI. According to statistics from NTU at the beginning of 2007, the sum of royalties have accumulated to NT$600 million, and the commercialization of the technologies has created much more than NT$10 billion.

*Commercialization of MPEG-4 codec* Dr Liang-Gee Chen's first technology commercialization was triggered from industry, not from himself personally. Company A sought technologies they wanted and found from an international journal that the digital signal processing integrated circuit (DSP IC) they needed was Dr Liang-Gee Chen's invention. Dr Liang-Gee Chen's technology was rare then but was also a pioneer technology. This invention had a huge range of applications, including digital cameras and image transmission on the Internet.

Due to the hurried Time-to-Market in the industry, Dr Liang-Gee Chen was at first hesitant about this business cooperation. Moreover while Dr Liang-Gee Chen had full confidence in the R&D results of his lab, he knew clearly that there were differences between the conditions in industry and in his lab. He doubted if his invention could meet the demands of the industry without any problems. In order to lighten the load of Dr Liang-Gee Chen, Company A only asked him to make a small necessary modification. Dr Liang-Gee Chen finally was persuaded to provide his assistance in response to the enthusiastic attitude of the company.

*The industry-academia collaboration model* Based on this successful technology commercialization, Dr Liang-Gee Chen derived some key factors from his experience, and developed them into his model on technology transfer and collaboration. In the above case, Dr Liang-Gee Chen found that bugs cannot be avoided when the R&D results from the lab are converted into business uses, and Company A fully understood this reality. Therefore the company asked Dr Liang-Gee Chen to build up the core technology, and make sure that the technology could work on the sample products. Most important of all, Company A did not request Dr Liang-Gee Chen to guarantee that there would be no technical problems after this technology transfer. Company A only asked Dr Liang-Gee Chen not to disclose the possible technical problems and solutions to its competitors. This collaboration model won Dr Liang-Gee Chen's trust and led to a satisfactory achievement.

The origin of the commercialization was Company A being attracted to Dr Liang-Gee Chen's invention. Company A decided on this invention partly because the technology had been almost developed completely by Dr Liang-Gee Chen. This presented an incentive for both the industry party and the academia party to embark on commercialization without R&D costs. Using this experience, Dr Liang-Gee Chen has led his lab to develop technologies for about five years before they are broadly known and accepted by the market. As soon as the economic value of the technology emerges, Dr Liang-Gee Chen's lab can provide not only know-how and patent licensing but also human resources. This model catalyzes the collaboration between industry and academia, and provides career opportunities for students.

*Successful factors of commercialization* In addition to the lessons learned from the above case, there are some crucial points about Dr Liang-Gee Chen himself. First, he has a personality that is happy to step out of the lab and cooperate with the industry. There cannot be technology commercialization if an inventor does not let his invention be verified on the market. Second, Dr Liang-Gee Chen's exceptional competence in R&D means that there are many companies that are willing to commercialize his inventions, and excellent students to participate in his team. Third, the domain of technology Dr Liang-Gee Chen chose corresponds with the industry needs in Taiwan. Fourth, Dr Liang-Gee Chen knows the concerns and language of the industry. The above factors finally constituted this success story.

### 8.3.5 Prospects and Recommendation

For NTU, there are some challenges in the future. The most urgent problem is finding a qualified professional, especially when it is planning to reorganize its R&D Office. NTU requires a person with industrial experience and legal knowledge as well as technology background to be a Tech Transfer Manager. Few of the college faculty members in Taiwan would be suitable candidates for the position.

In Taiwan, technology transfer and licensing is not yet a profession like attorneys or accountants. An experienced professional may not be interested in working in the university even if a high salary is offered. The lack of future development is one of the barriers to NTU attracting a suitable Tech Transfer Manager. This is the reason that NTU does not require a PhD degree for their Tech Transfer Manager. Despite all these disadvantages, NTU is still optimistic about finding a qualified Tech Transfer Manager because of its academic reputation and achievements.

Moreover the lack of input is attributed to the rigid system in NTU. NTU experiences a difficult human resource problem and has problems in engaging new staff to join the R&D Office. In addition NTU cannot refuse cases with small funds due to the special industrial environment of Taiwan, which is formed by small and medium-sized enterprises and totally different from the USA and Europe. Companies in Taiwan are not accustomed to paying royalties. Instead they have more interest in commission-to-research cases. They are used to paying for commissioned research and owning the technologies or patents arising from the research. Thus NTU cannot enlarge its scale with its royalty returns. The most urgent problem is the dramatic growth of patent filing, which will cost more and require more staff to deal with the additional operations. Finally, talented marketing people are also relatively scarce in Taiwan, especially those with knowledge of technology. In the experience of NTU, people with technology background will put working with high-tech companies as their top priority. To solve this problem, NTU is considering engaging people who have been working in ITRI. ITRI has many experienced people in technology transfer and licensing. NTU hopes the outstanding technology and greater chance to spin-off a new start-up will help it to attract elites in the field to join the university.

## 8.4 FOCAL UNIVERSITY: NATIONAL TAIWAN UNIVERSITY OF SCIENCE AND TECHNOLOGY (NTUST)

### 8.4.1 Brief Profile of NTUST

The whole office of NTUST appears clear and new, just like a business place. Every member of staff has their individual phone number, and can negotiate technology transfer and licensing independently. The always-occupied phones, documents piled on the tables and the typical atmosphere of the official bureaucrat are never seen in this office. Currently there are four full-time managers responsible for patent licensing, and the number is expected to climb in the future. In the incubator of NTUST, there are currently three staff members and the number is expected to double. NTUST has rented a new space in Taipei Neihu Technology Park from the Taipei government, about 90 percent of which has already been booked by start-up companies. It is proposed that there will be ten companies there, one of which is now operating. If possible NTUST wants to strive for another place in Tu-Cheng, to build its third incubator.

As the head of the technical schools in Taiwan, NTUST has been accumulating high quantity and high quality R&D output, thanks to the continuous endeavors of its faculties. On technology commercialization, NTUST takes the lead in the achievements of industry-academia cooperation among professional schools, and has become a model for its followers.

### 8.4.2 Indicators of Education, Research Output and Technology Commercialization

In addition to the number of cases and monetary indicators (Table 8.4), 'penetration' is another indicator to assess the commercialization achievements of teachers. 'Penetration' is defined as the ratio of the teachers who have filed patents by the teaching faculty.

*Table 8.4 NTUST indicators of education, research output and technology commercialization*

| Performance index | 2005 | 2006 goal |
|---|---|---|
| **Admission** | | |
| 1. Undergraduate students enrolled | 1638 | 1812 |
| 2. Admission approach improvement | 30% admissions are waived of the entrance examination | 40% admissions are waived of the entrance examination |
| **Internationalization** | | |
| 1. Number of foreign students | 78 | 20% growth per year |
| 2. Number of international exchange students | 10 | 20% growth per year |
| 3. International cooperation with contracts and funds | 4 | 15% growth per year |
| 4. Programs in foreign language | 103 | 10% growth per year |
| 5. Hosts to major international conference | 2 | ≥2 |
| 6. Visits of foreign scholars | 30 person-time | 20% growth per year |
| **Faculty and Research Results** | | |
| 1. Growth rate of international essays[a] | 335 | 5% growth on the basis of past five-year average |
| 2. Growth in citation of essays | – | 10% growth on the basis of past five-year average |
| 3. Editors of major international journals | 20 | 5% growth on the basis of past three-year average |
| 4. Accumulated number of major awards | 20 person-year | 5% growth on the basis of past three-year average |
| **Teaching** | | |
| 1. Number of undergraduate students taking multi-discipline programs[b] | 1% | 1% growth per year |
| 2. Undergraduate students participating in pragmatic programs | 100% | Raising the quality year by year |
| **Industry-academia cooperation** | | |
| 1. Cash flow from industry-academia cooperation | 3.4 billion NT dollars | 5% growth on the basis of past three-year average |

*Table 8.4* (continued)

| Performance index | 2005 | 2006 goal |
|---|---|---|
| **Industry-academia cooperation** | | |
| 2. Number of patents | 50 | 5% growth on the basis of past three-year average |
| 3. Number of technology transfers | 30 | 5% growth on the basis of past three-year average |

*Notes:*
a. International essays are SCI, SSCI and A&HCI essays.
b. Multi-discipline programs include double major and minor.

*Source:* Office of Research and Development, National Taiwan University of Science and Technology, 2007.

### 8.4.3 Technology Commercialization Institutions

NTUST is integrating its organization framework for technology commercialization. Beside this, NTUST attempts to have a 20 percent growing accomplishments per year. This is planned to be one mission of mentors. Mentors will be paid NT$10 million per year and asked to commit to attaining this goal and bringing in income in excess of payments. The operation is totally self-sufficient. The background of mentors includes knowledge management, construction and so on. Reciprocally, the university gains royalty income, credit and the achievement of bringing to prominence the technology of professors. In practice, taking consultancy service, for example, there are no executive issues, so there must be some surplus. This kind of cash balance could be counted as part of salary to make the total remuneration package more attractive. The mentors will not be hired again if they do not cover their cost and expenses. In the first one or two years, however, the performance may not be monitored too strictly, to give some buffer to new mentors. The mechanism and elastic thinking above show the ambition and animation of NTUST.

NTUST has introduced Tech Transfer Managers, who embrace abundant industrial experiences and good networking, into the team of industry-academia cooperation. Tech Transfer Managers connect the external resources, including enterprises, governments and research institutes, with the R&D efforts of NTUST and carry out technology commercialization.

The professional managers in NTUST are called industry-academic mentors. They are hired from alumni as consultants. This systematic model is very general over the world, but still rare in Taiwan. These alumni are sought from chief executive officers (CEOs), presidents or other high-level directors of listed companies. They can be persuaded to share their experiences and knowledge because of this precious alumni relationship. In one year, NTUST has found out about 40 alumni, including Electrical Engineering (EE), Mechanical Engineering (ME), Business Administration and so on. NTUST has conducted over 20 meetings, to let research staff report to mentors and decide which inventions have industrial value and how to commercialize them.

NTUST relies on the industrial experiences and management knowledge of Tech Transfer Managers, and authorizes them to lead the R&D team. Tech Transfer Managers will act as entrepreneurs, proceeding market analyses, resource allocation and R&D planning, to drive the industry-academia cooperation of NTUST.

Further, NTUST has a mature industry-academia cooperative mechanism in its industry-academia center, technology transfer center and incubator, and makes good use of this advantage. Moreover NTUST is striving to construct its Office of Research and Development as the sole unit on technology commercialization.

NTUST puts equal emphases on both essential technology development as well as industrial commercialization, to meet various demands from enterprises and thus to enhance IP competence and value. Through the cooperation from multi-discipline teams, NTUST will excel in its R&D achievements and marketing profession.

#### 8.4.3.1 The idea of knowledge transfer

In the opinion of NTUST, knowledge transfer is more important than technology or patent transfer. Based on Professor Jeng's working experiences in the TECO Group and its other conglomerates, he thought of this idea to rebuild the commercialization of technology in NTUST. For instance, the staff of NTUST TLO were tasked to guide the R&D scientists of client companies to analyse the patents of competitive companies. By doing this, NTUST could see if the products or technologies of client companies infringe the patents owned by competitive companies. This work using much detailed confidential information of client companies, however, should be on the basis of full trust and commitment.

After reading the patents owned by competitive companies, R&D scientists could understand how many inventions are patented and how to write the claims of patents. This would raise the patent applications of the companies which R&D scientists belong to. On the other hand, this is also beneficial to R&D scientists themselves. In addition to the reward from their companies for increasing patent quantity, as the inventors of the patents, scientists can also fulfill their achievements in their resumes. Moreover the scientists can retain their positions when there are mergers and acquisitions (M&As) of their companies.

NTUST has another case of knowledge transfer. Professor Jeng has helped a company to enhance the efficacy of a piece of laser equipment to 4/3 times compared to its original use. This will reduce the total amount of equipment the company needs to buy, with the price of the individual equipment being 200 million yen. Thus the value of this knowledge transfer is obvious. In this case, Professor Jeng successfully reduced the number of pieces of equipment the company planned to buy by two. That is, the industry-academic revenue in this case amounted to 200 million yen. This money could be paid to NTUST under the account of technology transfer or industry-academia mechanism.

Knowledge transfer could also help resolve the incentive problem of non-patent technology transfer. NTUST intends to create a new account named 'knowledge transfer royalty', in addition to the former 'technology transfer royalty' account. The new account is not necessary to cover the application or maintenance fees of patents. Therefore NTUST can increase the salaries of directors from a fixed NT$30000 per month by adjusting the royalty release ratio between the university and the director.

#### 8.4.3.2 The one-stop of patent application for universities countrywide

NTUST wants to be a one-stop center to handle all the patent applications and maintenance of universities in Taiwan. This idea came from the special circumstances of Taiwan, where almost every university only files a maximum of five patents per year. This causes a low economic scale, and a huge waste of resources in many of the universities. Therefore NTUST wants to run a model to file all the patent applications for universities' needs countrywide.

NTUST plans to present its proposals to the Department of Education, to apply for a package budget in the first three years, to subsidize inventors to file their patent applications. If this is approved, NTUST may create an incentive for inventors to habitually use their services. Based on profession, NTUST will hold a commission to determine which invention is suitable for patenting, across universities all over the country. If the conclusion is to omit the patent application, the commission can discuss with the inventor if he or she intends to file the patent by themself. This type of service is not only for NTUST professors but also for outside cases. RCT Corp., a US company, provides a similar service, including valuation and application. As such this business model has been verified to be profitable.

*Source:* Shot by Michael S.S. Huang.

*Figure 8.4* *Steel beam-column connection model*

### 8.4.4 Technology Commercialization Cases at NTUST

#### 8.4.4.1 Dr Gin-Shan Chen's steel beam-column connections

*The great achievement behind Taipei 101* A few years ago when Discovery began to show the great achievement of the world's giant buildings, these buildings were shown to be real monsters, possessing astonishing height. The Taipei 101 building (also known as Taipei Financial Center) is a landmark that each dweller can spot from every corner of Taipei. It is also a famous scenic spot introduced by Discovery. To construct these giant buildings requires many different technologies as support. Located on the seismic belt, Taipei 101 must possess a strong structure. One of its many technologies is the 'steel beam-to-column connection' technology, which was developed by Taiwan's famous Architecture Professor at NTUST, Professor Gin-Shan Chen (Figure 8.4).

*The design is filled with the philosophy of 'Tai-Chi'* Another description to this technology is 'a jerry-built H-type steel bar'. The H-type steel bar is commonly used at high-level steel-structure buildings, and this kind of construction possesses several merits such as stiffness, ease of construction and outstanding shock-absorbing capability. However its cost is substantially higher. As the Taipei basin is located upon the seismic belt, this design is suitable for application to steel structure construction. While encountering vigorous shaking, conventional steel construction centralizes the physical stress on the beam-to-column joint, resulting in serious fracture. By adapting Professor Chen's 'beam-to-column connection', the building's structural tolerance for physical stress is three times better than conventional methods.

Interestingly, unlike conventional methods, this invention is not a head-to-head method to overcome physical stress. Instead it ingeniously shunts the force in the other direction. It is a little bit like the 'Chinese Tai-Chi'. Normally the way to increase the shock absorption of buildings is by strengthening its structure, hoping that the stiffer the structure, the better it can withstand sudden earthquakes. However Professor Chen did not adapt this method. Contrarily, he created two breaches upon both sides of the flanking plate. By carefully calculating and cutting off unnecessary portions, the structure can precisely and evenly spread the stress onto the region previously selected.

*Low-technology cost and high utility* High-technology does not necessarily come with high cost and entry barriers. In the perspective of utility, this beam-to-column connection technology is three times better than conventional technology. The benefit is obvious. From another point of view, however, this technology involved merely removing a portion of material. For an H-type steel bar manufacturer, there is no need to acquire extra technology. Therefore the manufacturer does not have to place extra input, and at the same time saves more material cost.

*Thorough and effective steps complementary to commercialization* The specification notified the inventor, assignee, qualified applier and other information needed, and certified the rules for calculating royalties as follows: each construction is valued at least NT$300 000. If the construction cost is NT$500 million, the royalty is one thousandth (1/1000); if the cost is between NT$500 million and NT$2 billion, the royalty is five ten-thousandth (5/10 000); if the cost exceeds NT$2 billion, the royalty is one ten-thousandth (1/10 000); if the cost is below NT$50 million, the royalty is preferential for 150 000 NTD. Compared with other technology licensing offices, NTUST's TLO is now doing booming business with this technology. According to documents provided by NTUST's TLO, the invention 'beam-to-column connector' generated NT$22 million in revenue from licensing in only three years, from 1994–96.

After a careful evaluation, most constructors interested in the technology would contact NTUST's office directly. The TLO of NTUST also stated that, after these many years, the office is now experienced in licensing affairs. They also mentioned that the prosecution process is well established; two parties only have to proceed in written document between each other, without spending too much effort on promotion, negotiation and dealing.

*Conclusion: successful technology licensing relies on coordination between technical and managing aspects* The success of this invention, in both royalty and commercialization aspects, brings constructors and dwellers a better and safer building structure. It helped in constructing the famous Asian landmark tower – Taipei 101 – and at the same time promoting images of Taiwan's architectural capability, establishing the cornerstone for Taipei to become a global city. From each aspect of economic, social security and city globalization, the success of this invention is truly a great achievement. This extraordinary achievement resulted not only from the technical advantages the technology possessed; NSC's efforts in technology licensing and commercialization also played significant roles. The coordination and integration model is the best example for other inventors or technology licensing/managing groups.

### 8.4.5 Prospects and Recommendation

In the opinion of NTUST, a perfect technology transfer should contain industry-academia cooperation, patent filing and the transfer of the technology and its know-how. Through the demands disclosed in industry-academia cooperation, the academia can concentrate their intelligence and research capabilities on what enterprises exactly need. Additionally, under the patent protection, the cooperative enterprises would be more willing to be involved in technology transfer.

## 8.5 CONCLUSION

From providing funds on R&D and industry-academia collaboration, to releasing the ownership of R&D results from the government to universities, delegating responsibilities to government agencies to direct and conduct the related policy, to proposing the 'Adding value to joint innovation from industry-academia (research sector) cooperation' project to ensure the universities make good use of the above resources, the Taiwanese government has positively encouraged technology transfer and licensing from universities to industry for years. In order to improve industrial performance and upgrade the overall technology level, the Taiwanese government plays an important role in the promotion of technology commercialization and industry-academia collaboration.

On the other hand, NTU and NTUST, which are taken as cases in this chapter, both have their visions, advantages, organizations, strategies and their own successful technology commercialization stories. Comparing and summarizing these two cases, they both receive much contributions from the government, and have set up indicators to verify that they make excellent use of the resources. In addition, following government policy, the two schools are trying to integrate their industry-academia center, technology transfer center and incubator into a single unit, and hire professional managers embracing working experience in the industry. The successful stories of NTU and NTUST, however, have their specialties and merits. These different models provide good experiences for technology commercialization.

Currently technology commercialization is emphasized in Taiwan. There are more than 80 incubation centers in universities to do the technology transfer, generating more than 500 cases of technology transfer from universities to industries each year. On the other hand, the ITRI, the largest research institute in Taiwan, has done a remarkable job of commercializing its research results. It produced about 1000 products in 2004 and the income from IP rights amounted to more than US$35 million. In 2006 it auctioned 480 cases of its patents to industries and had 470 licensing cases.

## REFERENCES

Chen, D.Z. (2008), 'On the socio-economic environment and policy for industry-academia collaboration in Taiwan', Association of East Asian Research University International Symposium: Collaborations with Industry, City and Public, Sendai, Japan, 27–28 March.

Higher Education Evaluation and Accreditation Council of Taiwan (2008), *University Industry-Academia Collaboration Evaluation (2007)*.

# 9. University technology transfer and commercialization in China: using Tsinghua University as a case

**Yixin Dai and Lan Xue**

## 9.1 INTRODUCTION

As a dedicated research-intensive university, Tsinghua University obtained tremendous achievements in research, technology transfer and commercialization. In the year 2007, for example, the university received research and development (R&D) funds of 1.49 billion yuan, a 0.1 billion yuan increase compared with 2005, out of which 0.5 billion yuan came from industry sectors. In the same year 568 out of 875 patent applications were issued, with 90 percent being utility patent applications. The university gained more than 200 foreign patents. The number of academic articles has reached a historical high, with 1768 Science Citation Index (SCI) recorded articles, 3242 Engineering Index (EI) recorded articles and 1768 Index to Science & Technology Proceedings (ISTP) recorded articles.

Tsinghua University designs its technology transfer and commercialization management system based not only on its experience over the past 50 years, but also reflecting the administrative model of Chinese research universities. Its commercialization therefore serves as a representative sample of Chinese universities who try to make their research relevant and spread knowledge to increase national innovation capability. This chapter starts by introducing a general model of Chinese research universities adopted in technology transfer and commercialization. Further, the technology transfer and commercialization system in Tsinghua University has been elaborated emphasizing the long-term relationship the university has been trying to build through institutional design. The chapter then presents a successful commercialization case in the university and concludes with key factors in this process followed by a certain degree of generalization.

## 9.2 LINKAGES BETWEEN TSINGHUA UNIVERSITY AND INDUSTRY SECTORS

As with most other research universities in China, Tsinghua University builds multiple connections with industry partners, through which it conducts its technology transfer and commercialization initiatives.

It should be recognized at the outset that the most important form of university-market linkage is the flow of university graduates to the market, as well as the flow of new knowledge generated by university-based research projects. Tsinghua University is no exception.

Additionally five major forms of university-industry linkages are adopted nationwide: technology contracts; intellectual property transfer and licenses; university-owned enterprises; joint research centers; and university science parks. Among all forms of linkages, technology contracts are probably the most commonly used and with the most flexible format. While the number of university licenses rose recently, they are limited to schools and departments which have strong patenting capacity. Joint research centers, a newly emerged format, are making their appearance strongly as a mechanism for cooperation between universities and firms, both domestically and internationally. This has gained considerable support from the Chinese government, with the Law on Promoting the Transformation of Scientific and Technological Achievements that encouraged different means of joint research being launched in 1996 (CPPCC 1996). University-based science parks have gained popularity in recent years but have been built mostly in major cities with dynamic entrepreneurial activities. The following sections of this chapter discuss these types of university-industry linkages in detail.

### 9.2.1 Technology Contracts

Of all the forms of university-industry linkages, technology contracts have been the largest source of research funding for universities. Technology contracts in China are usually composed of categories such as technology development, technology transfer (non-patent technology transfer and patent licensing), technical services and technical consultancy (MOST 2001–04).[1]

As seen throughout the country, technology development has also played a dominant role in Chinese universities' efforts in technology transfer. Since the 1990s, industrial funding has claimed over half of university R&D investment, a majority of which was realized through technology development.

*Table 9.1 Technology contracts of Tsinghua University (1992–2006)*

| | Technology development (%) | Technology transfer (non-patent) (%) | Patent licensing (%) | Technical service (%) | Technical consultancy (%) | Total (million yuan) |
|---|---|---|---|---|---|---|
| 1992 | 69.03 | 24.64 | 0 | 5.83 | 0.49 | 5.168 |
| 1993 | 86.55 | 7.22 | 0 | 5.19 | 1.05 | 6.621 |
| 1994 | 94.09 | 1.58 | 0 | 4.1 | 0.24 | 12.789 |
| 1995 | 83.56 | 5.74 | 4.51 | 5.19 | 1.01 | 8.866 |
| 1996 | 75.06 | 5.61 | 0 | 17.97 | 1.36 | 12.304 |
| 1997 | 76.2 | 13.33 | 0 | 9.14 | 1.33 | 13.1 |
| 1998 | 79.16 | 6.02 | 4.43 | 8.05 | 2.34 | 17.52 |
| 1999 | 79.11 | 7.26 | 0.09 | 11.21 | 2.32 | 25.287 |
| 2005 | 70.07 | 0.99 | 0.66 | 11.07 | 3.9 | 489.683 |
| 2006 | 61.7 | 2.33 | 3.07 | 21.9 | 11 | 493.631 |

*Source:* Compiled by authors according to data from the Contract Office of the Department of S&T Development, Tsinghua University.

Technology contracts have consistently accounted for the biggest share of Tsinghua University's technology transfer. As shown in Table 9.1, almost every year over 70 percent of its technology contracts were signed in the form of technology development contracts. This share sometimes rose to more than 90 percent, with technology service contracts having the second-biggest share. The share of technology transfer contracts though experienced a great increase as a percentage of all the contract research from 2003 to 2006 (Tables 9.2 and 9.3; Research and Development Management Department report, Tsinghua University 2006). However, the majority of technology transfer contracts are in the form of non-patented technology contracts (Table 9.3), which also accounts for an important part of Chinese university technology transfer. Charging a lump-sum fee in the contract rather than patenting and licensing-out the technology is another feature of many university-based technology transfers. This can mostly be attributed to the inadequate protection of intellectual property rights, as well as to lack of trust between universities and their business partners.[2]

Compared with other forms of technology transfer, technical consultancy and service are more flexible. They mainly involve technology information providing and technical training. This is very similar to technology contracts, but less formal and often based on personal trust between the two parties.

*Table 9.2 Distribution of contract research in Tsinghua University (2005–06)*

| | 2006 | | 2005 | |
|---|---|---|---|---|
| | Contract amount (million yuan) | Percentage of all contract research (%) | Contract amount (million yuan) | Percentage of all contract research (%) |
| Technology development | 400.1438 | 61.7 | 431.6520 | 70.07 |
| Technology transfer | 350.56 | 5.40 | 101.450 | 1.65 |
| Technology service | 142.0054 | 21.90 | 681.761 | 11.07 |
| Technology consultancy | 71.3084 | 11.00 | 240.322 | 3.90 |
| Self-formed enterprise | 960 | Note: included in the total amount of contract research | 820.000 | 13.31 |

*Table 9.3 Technology transfer contract statistics (2003–06)*

| | | Non-patent license | Patent license | Total | Percentage of the total technology transfer (%) |
|---|---|---|---|---|---|
| 2003 | Number of contracts | 29 | 8 | 37 | 4.41 |
| | Contract amount (1000 yuan) | 1343.7 | 86 | 1429.7 | 4.78 |
| 2004 | Number of contracts | 46 | 4 | 50 | 5.59 |
| | Contract amount (1000 yuan) | 1212 | 205 | 1417 | 3.84 |
| 2005 | Number of contracts | 38 | 9 | 47 | 4.34 |
| | Contract amount (1000 yuan) | 613.5 | 401 | 1014.5 | 1.65 |
| 2006 | Number of contracts | 38 | 23 | 61 | 5.72 |
| | Contract amount (1000 yuan) | 1512 | 1993.6 | 3505.6 | 5.4 |

### 9.2.2 Intellectual Property Transfer and Licensing

A fully developed technology can be transferred from the university by two methods: patent licensing and non-patent technology transfer.

Compared with the USA, where patent licensing and sales are among the most important methods for universities to transfer technology, patenting activity in Chinese universities has been relatively rare. According to a 1998 survey, the total revenue generated by American universities through patent licensing was US$576 billion, greatly outnumbering the 2.2 billion yuan technology transfer revenue generated by Chinese universities in 2005 (Center of Technology Development, MOE 2006).

Shao (2002) analysed factors that prevent university patenting and licensing including: inadequacy of patent awareness; the shortage of patents with commercial value; lack of original innovation capability; and lack of government support for university technology commercialization. For example, 1653 patents were granted to domestic applicants in China in 2001, as compared with 125 704 in Japan and 35 900 in Korea.[3]

Although the absolute number is not very high, patent licensing and sales by universities in China have great influence on the patenting activities of the national innovation system (Chen 2004). While enterprises are the most active participants in patenting activity, Chinese universities are nevertheless important players. This is especially true for the application of invention patents. Table 9.5 shows that service invention patents granted to universities have exceeded one-fourth of the nation's total.

As can be seen in both Tables 9.4 and Table 9.5, since the early 1990s income generated through patents by universities has increased substantially as compared with the 1980s, while the number of patents increased modestly. This can be partly attributed to the fact that many universities have begun to pay more attention to intellectual property and learnt to better exploit the commercial benefits of their patents. Many universities have set up technology transfer offices or offices of intellectual property rights to promote the commercialization of intellectual properties generated by universities.

### 9.2.3 Joint Research Centers

Joint research is a relatively new method for enterprises to develop technologies in cooperation with other enterprises, universities or public research institutions. It can help enterprises make full use of internal and external resources, lower the cost of technology development and reduce potential risks in technology development. Among the available partners, universities have become favorite choices for enterprises due to the high

*Table 9.4 Income generated by Chinese universities through licensing and sales of patents (1985–2002)*

| Year | Number of patents | Income produced (1000 yuan) | Year | Number of patents | Income produced (1000 yuan) |
|---|---|---|---|---|---|
| 1985 | 89 | 6135 | 1994 | 326 | 25 345 |
| 1986 | 215 | 9024 | 1995 | 364 | 36 641 |
| 1987 | 189 | 7033 | 1996 | 367 | 39 584 |
| 1988 | 259 | 9117 | 1997 | 362 | 36 532 |
| 1989 | 194 | 5196 | 1998 | 371 | 55 369 |
| 1990 | 336 | 14 410 | 1999 | 298 | 70 096 |
| 1991 | 331 | 18 470 | 2000 | 299 | 125 396 |
| 1992 | 486 | 39 045 | 2001 | 410 | 185 967 |
| 1993 | 390 | 49 831 | 2002 | 532* | 150 097* |

*Note:* * Number of contracts reached and the income generated.

*Source:* http://www.cutech.edu.cn/.

*Table 9.5 Service invention patent granted to Chinese universities (1996–2003, numbers, percentage)*

| | Service invention patents to university | Service invention patents | Invention patents |
|---|---|---|---|
| 1996 | 228 (27.6) | 825 (59.7) | 1383 |
| 1997 | 256 (28.1) | 912(59.5) | 1532 |
| 1998 | 243 (25.5) | 954 (57.6) | 1655 |
| 1999 | 425 (25.2) | 1685 (54.4) | 3097 |
| 2000 | 652 (23.1) | 2824 (45.7) | 6177 |
| 2001 | 579 (22.1) | 2614 (48.5) | 5395 |
| 2002 | 697 (22.2) | 3144 (53.6) | 5868 |
| 2003 | 1730 (25.1) | 6895 (60.5) | 11 400 |

*Note:* Although invention patents are granted to both Chinese nationals and foreigners, this table presents only those granted to Chinese applicants. Figures in brackets indicate the ratio of the number in the same column to the number of the next column.

*Source:* http://www.sts.org.cn.

quality of talents and solid research capacity. Many enterprises regard joint research centers as an effective way to carry out such collaboration (Xue and Zhou 2007).

In 1999 the Ministry of Education formally launched a national

*Table 9.6 S&T research centers in China by type of university (1999–2002)*

| | 1999 | 2000 | 2001 | 2002 |
|---|---|---|---|---|
| Total | 1456 | 4432 | 4599 | 4842 |
| Comprehensive universities | 243 | 1122 | 1073 | 1099 |
| Engineering universities | 541 | 1972 | 1958 | 2117 |
| Agriculture universities | 193 | 420 | 452 | 481 |
| Medical universities | 381 | 635 | 604 | 579 |
| Normal universities. | 94 | 390 | 434 | 476 |
| Others | 4 | 73 | 78 | 90 |

*Source:* S&T Development Center, Ministry of Education.

program: 'Invigorating Education Towards the 21st Century', in which the 'project to stimulate the development of high-tech industry in universities' was a key element. The project encouraged universities to interact with enterprises and research institutes by various means, particularly by jointly setting up engineering research centers and productivity promotion centers in universities. Ever since then, science and technology (S&T) research centers in China's universities have undergone a rapid increase (MOE 1999). As shown in Table 9.6 S&T research centers in China's universities witnessed a big increase in the year 2000, nearly tripling in number from the year before, and then maintained a stable growth in the following years. The importance of research centers in universities has attracted much attention from government and enterprises. According to statistics from the Ministry of Science and Technology, over 1000 research centers for technology development in universities have developed a very close relationship with both public and private enterprises.

Meanwhile, universities benefit from joint research centers through: (1) accelerating technology commercialization; (2) obtaining adequate funding support for S&T research; (3) gaining opportunities for students to have practical training; and (4) enriching the university's experience in technology management. Meanwhile, multinational companies also admit that intangible assets of a well-known university help the company in expanding its market share in China (Ma et al. 2003; Fan and Fang 2004; Zhou and Zhu 2004)

Tsinghua University is a good example. In 1992 the university set up its first joint research center with the Japanese company Panasonic. Over the next decade, the number as well as scale of joint research centers increased dramatically. By the end of 2002 multinational companies from nine countries and regions had set up 48 joint research centers with Tsinghua

*Table 9.7 International joint research centers in Tsinghua University*

| Year | Number of new centers | Number of new centers with Fortune 500 Companies | Total contract amount (10 000 yuan) | Contract amount with Fortune 500 companies (10 000 yuan) |
|---|---|---|---|---|
| 1992 | 1 | 0 | 80 | 0 |
| 1993 | 1 | 0 | 100 | 0 |
| 1994 | 5 | 2 | 3531.19 | 250.22 |
| 1995 | 3 | 2 | 3018 | 540 |
| 1996 | 4 | 3 | 1192.5 | 1102.5 |
| 1997 | 4 | 1 | 647.7 | 320 |
| 1998 | 5 | 1 | 1653.28 | 397.8 |
| 1999 | 5 | 1 | 1777.88 | 600 |
| 2000 | 7 | 4 | 1933.61 | 1025.81 |
| 2001 | 9 | 3 | 5268.92 | 812.34 |
| 2002 | 4 | 1 | 147 250 | 1652 |
| Total | 48 | 18 | 166 453.08 | 6700.67 |

*Source:* S&T Department of Tsinghua University.

University, 18 of which were with Fortune 500 companies (Tables 9.7 and Table 9.8; Ma et al. 2003).

Within a decade over 1.6 billion yuan have been invested in 12 research areas such as information technology, machinery, automation, automobiles and biology. Overall, Tsinghua University has a 29 percent share in these centers. Multinational companies own 4 percent, with the rest being jointly owned by both sides (Table 9.9).

Cooperation during the past decade in these centers generated 36 domestic patents, 23 international patents, over 500 published papers and 12 commercial products. Among the 145 projects undertaken by these centers, 77 were jointly developed by both sides. In total, over 5000 students have benefited from such joint research (Ma et al. 2003).

### 9.2.4 University Science Parks

China's first university-based science park – Northeast University Science Park – was established in 1989. Since then, university-based science parks have become new avenues for commercializing university technologies. In 2007 there were 62 university-based science parks at the national level throughout China. There are also a large group of university-based

*Table 9.8 Regional distribution of international partners in Tsinghua University*

| Country/region | Number | Total sum (10 000 yuan) |
|---|---|---|
| USA | 23 | 142 584.8 |
| Japan | 10 | 6337.52 |
| Germany | 5 | 734.8 |
| Canada | 2 | 1136.9 |
| South Korea | 2 | 3057.82 |
| Taiwan (China) | 2 | 10 619.5 |
| Hong Kong (China) | 2 | 173.4 |
| Britain | 1 | 1652 |
| Swiss | 1 | 156.34 |
| Total | 48 | 166 453.08 |

*Source:* S&T Department of Tsinghua University.

*Table 9.9 Ownership of research centers (percentage)*

| | Tsinghua University | International Co. | Jointly owned |
|---|---|---|---|
| Research center | 29 | 4 | 67 |
| Project | 29 | 4 | 67 |

*Source:* S&T Department of Tsinghua University.

science parks launched by local governments or independently managed by universities.

China's university-based science parks were partly inspired by the legends of Stanford Science Park, Cambridge Science Park and many others. These parks are typically created through the joint effort of local government and university administration. Most of them are located on or adjacent to the university campus and administered by a commercial entity established by the university, or through a joint venture between local government and the university.

Science parks are to make full use of the advantages of universities in terms of innovative capability, talent and solid research foundations. Instead of a place for mass production, university-based science parks have been launched mainly for high-tech entrepreneurship. University-based science parks have played an important role in three areas in particular: (1) incubating spin-offs created by faculty or students from

*Table 9.10 General statistics of university-owned enterprises in science parks (in 2004, billion yuan)*

| | Enterprises in science parks | Total university owned enterprises |
|---|---|---|
| Number | 1121 (24.57%) | 4563 |
| Income | 58.26 (60.10%) | 96.93 |
| Profit | 3.24 (64.95%) | 4.99 |
| Tax paid | 2.39 (49.26%) | 4.87 |
| Income to university | 0.45 (25.56%) | 1.75 |

*Source:* S&T Development Center, Ministry of Education. Figures in parentheses indicate the ratio to the total number of the year.

universities; (2) widening and increasing the channels to integrate and commercialize all kinds of scientific outcomes from the local market as well as other places (in addition to providing incubators for spin-offs, these parks have also become a magnet to attract other high-tech start-ups, including those created by expatriates); and (3) providing services to enterprises located in the science park, which means not only managing the real estate of the park but also providing a sound environment for innovation, ranging from fundraising to legal counsel as well as talent absorption (He and Zhang 2005).

Table 9.10 shows the function of university-based science parks as important incubators for university-owned spin-offs. In 2004, of the 4563 enterprises affiliated to universities, over 24 percent were located in the science parks. Although relatively small in number, these enterprises have operated far better than those outside science parks in terms of income, profit and tax paid, accounting for over 60 percent, 64 percent and 49 percent, respectively, of the total figures generated by university-owned enterprises.

While these science parks have developed very rapidly over the past few years, several problems have haunted their further development: (1) many parks are not very clear of their mission and strategy and so, seeking short-term economic achievements, accepted all kinds of enterprises without careful selection – only when the parks decide on the specific area for development according to the strengths of the university will they obtain comparative advantages and long-term development; (2) multiple layers of administration from the district, municipal to provincial government level, burden the development of the parks; and (3) lack of human capital for technology assessment, technology transfer and capital operation (He and Zhang 2005).

*Table 9.11 Growth of university-owned enterprises (billion yuan)*

| Year | Number | Sales | Profit | Tax paid | Income to universities |
|---|---|---|---|---|---|
| 1997 | – | 29.55 | 2.72 | 1.23 | 1.58 |
| 1998 | 5928 | 31.56 (6.8) | 2.59 (–5.6) | 1.35 (9.7) | 1.50 (–5.1) |
| 1999 | 5444 | 37.90 (20.1) | 3.05 (18.0) | 1.66 (18.6) | 1.59 (6.0) |
| 2000 | 5451 | 48.46 (27.9) | 4.56 (49.5) | 2.54 (53.3) | 1.69 (6.2) |
| 2001 | 5039 | 60.30 (24.4) | 4.81 (5.5) | 2.84 (11.8) | 1.83 (8.3) |
| 2002 | 5047 | 72.01 (19.4) | 4.59 (–4.6) | 3.63 (27.8) | 1.72 (–6.0) |
| 2003 | 4839 | 82.67 (14.8) | 4.29 (–6.4) | 3.87 (6.61) | 1.8 (4.7) |
| 2004 | 4563 | 96.93 (17.3) | 4.99 (16.3) | 4.87 (25.8) | 1.75 (–2.8) |

*Source:* S&T Development Center, Ministry of Education, 2004. Figures in parentheses indicate growth rate (%).

Chinese universities regard science parks as being an important symbol of public service achievement, which will help develop the regional economy and boost technology innovation.

### 9.2.5 University-owned Enterprises (UOEs)

Many times, universities start their own companies to commercialize technology (Chen 2004). While there is no formal definition, university-owned enterprises (UOEs) refer to those enterprises that are still in one way or another controlled by a university. In some other cases, enterprises are willing to submit the control power back to the university so that they may access intangible benefits.

UOEs are not new for Chinese universities. Many, particularly the engineering and science-based universities, have had university-owned factories since the 1950s as part of their training facilities. They were mainly used for students to have short-term internships or apprenticeships. Moreover under the 'work unit system' (a self-sufficient organizational system for enterprises, universities and other social institutions in China after the founding of the People's Republic of China), many Chinese universities had their own service providers such as print shops, publishers, guest-houses and so on (for a detailed discussion of the system, see Lu 1989). Under the market environment, UOEs are presenting more complicated relationships with their parent universities.

In 2004 there were 4563 enterprises affiliated with regular Chinese universities (MOE 2004). Table 9.11 presents an overall picture of the development of UOEs. As can be seen, over the past several years and in particular between 1998 and 2000, UOEs have maintained their growth momentum in terms of sales, profit and tax paid. Since 2001, however, the growth rate has slowed down to some extent, with the profit and income to universities even decreasing in 2002 and 2003. This slowing of the growth rate partially reflects the effect of the de-linking effort, since some firms are not calibrated as UOES anymore.

While there are many UOEs in China, only a very small proportion of them are successful. These successful and influential UOEs are concentrated in a small number of selected universities and cities around the country. With respect to income in 2004, for instance, the top 13 universities that achieved higher income accounted for over 84 percent of the total amount of the year, leaving the remaining universities with less than 16 percent of total income.

In 1959, echoing the country's call for 'education serve for production', Tsinghua University added a production function to the Science and Research Institution (SRI), which gradually formed the UOE system in the university. To date, the development of commercialization activities in Tsinghua University has experienced three stages, echoing the national development stages of UOEs.

The first stage (1949–1970s) saw the formation of several UOEs, such as Tsinghua Computer Manufactory, Tsinghua Power Station, Tsinghua Auto Manufactory and others that provide practical training for relevant courses.

The second stage dates from 1980 with the operation of Tsinghua Technology Service Company, which was regarded as a milestone action by using modern enterprise management in UOEs with technology transfer and commercialization as the ultimate goal. In the next 15 years that followed, numerous enterprises by faculty members, students and alumni popped up under the umbrella of Tsinghua University. Poor management and short-term profit seeking of these enterprises raised controversies as to whether the government should regulate UOEs. As a result, the State Commission of Education, State Science and Technology Commission and Investigation Office of the General Office of the Party formed a joint investigation team in November 1990 and proposed a report that endorsed the development of UOEs (Xue 1997).

Echoing the national trend, Tsinghua University decided to establish a unified firm to manage its spin-offs for better quality control, which resulted in the establishment of Tsinghua University Enterprise Cooperation in 1995. This started the operational reform of all the firms related to Tsinghua University. One of the efforts was to introduce investment business into the system, which later brought several successful commercialization cases up to the record, such as the NUCTECH case (see Section 9.4 for details).

Tsinghua University owned 2.71 billion yuan in valued property as at 31 December 2001 within 42 independent or joint investment companies. Under this effort, sales of all Tsinghua-owned enterprises reached 10 billion yuan in 2003 for the first time in its history with a couple of firms being successfully listed on the stock market.[4] At this stage universities still executed major control power over UOEs, which led to two operational paradoxes. On the one hand, it was inevitable for the university to exercise too much control over the UOE's daily operation since the administrative structure of the university was not designed for business running. On the other hand, the university and the UOEs shared the same accounts book in financial management, which exposed the university's property – which literally belongs to the nation – to too much risk, especially when the value of the UOEs was decreasing.

The early 1990s marked another round of rapid growth in UOEs lasting from 1991 to 2000. In 1991, China's State Council issued its endorsement of UOEs in a document submitted by the Commissions on Education and Science and Technology to provide guidelines for administering UOEs. Meanwhile in 1993 another document submitted by the Commission on Education was again endorsed by the State Council to expedite the reform and development of higher education. This document prescribed that university-affiliated enterprises of a high-tech nature should be actively developed. Since then, particularly after Deng Xiaoping's southern tour in 1992, UOEs have been developing at an accelerated speed. In 1992 sales of such enterprises jumped to 2.9 billion yuan from 1.76 billion yuan in 1991. By 1999 this number had reached 37.9 billion yuan (SSTC 1992; STS 1999).

The third stage started around the year 2000, when new controversies began to surface over the appropriateness of universities becoming involved in running enterprises. The country introduced its first 'enterprise law' in 1997, which requires UOEs founded before this law to adjust their operational rule and property rights to fulfill their legal legitimacy (CCPIT Patent and Trademark Law Office (2001, p.13). Concerns of potential financial risks, together with the desire of UOEs to operate as real commercial enterprises, drove the 'de-linking' process starting from 2000. Universities were required to separate UOE management from university administration.

In order to give UOEs more flexibility in administration and to protect university property at the same time, nine central governments jointly published the *Guidelines over Regulating the System of University-owned Enterprises* – using Tsinghua University and Beijing University as pilot cases in November 2001. In Tsinghua University this directly led to the third round of UOE reform (General Office of the State Council of the People's Republic of China, 2001[5]).

## 9.3 COMMERCIALIZATION AND TECHNOLOGY TRANSFER SYSTEM STRUCTURE IN TSINGHUA UNIVERSITY

As shown in Figure 9.1, Tsinghua University has a large administrative organization to manage the research output of 6945 faculty members from 55 departments.[6] The university has executed its technology transfer and commercialization activities for more than 50 years mainly through two channels: Tsinghua University Science and Research Institution is the official organization for university-wide technology transfer, while Tsinghua Holdings Co., Ltd is a totally commercialized enterprise that takes charge of university commercialization related activities. The function of these two sectors only intersects a little, at the International Technology Transfer Center where foreign technologies are connected with domestic market calls using university-invested capital to finalize the commercialization process. This section introduces the development history and administrative structure of the two organizations, and lists relevant policies at the university level to fulfill commercialization and technology transfer goals.

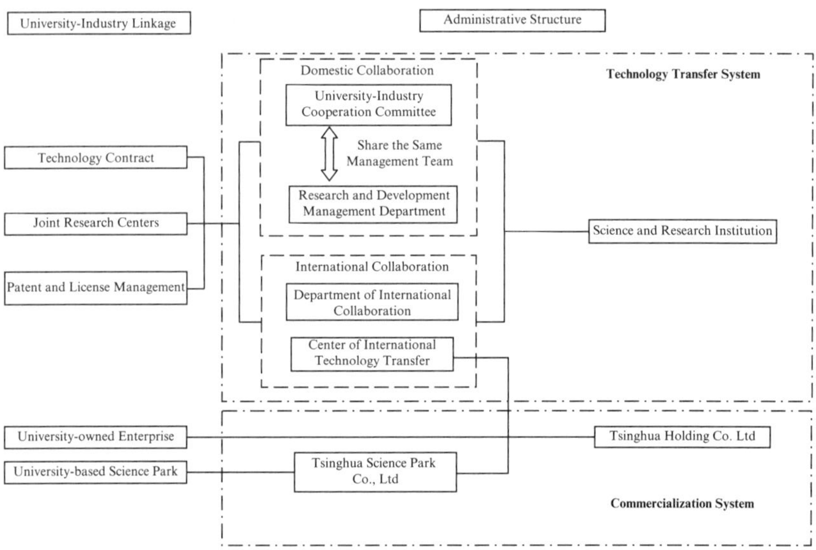


*Figure 9.1 Administrative structure for Tsinghua University technology transfer and commercialization*

### 9.3.1 Administrative Structure for Technology Transfer in Tsinghua University

The initiative of Tsinghua University Science and Research Institution (SRI) started in 1956, when the country was developing its first long-term science development plan – 'Guideline of the Long-term Plan for Science and Technology Development: 1956–1967', which was also called the 'Twelve-year Science and Technology Plan'. Although Tsinghua University only had a very small number of scientific research projects at that time, it still established an official institute for future R&D management with the purpose of strategically adding scientific research as part of the university mission. Until 1978 SRI gradually developed its research management and patent management branch that enhanced academic knowledge exchange and technology transfer. The university formed its organizational base for academic research exchange and technology transfer and continuously expanded its functions to achieve these goals.

Tsinghua University developed its SRI structure according to customer type in order to better serve the university's goal of providing solutions to the economic and social development of the nation. Figure 9.1 shows the current technology transfer system in Tsinghua University, among which two institutes are responsible for the majority of technology transfer and commercialization services: the Research and Development Management Department (RDMD) serves domestic collaborations between the university and industry sectors; while the Department of International Collaboration (DIC) helps with technology transfer activities and other types of collaborations with foreign entities. Although designed to help different customers, these two branches work collaboratively most of the time. Additionally SRI also contains the University-Industry Cooperation Committee (UICC) as a collaboration club serving both domestic and foreign enterprises, while the International Technology Transfer Center (ITTC) is one example of many investment companies owned by Tsinghua University to help cross-national technology transfer and commercialization. The following sections will discuss these organizations in detail.

#### 9.3.1.1 Research and Development Management Department (RDMD)

Developed from the research management branch in the 1950s, the Research and Development Management Department (RDMD) was founded in 1983 reflecting the university's long-term effort in supporting R&D and its application in industry sectors. RDMD has a broad range of authority, including technology contract management, patent application and licensing management, and joint research institute management. Authorized directly by the president of the university, RDMD builds stable and close ties with many provinces, municipality governments and local industries aiming to transfer university research outputs to a certain region. For example, Tsinghua has built stable research collaborations with the Beijing municipal government, having annual contract values of up to 0.1 billion RMB.[7]

RDMD achieves its technology transfer goals through four models: (1) university-industry/local government collaboration offices; (2) joint research collaboration foundations sponsored by the university and provincial governments, supporting university-industry research collaborations;[8] (3) a UURR collaboration model that ties the domestic university together with foreign universities, domestic localities and foreign localities to pursue research collaboration; and (4) regional research and development institutes, a joint research foundation, sister universities or joint research laboratories, which combine the university's knowledge and human capital with industry research requirements and funds. Box 9.1 illustrates different platforms adopted by RDMD to execute its goals.

Continuously being the largest institution managing domestic technology transfer issues, RDMD inevitably shares managerial functions with the UICC while serving domestic technology collaboration and transfer.

#### 9.3.1.2 The University-Industry Cooperation Committee (UICC)

The university continually strives to promote technology commercialization, to participate actively in national and regional development, and to contribute to society. Starting operations in 1995, UICC is designed as a broad platform for accelerating Tsinghuas University's technology transfer and as a specific tool for strengthening collaboration between the university and industry. On the one hand, UICC encourages in-depth collaborations between the university and the industry sector with the belief that connecting technology and economic development is essential to improve the national innovation capacity. The university has the talent to forecast future trends in technology development and bring about applications of technologies. On the other hand, UICC uses a membership club to strengthen the ties between firm demand and university technology supply. For example, domestic industry accounted for more than 800 contracts with a total value of 0.4 billion yuan in 2007. Technology consultancy, contract research projects and training were provided to over 180 industry members through the UICC system in the same year.

#### 9.3.1.3 Department of International Collaboration (DIC)

In Chinese engineering universities, practical skills have traditionally been ranked high in talent training colloquiums. Thus the technology transfer service was originally designed to work with domestic partners. However,

## BOX 9.1 RDMD BUILDS DIFFERENT PLATFORMS FOR REGIONAL DEVELOPMENT

There are four platforms that help RDMD in Tsinghua University to help certain regions in their social and economic development. First and foremost is the Office of University-Industry-Research Institution Collaboration that aims at connecting the university with local government. Functioning as technology transfer offices, these offices are headquartered in RDMD but deal with issues related to university-industry collaboration with a special focus on certain municipalities. Their responsibilities include – but are not limited to – duties such as: building technology information exchange platforms between the university and the localities to find the most suitable means of university-industry collaboration; bridging the locality with the most advanced technologies either from Tsinghua University researchers or from other countries who want to transfer their technologies in China; organizing expert consultancy; and enhancing joint training programs both in the university and at the local level.

The Technology Development Collaboration Foundation is an agreement signed between Tsinghua University and a local government. Tailored by local technology needs, these foundations bring Tsinghua researchers and technologies to the region for technology consultancy, contract research and technology commercialization. For example, the He Bei Province-Tsinghua Technology Collaboration Development Foundation mainly focuses on sponsoring high-tech firms under the provincial supporting list with an annual budget of 5 million yuan. There were nearly 100 joint projects underway at the end of 2006, with a contract value of over 0.3 billion yuan.

The UURR model is also called the "Four-Party-Collaboration" model, referring to collaboration relationships among Tsinghua University, foreign universities, a region in China and regions in other countries. This is a multilateral open collaboration model developed in line with globalization trends, where Tsinghua integrates possible resources for all four parties. Undergoing projects include 'Tsinghua University, Iwate University, Zhejiang Province, Iwate county in Japan' and 'Tsinghua University, Fraunhofer Institute, Jiangsu Province, Nordrhein-Westfalen in Germany'.

> Joint research laboratories are the fourth platform RDMD uses to transfer Tsinghua's technologies to regional enterprises. Tsinghua collaborates with a variety of partners to build these laboratories from local firms to foreign firms. Between 1999 and 2005 there were around 80 joint research institutes built within the university. As mentioned above, this format of university-industry collaboration not only provides direct technology support to the enterprise, but also forms a training base both for university students and for technicians in the firms.

along with the economic reform starting from 1978, a significant amount of technologies were transferred from foreign countries. Chinese universities had more technology contracts and patents with foreign partners. For example, in 2006 the DIC recorded 472 newly signed international technology contracts with research funds of up to 237 million yuan. This accounted for 36.5 percent of total research funds and for 26.6 percent of the total research contract amount (RDMD report 2008).

As the organization in charge of foreign research collaboration, DIC was founded in 1996 with a focus on: (1) encouraging communication between Tsinghua University and international counterparts such as research institutes, universities and enterprises; (2) connecting the university with foreign enterprises in technology innovation collaboration and commercialization; (3) connecting domestic enterprises with foreign enterprises through UICC for potential technology transfer and collaboration; and (4) commercializing international technology at local firms. As the counterpart of RDMD in serving foreign partnership and collaboration, DIC duplicates RDMD's managerial structure.

#### 9.3.1.4 International Technology Transfer Center (ITTC)

The fourth branch of Tsinghua's technology transfer system is ITTC, whose major goal is to disseminate advanced international technology to local industry sectors. Sponsored by the Ministry of Education (MOE) and the Ministry of Science and Technology (MOST), ITTC is a state-authorized technology transfer center. Tsinghua University helps ITTC by exploring new technologies, targeting cutting-edge technologies, executing the R&D, and disseminating and commercializing technology through two commercial entities (Coway International Tech Trans Co., Ltd and Beijing Tsinghua Luyin Science and Technology Development Co., Ltd). As the biggest shareholder, Tsinghua University initiated and founded these two entities in 1998 and 2002, respectively, in order to carry out technology commercialization and technology management for ITTC, with a view to developing and promoting international technology transfer and commercializing Tsinghua University's scientific and technological achievements. As the current policy emphasizes absorbing imported technologies, ITTC provide services ranging from purely technology intermediate services to a basket of services including technology targeting, technology importing, absorbing and redevelopment of imported technology with the purpose of commercialization.

### 9.3.2 Commercialization in Tsinghua University

Tsinghua Holding, Co., Ltd, although a separate commercial entity from Tsinghua University, still has a close relationship with the university in research and shares with the university the goal of technology transfer – 'to build enterprise-centered university industry technology transfer system and innovation system based on market economy'. Not only has the company invested in several star technologies from Tsinghua University and successfully commercialized those technologies, but the company also signs 20–25 percent of the university's technology contracts.

Tsinghua Holding, Co., Ltd has been listed among the 500 biggest state-owned enterprises from 2003–06, and has been rewarded as one of the 500 enterprises with the strongest innovation capacity.

The 2001 policy of de-linking UOEs from Tsinghua University led directly to the formation of the Tsinghua Holding, Co., Ltd in 2003, combining former Tsinghua Enterprise cooperation, Tsinghua University Technology Incubator and Unisplendour Corporation Limited (UNIS). Four management reforms have been made. Firstly, this reform completely separated UOE property from university property through a professional auditing process. The new company liquidated all the operating assets of the university and transferred them into the holding company. Secondly, from 2001 onwards the university administration withdrew from both the administrative team and financial team of UOEs. For example, all university administrators resigned from UOEs, where they used to be the chief executive officers. This wave of position adjustment ensured future separation between company decisions at the business level and university decisions at the policy level. Physically, all UOEs moved out of campus and employees were no longer jointly appointed. Thirdly, the 42 companies that were then owned by Tsinghua University were either merged into Tsinghua Holding Co., Ltd or dismissed. Last but not least, no UOEs were allowed to include 'Tsinghua University' in their names for further intellectual property rights and reputation protection. The merger dealt with capital merger and institution reform. Tsinghua

University technology commercialization was then formally set as a modern enterprise format.

The Executive Board of Tsinghua Holdings directly reports to the Tsinghua University Operating-Asset Management Committee, which was formed within the university management system as an independent legal body representing the university to appoint board members and the supervisory board. Under the leadership of the president, Tsinghua Holdings provides capital investment, property management and consulting services. Tsinghua University is the technology backbone as well as the biggest shareholder of the company.

The property value of Tsinghua Holdings rose from 18.4 billion yuan to 2.8 billion yuan in 2006, an increase of 52 percent. Its sales almost doubled from 2003 to 2006. Most importantly Tsinghua Holdings and companies initiated 206 million yuan worth of research projects with Tsinghua University. This contributed 14 percent of the university's total R&D funds.[9] Tsinghua Holdings has become an important player not only in Tsinghua University's technology transfer but also for local economic development (Song and Wang 2005). For example, Tsinghua Science Park, Co. is jointly invested by Tsinghua Holdings and other companies. It is not only an incubator to start up companies, but also provides R&D resources, training programs for high level management, and domestic and international technology transfer activities. Four years of experience helped Tsinghua Holdings Co., Ltd to rank 143th among the top 500 Chinese companies with a sales income of 21.5 billion yuan in 2006.

## 9.4 CASE STUDY OF COMMERCIALIZATION IN TSINGHUA UNIVERSITY: NUCTECH

NUCTECH is one of the university-owned companies in which Tsinghua Holdings Co., Ltd invested. X-ray technology is used for a quick inspection of large-sized containers or vehicles with large containers without opening them. The product line has been extended from a single product – a stable security inspection system – to a small container inspection system, liquid inspection system, mobile inspection system and other municipal security inspection systems. In 2008 the company had 1200 employees with sales of 1.2 billion yuan in 2005 and 0.12 billion yuan profit.

The core technology developed from a joint research project among three departments in Tsinghua University: the Department of Engineering Physics, the Department of Mechanical Engineering and the Institute of Nuclear and New Energy Technology. It took eight years from research

*Table 9.12 Technology development and commercialization timetable in NUCTECH*

| Time | Technology Development/Commercialization Improvement |
|---|---|
| 1991 | Professors from three departments initiated the research project and started looking for research funding |
| December 1991 | The research project obtained funding from the 'eighth five-year technology plan', with Tsinghua University as the Primary Investigator |
| March 1992 | Research project officially started |
| 1993 | One-quarter-sized prototype is successfully launched in the laboratory |
| January 1996 | National funded research project completed |
| January 1996–November 1996 | Due to the lack of further financial support, technology development process idled for nearly a year |
| November 1996 | With Vice Premier Li Lanqing's intervention, Chinese Customs invested 2 million yuan as development funding. NUCTECH was founded in 1997 |
| November 1996 – January 1998 | Security inspection system finished commercialization and passed the national qualification test |
| January 1998 – June 1998 | Not enough orders for further commercialization |
| June 1998 – September 1998 | With the help of the central government, Chinese Customs set up a special project to order a large number of Tsinghua University's inspection system |
| September 1998 | Tsinghua University adopted the 'visual manufacture' method to meet the large order |
| December 1998 | Developed a mobile security inspection system according to Chinese Customs' request |
| July 1999 | Sales reached 1.1 billion yuan with 40 sets of products installed in Chinese customs and customs in other countries. This reflects the success of the commercialization. The company continued developing on a good track |

*Source:* Zhang (2002), Table 2.2. p. 18.

project orientation until the first commercialized product was successfully sold. Table 9.12 shows the technology development and commercialization timetable of NUCTECH. After 1997, with the investment of Tsinghua Holdings Co., Ltd, NUCTECH reached its boom period. So far more than 210 THSCAN™ Systems have been exported to over 50 countries and regions in Europe, America, Asia, Oceania and Africa. As one of the three biggest suppliers of high-energy security inspection systems, NUCTECH held the largest market share in the world in 2008.

### 9.4.1 Brief History of NUCTECH

In the late 1980s Chinese Customs was considering purchasing a customs inspection system from the UK or Germany. One of the NUCTECH co-founders, a professor in the Department of Engineering Physics in Tsinghua University, was invited as a technology consultant. The professor found that, surprisingly, Tsinghua University had all three core technologies – accelerator technique, X-ray inspection technique and radiation imaging technique – for developing a security inspection system independently. This idea was not executed until the late 1980s.

The Department of Engineering Physics which heavily relied on military–related research projects, experienced a financial dip due to a sudden decrease of research funds from the government. Academic research slumped along with the decrease of research funds. The department had to find new research directions that could either make a theoretical breakthrough which would allow it to obtain sufficient research funds, or provide civil applications by technology transfer or commercialization. Thus, on receiving the news that the first security inspection system was launched at a French airport in 1991, the evaluating professor immediately started to organize a project to produce the system. A group of professors from different departments sat together for a feasibility discussion. Zhang (2002) recorded the conclusions reached in the discussion. The first conclusion shows great confidence among participants about the potential market for the security inspection system, both in China and all over the world; the discussion then affirmed the prominent academic value of putting three different techniques together to integrate a new system. For the rest of the discussion, the group estimated that the development period of the product would not be very long due to the mature techniques Tsinghua University already had. The combination of academic value and commercial value made this project very attractive, and all the participants estimated that limited requirements of people and resources would be needed to produce a prototype.

After the feasibility discussion conference, a joint research proposal seeking administrative and financial support was written by relevant departments to the university administration. It was relatively easy to gain administrative support for starting a joint research group across departments; however faculty members had to apply for national technology innovation research funding for financial support in applied technology development. In the early 1990s, the National Research Foundation was a major resource for university research, emphasizing pure basic research or applied research, but rarely technology development or product development. Only a small amount of research funds were issued to the joint research group in 1992, the official starting date of the project. With prominent researchers from different departments, the academic part of the project went fairly well: the first one-fourth-sized prototype was successfully launched one year after they started, and multiple efforts had been made to turn it into a product.

## 9.4.2 Key Factors in the NUCTECH Commercialization Process

### 9.4.2.1 Financial support and venture capital

The joint research group encountered the first problem of a funding shortage in 1996 when the national research project ended after a three-year period. Since governmental funds only support basic and applied research, the joint research group received hardly any national research grants for product development. With no subsequent funds, the project lingered until around 1998 with no essential progress. Some faculty members withdrew from the project, leaving it with no foreseeable future. Financial investment, therefore, was the highest priority for the Department of Engineering Physics and the research group.

Back in 1996 China had no mature investment banking system to support high-tech business at an early stage, not to mention investing in research projects. Potential financial support came from private firm investment or from national bank loans. Obtaining this was a task beyond the capability of university researchers or department administrators for two reasons. Firstly, the product was still in the process of laboratory experiment within the academic department; there was therefore no legal person to take responsibility for the financial return. Secondly, investors regarded high-tech products as a high-risk investment and were hesitant to support a product at such an early stage. The department tried to communicate with Chinese Customs, the potential buyer of the product, for a small grant as down-payment to use as a development fund. However this attempt failed due to the high level of investment risk.

The government and governmental officials played a significant role in financing the product development. Vice Premier Li Lanqing was informed about the project and visited Tsinghua University four times about this project. His communication with related organizations finally pushed Chinese Customs to agree to fund the project with 2 million yuan as the down-payment for future orders. When Tsinghua Holdings Co., Ltd, one of the Tsinghua's largest UOEs, went to market in 1997 it agreed to invest another 20 million yuan in the project based on the suggestion of the Vice Premier and the university administrators. NUCTECH was officially founded in 1997.

Financial support for technology commercialization should come from market mechanisms such as venture capital, industry investment or technology intermediary organizations. Government officials may push investment decisions when the degree to which people are willing to take risks is low. Along with the development of the financial market, the opportunity for UOEs to obtain angel funding or other forms of early investment are now increasing.

#### 9.4.2.2 Balancing teaching, research and technology commercialization within the department

The Department of Engineering Physics adapted a special mechanism to solve the problem of balancing its teaching and research roles with technology commercialization: the so-called strategy of 'Transplanting with the soil first, reward the nursery later'. 'Soil' refers to faculty members, equipment and other resources in the department.

Founded in 1956, the department offered courses for undergraduate, masters and doctoral students. The teaching load as well as research duties and public service did not leave enough time and energy for technology commercialization, while commercialization required scientists to work full time on the project. On the one hand, the traditional teaching load and the additional load of technology development generated human capital scarcity given the unchanged number of faculty members. On the other hand, the commercialization activity challenged the university evaluation system that regarded academic improvement such as publications, student training and knowledge publicity as essential outcomes. The majority of the participating scientists were tenure track faculty members in the Department of Engineering Physics. They had to sacrifice a certain amount of publication opportunities to technology development and commercialization without being rewarded within the existing evaluation system. Similarly faculty members who did not engage in this project but taught extra courses in the department and provided more public service could not be rewarded in the evaluation system either. Recognizing both groups and making full use of the limited human capital in the department was a huge challenge.

As a solution the department raised a plan called 'transplant with the soil first, reward the nursery later'. During the first stage, the department allowed some faculty members to work full time in NUCTECH for knowledge and experience dissemination; this was during the fast expansion period of NUCTECH. While these faculty members worked in NUCTECH, they retained their positions in the department and had their teaching load and public service requirement eliminated. In return, a certain percentage of sales income was given back to the department as a reward to faculty members who offered more classes and to support other research projects in the department. In this way the department and NUCTECH reached a win-win equilibrium. Faculty who worked full time in NUCTECH during its fast-growth period still contributed to the department in addition to the revenue repayment: although they did not need to teach or serve public duties, they did supervise students on theses and conducted research work related to the project. This generated publications as well as providing students with the opportunity to access the most updated technological and practical experience. These faculty members also advised the technicians who took over the technical know-how in the company after the faculty members withdrew. Nowadays, with over 60 research scientists in the company, less than ten researchers are jointly affiliated with the department. This forms a clear separation between the academic research in the department and R&D in the company. Besides a few adjustments adopted in the early stage, this model did not change the academic evaluation system in the department or at the university level.

## 9.5 SUMMARY AND CONCLUSION ABOUT NUCTECH

As the most successful technology commercialization case in Tsinghua University, NUCTECH has been studied through many lenses, such as its technology transfer model, its role in the national innovation system and so on. Although NUCTECH may not work as a universal model for university commercialization in China due to many unique opportunities along its development path (Zhang 2002), it does represent a development path along which Chinese universities and UOEs explore innovative ways of technology commercialization. Institutional reform, such as the separation of administrative authority from financial management, pushes university commercialization towards a market-oriented mechanism; a good combination of government support and mature market rules are expected to guide university technology transfer and commercialization to a better chance of growth.

### 9.5.1 The Importance of Financial Resources in University Technology Transfer and Commercialization

UOEs are still having a hard time attracting venture capital with many obstacles. Firstly, the investment market in China is relatively small and at an early stage. According to Dow Jones Venturesource, Chinese venture capital investment hit its three-year peak in the third quarter of 2008 with US$ 0.96 billion. However it is a small scale compared with the US$ 7.1 billion market in the USA during same period. Secondly, the risk-aversion logic adopted by Chinese venture capitalists prevent university technology from being commercialized (Xue et al. 2005). In NUCTECH's case, Tsinghua Holdings gave 20 million yuan to NUCTECH in 1997 mainly because the vice premier persuaded it to do so. This investment was regarded more as an input aimed at maintaining a good relationship with the government than for financial return. The project was evaluated as very risky according to one of the managers in Tsinghua Holdings, who also admitted that if they had realized how much value NUCTECH would generate back in 1997, they would have invested much more.

At the same time, end users are also risk-averse. Chinese Customs would not place an order without seeing the mature product certified by the national qualification standard, and no private companies were willing to invest in the project before they saw the order contract from Chinese Customs which would guarantee success of the product. The financing process then fell into a deadlock (Zhang 2002).

The lack of market support for technology commercialization, especially the lack of a venture capital mechanism, forces universities to find other ways to fill the financial gap. It is easier for them to obtain funding from government sectors due to the fact that Chinese universities maintain a good relationship with the government. Financial support then is not determined by evaluations regarding the degree of technology maturity, potential market value or rate of return on the project; rather it is determined by how close a university or university researcher is to a government sector. This closeness varies from university to university, from researcher to researcher and from project to project. Each researcher has to try out their project through this system and there is no guarantee that enough financial support will be given in time. This may cause the termination of many technology transfer projects and commercialization projects.

### 9.5.2 Human Capital Accumulation in Technology Transfer and Commercialization

With talent training being one of the most important responsibilities of Chinese universities, human capital accumulation in technology transfer and commercialization has always been given a high priority. The university serves as the backbone of technology and human capital for UOEs, while UOEs provide research funds and the most advanced experiment labs to researchers and students in the department. This format of collaboration is widely adopted in Chinese universities and their UOEs. Qingdao MESNAC Co., Ltd is a software company that has built a very close relationship with Qingdao University of Technology. Located right next to the university, the company used to be a UOE and generated close ties on both research development and human capital sharing with the university.

To keep intellectual talents, the NUCTECH case provides an innovative solution in which faculty members balance their teaching and researching duties with commercialization. Even today, this evaluation system is not widely adopted by a majority of universities. As a public service unit, Chinese universities share a set of strict rules in human resource management. There was only a very short period of time in the late 1990s when universities encouraged faculty members and students to 'jump into the sea' by allowing a leave of absence for commercialization activities with no re-entry test. This trend ended fairly soon – around 2001 – when the government established firm monitoring of UOEs. Universities regard start-ups more as a natural market behavior than an extension of the research branch of the university. No special 'green light' is designed for commercialization behavior. If, for example, faculty members leave for commercial activity, they must resign from their position. When they want to come back to the university, they must wait for an opening and compete with other candidates under the same evaluation of academic achievement. No joint affiliation is officially allowed between UOEs and academic departments. Leaving a university position for entrepreneurship is therefore quite risky for faculty members.

In other words, the majority of university evaluation systems have not been modified to support university commercialization. Minor changes, however, have taken place during the past 20 years, such as including patenting as a research outcome in annual faculty evaluations.

### 9.5.3 Intellectual Property Rights Protection in University Technology Transfer and Commercialization

Both universities and faculty members have increased awareness of patent protection. They patent not only to protect the latest technology, but also to protect concepts, methods or theories using the patent network for their future benefit. One trend which accounts for a large degree of university technology transfer and commercialization is faculty members collaborating with or working in private companies who put great emphasis on intellectual property rights protection. In this process, faculty members are educated regarding the protection of intellectual property rights. The ownership of intellectual property rights varies from case to case. Prominent universities have more bargaining power in sharing or even gaining the ownership. Tsinghua University, for example, is quite strict, having a rule that all contract research with industry should lead to shared intellectual property rights. The only exception is research conducted through industry-sponsored university research labs, in which the private sector not only owns the patent but also decides the distribution of the patent map.

Licensing income is not a standard statistic in Chinese universities since its accuracy is hard to guarantee for two reasons: firstly, without professional help in the legal procedure of licensing out, university researchers usually find it time-consuming and energy-consuming. On the one hand, faculty members are not professional in licensing a patent out; on the other hand, the lack of law enforcement on intellectual property rights violation makes it more difficult to benefit from licensing. 'Faculty members usually get to know who uses his/her technology by media report rather than being informed by the user.'[10] Given this, the majority of researchers would prefer to license out technology via exclusive licenses in order to reduce the time and energy they need to spend. Thus licensing income accounts for only a very small amount of university income. Also, Chinese universities rarely have royalty sharing policies. In Tsinghua University rate setting for royalty sharing is based on case-by-case negotiations.

### 9.5.4 The Government's Role in University Technology Transfer and Commercialization

Due to the naturally close relationship between Chinese universities and different levels of government, the government has always been an important player within the university technology transfer and commercialization system. In recent years, local governments have broadened university-industry collaboration opportunities in local economic development. Many municipalities provide 'green channels' to finance university-industry collaborations as a method to adjust the local industry structure by attracting high-tech enterprises into the region. UOEs benefit from this policy as well. For example, municipalities around the Yangzi triangle area are making efforts to restructure the local industry portfolio – from labor-intensive manufactures to high-tech based industry with higher value added – in response to financial crisis. Geographically, several municipalities with similar degrees of economic development and social development are scattered in the area. The settlement decision for any high-tech company would therefore strongly depend on the differences between policy incentives among neighborhood municipalities. Municipal governments are competing rigorously with each other to attract high-tech industry. Enterprises with strong university relations are among the most attractive firms to these governments. The latest policy in some localities states that the local government will finance a good research project with up to 3 million yuan for start-up funds, with another maximum of 3 million yuan venture capital from a public investment firm. Local governments also build University Collaboration Parks for use by universities to transfer their technologies to the local firms or commercialize them by themselves.

As mentioned above, university-industry collaboration has also recently been emphasized at the central government level. Although the role government should play within the national innovation system, and how it could influence the university-industry relationship in technology transfer and commercialization, is still under discussion, universities are highly influenced by national innovation policy. Chinese universities do organize their technology transfer and commercialization aside from spontaneous activities.

Tsinghua University serves as the case in point. The university attributes its success in technology transfer and commercialization partially to the organization of its R&D in three aspects: (1) the university closely connects technology development with the national S&T development strategy by organizing relevant basic research ahead of time and optimizing human capital in technology development and commercialization demonstration; (2) the university builds close relationships with different layers of governments and firms in order to receive continuous support in terms of R&D resources and cutting-edge technologies; and (3) joint research institutes built by universities and local/foreign enterprises are the most successful format for connecting advantages from different partners in practice. This relatively stable format of jointly built organizations will build long-term relationships between universities and the industry sector, and will make real-life challenges the center of applied research on campus, resulting in relevant technology.

# NOTES

1. According to the Contract Law of the People's Republic of China (PRC): (1) a technology development contract refers to a contract concluded between the parties for the purpose of conducting research in, and development of, new technologies, new products, new processes and new materials as well as their systems; (2) technology transfer contracts include contracts on patent transfer, contracts on transfer of the right to apply for a patent, contracts on transfer of know-how and contracts on the licensing of patent exploitation; (3) technical consultancy contracts include contracts whereby feasibility studies, technological forecasts, technical investigations and analytical evaluation reports shall be provided in respect of specific projects; and (4) technical service contracts refer to contracts whereby one party undertakes to solve specific technical

problems by using its technical expertise for the other party, excluding contracts for construction projects and contracts for work.

2. Distrust towards enterprises has become a principle for universities when transferring technologies. During an interview, an employee of a certain university in charge of technology transfer stated: 'We try to charge the total fees in only one time in order to avoid future troubles. Even though about 20% of the total fees sometimes can't be taken back.' On the other hand, enterprises also complain that: 'Some of their technologies can't even be put into practice, however, they always keep on asking for the money.' There are quite a few quarrels between universities and enterprises over technology transfer issues, some of which even result in court appeals.
3. *The World Competitiveness Yearbook* (2001) International Institute for Management Development.
4. Firms that went to the market during 1995 to 2003 include Tsinghua Unis Co. Ltd, Tsinghua Tongfang Co. Ltd and Chengzhi Share Holding, Co. Ltd.
5. http://law.lawtime.cn/d435307440401.htm (accessed 29 May 2011).
6. This figure refers to 2007.
7. http://www.kfb.tsinghua.edu.cn/kfb/index.htm (accessed 29 May 2011).
8. Ibid.
9. Given that the total R&D fund in 2006 was 1.487 billion yuan (RDMD 2006).
10. Interview with Cao Jianguo, Director of RDMD, April 2008.

# REFERENCES


CCPIT Patent and Trademark Law Office (2001), *Patent Law of China*.

Chen, G. (2004), 'Research on the role of China's universities in the national innovation system', Doctoral dissertation, Zhejiarg University.

Chinese People's Political Consultative Conference (CPPCC) (1996), *Law on Promoting the Transformation of Scientific and Technological Achievements*.

Fan, X. and Y. Fang (2004), 'Three stages for China's universities to interact with other research entities in technological innovation', *Journal of Technology Development and Strategies*, **6**, 130–31.

He, J. and Y. Zhang (2005), 'Function of university-based science parks', *Chinese University Technology Transfer*, **8**, 27–9.

Lu, F. (1989), 'On work unit system', *China Social Science*, **9** (1), 71–88.

Ma, J., F. Yang, G. Li and M. Li (2003), 'Research on management model and development trend of university-affiliated International Joint Research Institute', *Journal of Chengdu University of Technology*, **11**, 56–8.

Ministry of Education (MOE) (2006), http://www.moe.edu.cu/publicities/business/htmlfiles/moe/moe_1597/200708/25711.html (accessed 29 May 2011).

Ministry of Science and Technology (MOST) (2001–04), *Annual Report on China's Technology Market*, Beijing: MOST.

Ministry of Education (MOE) (1999), *Action Scheme for Invigorating Education Towards the 21st Century*, Beijing: MOE.

Ministry of Education (MOE) Center of Technology Development (2004), *Analysis Report on University Affiliated Enterprise for Chinese Regular University in 2004*, Beijing: MOE

RDMD (2006), *Annual Report,* Research and Development Management Department, Tsinghua University, Beijing.

Science and Technology Statistics (STS) (1999), 'Technology activities of inde-

pendent R&D institutes nationwide in the fields of natural science and technology', available at http://www.sts.org.cn (accessed 5 April 2007).
Shao, Y. (2002), *Reports on China's Higher Education in the National Innovation System*, Post-doctoral research report Tsinghua University, Bejing.
Song and Wang (2005), 'Practice of industrial research interactions in high education', *Chinese University Technology Transfer*, **4**, 24–7.
SSTC (various years), *China Science and Technology Statistics Data Book*, SSTC.
State Council of China (2001), *Guideline on the Management Standardization of University-affiliated Enterprises in Both Peking and Tsinghua Universities.*
Xue, L. (1997), 'China's innovation system reform: a historical perspective', *Journal of Engineering and Technology Management*, **14**, 67–81.
Xue, L., J.Q. He and Q. Zhu (2005), 'University technology: Seeking development via challenges and opportunities', *Chinese University Technology Transfer,* **4**, 38–41.
Xue, L., and L. Zhou (2007), *University-Market-Government Interactions: The Case of China*, UNESCO report.
Zhang, H.Y. (2002), 'Rethinking about the success: a study of the innovation process of Tsinghua's container inspection system', Masters thesis, Tsinghua University, Beijing.
Zhou, L. and M. Zhu (2004), 'A study on transfer to China of multinational corporations' R&D facilities', *Academic Exploration*, **4**, 70–4.

# 10. Knowledge production and knowledge transfer: a study of two Indian Institutes of Technology (IIT Madras and IIT Bombay)

**V.V. Krishna* and Nimesh Chandra**

## 10.1 INTRODUCTION

Over the last decade, the links between academic research institutes (ARIs) and industry, generally known as university-industry relations (UIS), have attracted substantial research and policy attention in science, technology, innovation policy and knowledge management related literature. A number of factors have culminated in making ARIs in science and technology important actors in the national innovation systems (NISs), creating wealth from knowledge and contributing to the economy as a whole. The impact of globalization on trade, technology and economic growth generated a demand in industry for systematic knowledge in science and engineering, which in India is increasingly becoming the domain of the university system. The importance of human resource skills for a range of new technologies such as information and communication technology, biotechnology, nanotechnology, space and energy-related technologies, coupled with the rise of business enterprises and private industry in India including public sector enterprises, is placing more and more demands on academia, particularly in premier institutes such as the Indian Institutes of Technology (IITs). Together with the Massachusetts Institute of Technology (MIT) as a reference model institution, Humboldtian values of teaching and research excellence have remained quite important in the establishment and growth of IITs.

The IITs as a representative set of ARIs particularly in science and engineering education are known for their academic excellence,[1] particularly for producing high quality engineering, science and management graduates and post-graduates. Since their establishment, the primary objective for these institutions continues to be education and research in both pure and applied science, engineering, social sciences and humanities.

The linear model of innovation, which advances technology in several ways, has conventionally been considered as the basis for technological innovation. In the last decade, IITs have experienced increasing pressure from both government and industry to reorient and reorganize their institutional and organizational structures in order to promote transfer of knowledge to industry and for societal consumption at a faster pace. IITs in India are increasingly seen as the leading frontiers of innovation and technological change, particularly concerning science-based innovation. Creating and transferring intellectual property has assumed enormous significance. This requires institutionalizing proactive policies regarding patenting intellectual property and licensing and transferring it for generating revenues; it also requires policies promoting incubation and spin-off firms for generating revenues. Policies on these issues have assumed enormous significance over and above conventional forms of technology transfer such as consultancy and sponsored research from industry to universities. This trend gained momentum in the USA from the 1980s with the introduction of the Bayh-Dole Act. However in India, particularly in the case of IITs, trends in knowledge transfer have assumed significance since the 1990s with the rise of globalization. This study is basically concerned with the exploration of knowledge production and knowledge transfer in two IITs, namely, IIT Madras and IIT Bombay. The IITs are an important part of a larger university system and hence constitute a key actor or agency of the Indian NIS. Before we explore the case of the IITs, it is pertinent here to briefly explore the NIS in India with special reference to its university system.

## 10.2 INDIA'S NATIONAL SYSTEM OF INNOVATION (NSI) AND THE ROLE OF UNIVERSITIES

Over the last decade, a very influential group of scholars (Freeman 1987, 1995; Lundval 1992; Nelson 1993; Edquist 1997 among others) have drawn our attention to the dynamics of innovation within and across different national boundaries or economies by identifying various actors and agencies which interact and determine the process of innovation. Varying perspectives culminate towards underpinning the importance of what has come to be known as the national system of innovation (NSI). Freeman (1987, p. 1) was one of the earliest to define NSI 'as the network of institutions in the public and private sectors whose activities and interactions initiate, import, modify and diffuse new technologies'. Further, Lundval (1992, p. 12) and Nelson (1993) also underline the importance of organizations and institutions such as research and development (R&D)

departments, firms, technological institutes and universities which are crucial, as well as 'the elements and relationships which interact in the production, diffusion and use of new and economically useful knowledge'. These actors are either located within or rooted inside the borders of a nation state. In the next subsections, we shall briefly present the structure and main actors of India's NSI and focus on the role of universities.

### 10.2.1 Main Actors Constituting the Indian NSI

Publicly funded national laboratories under science agencies; universities and institutions of higher learning; business enterprises both public and private; public policies on science, technology and innovation; and civil society groups and institutions constitute the main actors of India's NSI. India's gross national expenditure on R&D (GERD) in 2007–08 was around 368 billion Indian rupees, of which the government accounted for 68 per cent (down from 80 per cent in the 1980s), business enterprises accounted for 30 per cent (up from 21 per cent in the 1980s) and other sectors accounted for the remainder. University and higher education R&D accounted for about 7.2 per cent of GERD in 2007–08. The R&D budget as a proportion of GDP fluctuated between 0.8 per cent and 1.13 per cent over 1992–93 and 2004–07.

#### 10.2.1.1 Laboratories under science agencies

As shown in Table 10.1, there are 12 main science agencies which constitute the main actors of India's NSI and continue to draw considerable support from the government. Growth in the R&D support given to these 12 main science agencies during the decade of the 1990s up to 2005–06 is shown in Table 10.1. The strategic science-related science agencies Defence Research and Development Organization (DRDO), Department of Space (DoS) and Department of Atomic Energy (DAE) witnessed an almost five- to seven-fold increase in R&D expenditures. Other civilian science agencies also recorded increases, with the exception of the science agency dealing with non-conventional energy sources. As per data available, 391 000 personnel were involved in R&D activities in both public and business enterprises.

#### 10.2.1.2 Universities

Higher education institutions witnessed considerable growth in the post-independence period after 1947. From 20 universities in 1947, the number increased to over 361 universities.

There are now 20 central universities, 217 state universities, 106 deemed to be universities and 13 institutes of national importance established

*Table 10.1 R&D expenditure of major science agencies (1990–91, 1998–99, 2000–01 and 2005–06)*

| Science Agency | 1990–91 | 1998–99 | 2000–01 | 2005–06 |
|---|---|---|---|---|
| DRDO (defence) | 6810 | 23002 | 20228 | 52000 |
| DOS (space) | 3862 | 15155 | 16423 | 25000 |
| DAE (atomic energy) | 2755 | 8367 | NA | 17500 |
| ICAR (agriculture) | 2762 | 8440 | 14988 | 17500 |
| CSIR (industrial research) | 2491 | 7133 | 8778 | 17500 |
| MOEF (environment & forest) | 1620 | 3780 | 8500 | 1900 |
| DST (all S&T) | 1198 | 2990 | 7314 | 10000 |
| DBT (biotechnology) | 413 | 945 | 1415 | 2000 |
| DOD (ocean) | 278 | 848 | 1688 | NA |
| ICMR (medical) | 445 | 862 | 1470 | 2000 |
| MIT (ICT) | 330 | 621 | 760 | NA |
| MNES (non-conventional energy) | 160 | 90 | NA | NA |

*Notes:* Figures in millions rupees.
Abbreviations: DRDO: Defence Research and Development Organisation, DoS: Department of Space, DAE: Department of Atomic Energy, ICAR: Indian Council of Agricultural Research, CSIR: Council of Scientific and Industrial Research, MOEF: Ministry of Environment and Forests, DST: Department of Science and Technology, DBT: Department of Biotechnology, DOD: Department of Ocean Development, ICMR: Indian Council of Medical Research, MIT: Ministry of Information Technology, ICT: Information and Communications Technology, MNES: Ministry of Non-conventional Energy Sources.

*Source:* DST, Ministry of Science and Technology Reports.

through central legislation and five institutions established through state legislation. The number of colleges increased from 500 in 1947 to 17 625 in 2005, indicating a 26-fold increase.

In the technical education sector there were about 1749 colleges comprising 1265 engineering and technology colleges, 320 pharmacies, 107 architecture and 40 hotel management colleges. In addition to these, there were 958 post-graduate management institutions and 1034 masters in computer application-based institutions by 2004.

Through the University Grants Commission, the government accorded special recognition of deemed university status to science agencies such as in space, atomic energy and other specialized agencies to train human resources. In May 2008 the government made budget provisions to set up four new Indian IITs, six Indian Institutes of Management and 14 more central universities. Among the new higher education institutions (HEIs), mention may be made of the Rajiv Gandhi Centre for Biotechnology

(RGCB), the Indian Institute of Science, Education and Research (IISER) and the Institute of Space Science and Technology (IIST). All three institutions were created in Kerala. These centres of higher learning are created to meet the increasing demand for high technologies and science-based innovation in space, nano- and biotechnologies and related areas.

Currently the gross enrolment ratio (GER) in higher education is relatively low at 12 (that is, 10.481 million students in HEIs in about 360 universities and 17 625 colleges affiliated to these universities in 2005–06). The target is to achieve GER of 15 in the 2007–12 XIth Plan period.

Historically R&D in science and technology has been concentrated in only about 15 per cent of the universities and partly their affiliated colleges. The top 20 per cent of the universities account for the bulk of R&D output from the university sector. There has been considerable increase in the overall publication of papers in all sciences, from 52 120 in 1999 to 89 297 in 2005. Universities produce more than 50 per cent of the total papers published from India, though they account for a mere 7.2 per cent of GERD. There has also been a considerable increase in the growth of science and technology (S&T) human resources, from 1.7 million graduates in 1980 to eight million in 2004. About 14 per cent of 391 000 R&D personnel are in the university sector.

The human resources for India's S&T system primarily come from the University Grants Commission (UGC) under the Ministry of Education. While general higher education is governed by UGC, technical education is governed and controlled by the All India Council for Technical Education (AICTE). The four layers of vocational and higher education consists of: (1) the UGC which has more than 381 universities and accredits private universities and deemed universities; (2) the AICTE which controls the technical, business management and other educational institutions such as hotel management and so on; (3) medical education which is governed by the Medical Council of India (MCI); and (4) vocational training through 5000 Industrial Training Institutes (ITIs) under the Directorate of Employment and Training in the Ministry of Labour, Employment and Training. Some trends in publications and human resources are shown in Tables 10.2 and 10.3.

Universities constitute an important part of the Indian NSI in the supply of highly skilled human resources and in knowledge production. However, the major challenge in the coming decade is to enhance the higher education enrolment ratio from 11 per cent to at least 15 per cent in the XIth Five-year Plan period 2007–12. One of the major weaknesses of the Indian NSI is the weak linkages between universities and industry; IITs are an exception which we shall explore in this chapter.

*Table 10.2 Growth of the total scientific and technical manpower 1970–2000. Figures in '000s*

| Fields | 1970 | 1980 | 1991 | 2000 | 2004** |
|---|---|---|---|---|---|
| Engineering Degree Holders | 185.4 | 221.4 | 546.7 | 969.5 | 1011.7 |
| Engineering Diploma Holders | 244.4 | 329.4 | 873.9 | 1456.0 | 1514.3 |
| Science Postgraduates | 139.2 | 217.5 | 482.0 | 767.1 | 795.6 |
| Science Degree Holders | 420.0 | 750.5 | 2430.3 | 3837.7 | 3978.4 |
| Agriculture Post-graduates | 13.5 | 96.5* | 168.4* | 231.2* | 237.5* |
| Agriculture Degree Holders | 47.2 | – | – | – | – |
| Medicine Degree Holders | 97.8 | 165.4 | 310.3 | 403.4 | 412.7 |
| Others | NA | NA | NA | 68.1 | 74.9 |
| Total | 1147.5 | 1780.7 | 4811.6 | 7733.0 | 8025.1 |

*Note:* * Includes first degree holders; ** estimation based on the decade (1990–2000) average.

*Source:* Department of Science and Technology, 1999 and 2002.

#### 10.2.1.3 Business enterprise

Even though the government currently dominates in both R&D effort (research activities) and expenditure to the extent of over 68 per cent, the share of business enterprise R&D has been increasing over the last decade. While it increased from 13.8 per cent in 1990 to around 21.6 per cent in 2000, it further increased to 30 per cent in 2007–08. In the business enterprise sector, R&D expenditure and performance is concentrated in the biotechnology, pharmaceuticals and chemicals, telecommunications and information and communications technology (ICT), electrical and electronics equipment, textiles, transportation and metallurgical industries. The innovation potential of the Indian business enterprise sector (both public and private) can be examined through its patenting activity in the USA. Some trends by business enterprise are shown in Table for the period ranging from 1999 to 2003–06.

Let us briefly look at two sectors. Pharmaceuticals have been one of the most dynamic sectors in India over the last two decades. According to estimates given by the Indian pharmaceutical industry and the World Trade Organization (WTO), drug production in India was barely US$10–12 million in 1974. This increased dramatically to about US$2.75 billion by 1994 and to US$6.5 billion in 2003 (see Jayaraman 2005).

Another important feature of the pharmaceutical sector system has been the evolution of technological capabilities. The first stage of technology support involved the development of an R&D base in national

*Table 10.3 Research papers published* from India during 1989–2005*

| Subject/Year | 1989 | 1990 | 1991 | 1992 | 1993 | 1994 | 1995 | 1996 | 1997 | 1998 | 1999 | 2005 |
|---|---|---|---|---|---|---|---|---|---|---|---|---|
| | Number of papers | | | | | | | | | | | |
| Agriculture | 11 660 | 11 382 | 11 014 | 10 903 | 11 264 | 11 025 | 11 575 | 11 739 | 11 700 | 12 782 | 11 702 | 16 526 |
| Biological Sciences | 8041 | 8012 | 8131 | 9244 | 10071 | 9987 | 9992 | 9537 | 9226 | 8880 | 8948 | 12 491 |
| Chemical Sciences | 11 670 | 11 372 | 12 449 | 13 510 | 12 964 | 13 125 | 12 569 | 13 448 | 13 467 | 14 237 | 13 384 | 23 668 |
| Earth Sciences | 1848 | 1602 | 1492 | 1399 | 1355 | 908 | 1390 | 1078 | 923 | 1102 | 890 | 1212 |
| Engineering | 3798 | 2778 | 2921 | 2421 | 3477 | 4292 | 3658 | 4540 | 4696 | 3755 | 4550 | 11 945 |
| Mathematics | 1337 | 1384 | 1463 | 1548 | 1352 | 1486 | 1841 | 2271 | 2247 | 2109 | 1318 | 1739 |
| Medical Sciences | 4273 | 5327 | 4396 | 4209 | 4068 | 4215 | 3988 | 4132 | 4490 | 4637 | 5633 | 12 142 |
| Physical Sciences | 5545 | 5192 | 5312 | 5778 | 5453 | 5652 | 5709 | 5655 | 5642 | 5725 | 5695 | 9574 |
| Total | 48 172 | 47 049 | 47 178 | 49 012 | 50 004 | 50 690 | 50 722 | 52 400 | 52 391 | 53 227 | 52 120 | 89 297 |

*Note:* * In international databases.

*Source:* Department of Science and Technology (DST) (1999, p. 42), DST (2002, p. 46) and R&D statistics at a glance, DST, New Delhi (2008).

laboratories. This capability enabled reverse engineering under the policy regime, which protected patents for about seven years. A second stage enhanced the capacity to exploit the innovation base for new drug discovery, development of vaccines and use of new technologies such as biotechnology knowledge for value addition in the production of high-technology pharmaceuticals and health-related industry products (see Krishna 2007).

Another sector which is a key component of the NSI is ICT software. From 1990 the ICT sector emerged as India's leading knowledge sector, registering a compound annual growth rate (CAGR) of more than 26 per cent for the last decade. It is contributing 5 per cent of GDP and currently employing about two million professionals, with demand projection of about 1 million for the next five years. In 2006–07 ICT software and services exports increased by 32 per cent, with the revenue aggregate growing nearly ten-fold in ten years. The Indian ICT (information technology (IT) services, business process outsourcing (BPO), IT-enabled services and engineering services and R&D products) sector's performance witnessed a ten-fold increase in the last ten years. From US$15.2 billion in 2004 the sector generated total revenues of US$36 billion in 2007, of which exports constitute approximately 78 per cent and the remaining 22 per cent was sales to the domestic market. The sector has set a milestone of US$54 billion by 2010.

## 10.3 BRIEF THEORETICAL AND LITERATURE OVERVIEW

The approach in this chapter builds on the view that the essence of the relationships between institutions and associated actors can be captured from the different concepts of innovation, including the 'triple helix' framework.[2] This framework is pertinent because 'triple helix' observes academic institutions to be playing a dominant role in the innovation system. 'Triple helix' has evolved gradually from a simple understanding of university-industry 'double helix' to trilateral reciprocal relationships between academia, industry and government, and lately to a more intricate adaptation of innovation and sustainability as 'triple helix twins' working together as a dynamic yin/yang pair that advance sustainable economic and social development.[3] Etzkowitz et al. (2000) argue that there is a widespread movement among ARIs to adopt a more complex entrepreneurial model – one that emphasizes the commercialization of knowledge and the fuelling of private enterprise in the local and regional economies.

Broadly speaking, this chapter builds upon the theoretical premise of Etzkowitz et al. (2000), Etzkowitz and Leydesdorff (1997, 2000) and

Lundvall (1992) wherein we examine how IITs as a representative set of ARIs, particularly in engineering and science, fit into the entrepreneurial model in the Indian context. Specifically, we attempt to fill in the knowledge gap by empirically examining how IIT Madras and IIT Bombay (two of the five IITs that have existed for nearly six decades) have sought to reform their institutional and organizational structure since the 1990s, and by considering the impact that such developments have had on the ARIs' changing role in the knowledge-based economic development of India.

There are limited studies in this area which focus on Asia, and even fewer undertaken in the Indian context. To name a few, there are studies by Fujisue (1998), Hicks (1993) and Kneller (1999) on the challenges facing academic institutions in Japan; Wong Poh-Kam's study of Singapore's NSIs (1996, 2007); Shulin Gu's paper (1999) on the implications of NISs for developing countries; Kim's book (1993) on the Korean experience; and Sharif and Baark's (2005) study on understanding Hong Kong's innovation system and innovation policies.

However many of these studies do not focus exclusively on the role of universities. Further, there are only a limited number of studies addressing the innovation system in India that focus on academic institutions. Chidambaram (1999) focuses on patterns and priorities in Indian R&D. Gupta and Dutta (2005) give a macro picture of the innovation system in the Asia Pacific region, Krishna (2001) looks at the changing status of academic science in India, Abrol (1983) addresses the issue of scientific research in Indian universities and Menon (2002) focuses on technology incubation systems in India. These studies have given little or no residual research attention to the subject of the changing role of universities. There are particularly few studies on IITs, and most of these have been written by decision makers and heads of institutions. Mention may be made of Indiresan and Nigam (1993) and Indiresan (2000) who address the challenges faced by IITs in general and in mobilizing resources. While Raju (1995) in his paper focuses on building bridges between IITs and Indian industry, Swaminadhan (1995) looks at how engineering education can be made relevant to industry. Sengupta (1999) looks at the aspect of marketing technological competence of ARIs in technology transfer. There are however some exceptions with a focus on IITs – such as the recent paper by Basant and Chandra (2007) looking at the role of educational and R&D institutions in city clusters, and another by Nath (1992) who shares the experience of IITs with industry and some others. Despite the growth of IITs and their significant position in the Indian economic and intellectual structure, very little attention is given by social science research in India to these institutions, except for some studies like Chandra and Krishna (2006) and Chandra (2007).

## 10.4 ORIGIN OF IITs AND THE IIT SYSTEM

The IITs originated in the late 1940s, an era which signified the end of the Second World War culminating towards the end of colonialism and its control over HEIs. It was indeed the beginning of an era when independent governmental efforts were devoted towards building science and engineering HEIs. The IITs were born at a time when the First Prime Minister, Jawaharlal Nehru's vision clearly held that S&T are key factors for the transformation and modernization of India. A policy discourse was initiated by Indian political and scientific leadership over the need for engineering and technical personnel for the economic development and growth of India. This led to the formation of an ad hoc committee under N.R. Sarkar in 1945. This committee consisted of scientists and businessmen with a view to advising on the provision of higher technical education facilities in India towards the end of the Second World War. In the meantime AICTE, which was also set up by the colonial government in 1945, upheld three primary objectives: to survey the whole field of technical education in India; to consider the desirability of establishing high-grade technical institutions in India along the lines of MIT; and to promote inter-provincial coordination in all plans regarding technical education.

The ad hoc committee submitted its interim report (Interim Report: Sarkar Committee 1946) recommending the establishment of not fewer than four higher technical institutions (HTIs) in different parts of India (east, west, south and north), which would serve as models of scientific and technical education, particularly of a high order. This document, known as the Sarkar Committee report,[4] was submitted in 1946 and is considered to be the blueprint for the establishment of IITs.[5] After independence the government established five IITs in Kharagpur, Madras, Delhi, Bombay and Kanpur.

Even though India inherited a number of colonial scientific organizations, it opted to emulate MIT as a 'reference model' for setting up a chain of five IITs. There were certain features in this 'MIT model' that inspired the planners of these Indian HTIs, most prominent of which were: a strong science base; course structure that integrated teaching and practical training; recognition of humanities and social sciences in the engineering curriculum; characteristics of a land-grant university committed to local/regional economic and social development; and most importantly cooperation with industry. The 'MIT model' was perceived as a solution to the lack of high-level engineering training in India (as has been noted by several scholars mentioned in previous sections), and possibly also as a means to promote an institutional research model in which science and engineering education and research would have a major impact on regional development.

### 10.4.1 The IIT System

The IITs have been in existence for over six decades, making significant contributions in terms of knowledge production and in creating highly skilled, well-educated graduates and post-graduates, particularly in engineering and sciences as well as in social science and management disciplines. After the creation of five IITs during the 1950s and 1960s, two more were established in 1994 and 2001 (at Guwahati and Roorkee, respectively) and three more IITs have been set up in 2008.[6] These institutions have played an important role in imparting quality teaching and undertaking scientific and engineering research. However the involvement of IITs in making use of their intellectual assets for economic development and also for generating revenue has gained considerable attention in the last decade or so.

The IITs are a unique system for two important reasons relevant to this study. First, they represent a subgroup of India's premier academic researchers, spanning a broad range of disciplines which serve as a model for other ARIs, in both knowledge creation and in knowledge transfer. Second, we presume that the IITs' mission (at the time of their establishment) was not explicitly focused on research, contrasting with the homogeneity of missions in the five IITs dispersed geographically. The IITs focused more on teaching, education and training in the initial years of their establishment. Keeping in mind the objective of producing highly skilled scientists and technologists, the emphasis on post-graduate education and research was rather subdued in the beginning until the 1970s.

More than any other aspect, IITs have earned a global reputation for producing highly skilled engineering graduates. The IITs as a representative set of ARIs particularly in science and engineering education are known for their academic excellence.[7] It is not surprising that admission into IITs is a highly competitive exercise in India with nearly 250 000 students competing for 5500 positions in all the IITs combined. The competition to get admission into an IIT is tougher than for any other prestigious institution of engineering and technology in the world (IITDAA 2004). This study shows that the acceptance rate into the undergraduate programme in IITs is one in 60 applicants, compared with one in seven in Stanford and one in ten in MIT. For post-graduate admissions, about 159 000 students sat for the General Aptitude Test in Engineering (GATE) in 2003, out of which 28 877 (18 per cent) were qualified to gain admission into the IITs.

*Table 10.4 Average student enrolment/graduation per year in IITs (1999–2005)*

| Enrolment | B.Tech | M.Tech | PhD | MSc | MBA | Dual Degree | Others | Total average/year |
|---|---|---|---|---|---|---|---|---|
| IIT Madras | 381 | 568 | 176 | 90 | 41 | 107 | 0 | 1362 |
| IIT Bombay | 335 | 536 | 191 | 112 | 64 | 172 | 13 | 1423 |
| IIT Delhi | 499 | 710 | 246 | 84 | 64 | 0 | 10 | 1613 |
| IIT Kanpur | 407 | 361 | 111 | 61 | 35 | 73 | 10 | 1057 |
| IIT Kharagpur | 431 | 540 | 197 | 55 | 55 | 134 | 26 | 1437 |
| Total | 2053 | 2715 | 920 | 401 | 259 | 485 | 58 | 6891 |

*Source:* Calculated from the *Annual Reports* of the respective IITs (1999–2005).

*Table 10.5 Increase in student enrolment (1999–2000 to 2004–05)*

| All five IITs | B.Tech | M.Tech/M.Sc/ MBA | PhDs | Others | Total |
|---|---|---|---|---|---|
| 1999–2000 | 2295 | 3506 | 541 | 18 | 6137 |
| 2004–05 | 2699 (17.6) | 3624 (3.4) | 1003 (84) | 115 (505) | 7441 (15.7) |

*Note:* Figures in brackets give percentage growth. B.Tech includes a five-year integrated MSc programme and a dual degree; M.Tech includes a two-year MSc, MBA, MS (Research) and M. Design; others include diploma and postgraduate courses.

### 10.4.2 Enrolment of Students

Table 10.4 shows a snapshot profile of student enrolment/graduation for five IITs from1999 to 2005. At the IITs there has been an overall increase in the enrolment of students and in the number of degrees awarded, which is reflected in the visible growth of the number of teaching disciplines and programmes of study.[8]

Looking at the data over two five-year periods (1999–2000 to 2004–05) as shown in Table 10.5, there has been higher growth in the number of post-graduate (PG) research students as compared with undergraduates (UGs). The ideal UG:PG ratio at the time of the establishment of the IITs was recommended by the Sarkar Committee to be 2:1 (Interim Report: Sarkar Committee 1946). The present UG:PG ratio computed for all five IITs is 1:1.72.[9] Thus the most striking feature of IITs in terms of student output is the reversal of the trend in the ratio of UG students to PG students in over five decades of the IITs' existence. Another outstanding

*Table 10.6 Faculty-student ratio at IITs (2004)*

| Institution | Faculty | Student | Student: Faculty ratio |
|---|---|---|---|
| Bombay | 410 | 4252 | 10:1 |
| Delhi | 468 | 4831 | 10:1 |
| Kanpur | 308 | 3104 | 10:1 |
| Kharagpur | 474 | 3797 | 8:1 |
| Madras | 355 | 4731 | 13:1 |
| All five IITs | 2015 | 20715 | 10:1 |

*Source:* IIT *Annual Reports* (2004).

change is seen in the enrolment of PhD students, which has almost doubled in just five years, from 541 to 1003 (Table 10.3).

### 10.4.3 Faculty Status at IITs

Faculty members and their academic stature constitute the core calibre of the IIT system and it is their intellectual value that drives output (IIT, Review Committee 2004). The number of total faculty in the period 1999–2005 has been stable and has not increased significantly. The number of professors is nearly equal to the total number of associate and assistant professors. The growth in the number of faculty members has been relatively insignificant when compared with the proportional increase in the number of students.

The IITs however face a critical challenge in maintaining their faculty base. There is a critical shortage of faculty in all the IITs.[10] The student:faculty ratio is given in Table 10.6 which shows that taking into account all five IITs for the period 1999–2005, there are ten students for each faculty member, with the highest ratio at IIT Madras.

### 10.4.4 Finance and Budgetary Provisions for IITs

By and large, IITs are now publicly funded institutions. The Indian government gives them grants through the Department of Higher and Secondary Education in the Ministry of Human Resource Development. The latest figures of 2007–08 (budget and revised estimates), seen in Table 10.7, show a significant rise in allocation to IITs as a proportion of the total allocation to the higher education sector given in the Union Budget 2008–09.

In 1993–94, with the introduction of a 'block grant system' for non-plan

*Table 10.7 Allocation to IITs in Union Budget 2007–08 (revised and budget estimates (BE) in million rupees)*

| 2007–08 | Plan (BE) | Non-plan (BE) | Total (BE) |
|---|---|---|---|
| All IITs | 11 117 | 4420 | 15 537 |
| New IITs | 800 | – | 800 |
| Total Technical | 29 290 | 9410 | 38 700 |

*Source:* Union Budget 2008–09: *Expenditure Budget, Vol. II* as shown in Tilak (2008, p. 54).

funding for IITs, a significant shift in the funding policy of the government was seen. This policy targeted reducing expenditure and encouraged internal resource generation. While the grant from government accounts for the largest share, earnings in 2002–03 from student tuition fees accounted for about 6 per cent of total expenditure, and the revenue generated through sponsored research and industrial consultancy ranged between 15 to 22 per cent of total expenditure at different IITs.

## 10.5 AN OVERVIEW OF IIT MADRAS AND IIT BOMBAY

### 10.5.1 IIT Madras

IIT Madras was opened on 31 July 1959 as the Southern Higher Technical Institute (SHTI) and located at Guindy, Madras. The government of Germany and its universities gave generous support to the institute during its inception, in the form of valuable equipment, instruments, books and powerful computers totalling a value of nearly DM 50 million, making it one of the largest German-aided educational projects in the late 1950s.

At SHTI too, the course structure was debated and the extent of adaptation of the German model was discussed. The first Director of the institute, Professor B. Senugupto, observed that while German students entered four- or five-year technical institutions at the age of 19-plus with adequate preparation in science combined with two years of practical experience, Indian students at IIT would come at the age of 16-plus with little workshop experience and inadequate preparation in basic sciences. He thus pointed out that consequently fundamental sciences and mathematics had to be taught more intensively in the first two years of the courses in IITs (Sebaly 1972, p. 75). Moreover the academic pattern emphasized course work on basic sciences during the first three years, while specialization through a system of electives would be offered in the fourth and fifth years.[11]

Apart from the equipment and expert support, German influence was seen in the course structure, where students could opt for courses in the German language and Indo-German politics and economic relations, particularly in the humanities department. Similar to the western HTI (that is, IIT Bombay) where problems of industries were considered, the Madras institute gave importance to problems peculiar to the southern region through its focus on tropical technology, agro industries and economics of development. Another important aspect of this institute was the research programme that emphasized technological development and industry cooperation, and included developing expertise in the erection of pilot plants, testing of products, provision of computer facilities and facilities for basic research, emphasis on practical training and guest lectures on industry-specific issues.[12] This feature perhaps explains a culture of large-scale industry cooperation and the sizeable number of start-up firms and technological development projects in the contemporary period in IIT Madras.

Table 10.8 provides a summary overview of the key changes that IIT Madras has undergone during 1999–2000 to 2007–08. In essence, this table shows that although there was only a moderate expansion in terms of conventional/traditional dimensions of education and research output, a more dramatic change can be observed in the new dimensions of entrepreneurship promotion, technology commercialization and collaborative R&D. This will be discussed in the upcoming sections.

#### 10.5.1.1 Skilled graduates/PhDs at IIT Madras

IIT Madras has witnessed a gradual increase in its number of PG and research students, which has enabled the institute to increase its research and innovation base. The number of UGs increased from 465 in 1999–2000 to 550 in 2004–05, and the number of PGs increased from 633 to 761 for the same period. The number of PhD degrees awarded at IIT Madras in 2004–05 was 255 compared with 115 in 2000.

### 10.5.2 IIT Bombay

IIT Bombay recently celebrated its golden jubilee in 2008. The planning for the institute began in 1957, and IIT Bombay (IITB) was established in 1958 with the cooperation and participation of the United Nations Educational, Scientific and Cultural Organization (UNESCO) and funding from the erstwhile Soviet Union (R&D Spectrum 2005, p. 20).[13] Indian Prime Minister Jawaharlal Nehru sought Soviet assistance to set up the institute in Bombay (now Mumbai), with Brigadier Bose appointed as

*Table 10.8 Key indicators of change at IIT Madras*

| Indicator | 1999–2000 | 2004–05 | 2007–08 |
|---|---|---|---|
| UG students enrolled | 465 | 550 | 518 |
| PG students enrolled | 633 | 761 | 690 |
| PhD students enrolled | 115 | 255 | 204 |
| Government grant to IIT Madras (Rs million) | 814 | 1100 | 1330 |
| Number of sponsored research projects | 42 | 79** | 139 |
| Value of sponsored research projects (Rs million) | 45.0 | 351.6 | 669.27 |
| Number of consultancy assignments | 833 | 452 | 549 |
| Value of consultancy assignments (Rs million) | 67.8 | 83.9*** | 177.26 |
| Number of research publications[a] | 317 | 555* | |
| Number of conference presentations | 391 | 881 | 881 |
| Number of patents filed | 1 | 12 | 23[b] |
| Number of patents issued | 1 | 2 | 14 |
| Number of spin-offs | 4 | 14 | 28[c] |

*Notes:* * In 2007 the publication count was 664 (~110 per cent increase since 1999), ** In 2006–07 the number of sponsored research projects increased to 112 valued at Rs 450 million; *** In 2006–07 the number of industrial consultancy assignments increased to 639 valued at Rs 220 million.

a. Thomson (ISI Web of Knowledge).
b. Intellectual property rights include two Patent Cooperation Treaty (PCT) applications and is computed for the calendar year 2007.
c. This includes 16 incubatee units from the TeNeT group and 12 from the Rural Technology Business Incubation Unit (RTBI) formed in 2006.

its first director.[14] As a fallout of the prevailing Cold War, the Americans offered to help to set up another IIT at Kanpur.

IITB is a residential campus with 2200 UG and 2000 PG students. The number of UG students increased from 434 in 1999–2000 to 500 in 2004–05, and the number of PGs witnessed a decrease from 815 to 699 for the same period (Table 10.9). However, PhD students increased from 105 to 252 for the same period. IITB also has schools in management (Shailesh J. Mehta School of Management) and information technology (Kanwal Rekhi School of Information Technology) on its premises. Despite a change in the name of the city, the IIT retains the original name.[15] As with other IITs, over the last decade IITB has received a good deal of encouragement and support from its alumni. It is said that the largest overseas concentration of IITB alumni can be found in the USA. It is therefore not surprising that the most prominent celebrations of IITB's

*Table 10.9 Key indicators of change at IIT Bombay*

| Indicator | 1999–2000 | 2004–05 | 2007–08 |
|---|---|---|---|
| UG students enrolled | 434 | 500 | 553 |
| PG students enrolled | 815 | 699 | 863 |
| PhD students enrolled | 105 | 252 | 263 |
| Government grants to IIT Bombay (Rs million) | 839.5 | 1024 | 1073.9 |
| Number of sponsored research projects | 140 (2001) | 200 (2003–04) | 180 |
| Value of sponsored research projects (Rs million) | 145.6 | 280 | 560 |
| Number of consultancy assignments | 506 | 950 | NA |
| Value of consultancy assignments (Rs million) | 52 | 100 | 167 |
| Number of research publications[a] | 238 | 552 | |
| Number of conference presentations | 233 (2001) | 514 | 604 (2006) |
| Number of patents filed | 9 | 12 | 13 |
| Number of patents issued | 1 | – | 14 |
| Number of spin-offs | NA | 7 | 33 |

*Note:* a. Thomson (ISI Web of Knowledge).

golden jubilee were organized in New York. At this event Romesh Wadhwani, founder of the Wadhwani Foundation and an electrical engineering IITB alumnus of the class of 1969, donated US$5 million for research and innovation in the area of biosciences and bioengineering and for setting up the Wadhwani Research Centre in Biosciences and Bioengineering at IITB.

## 10.6 KNOWLEDGE PRODUCTION

### 10.6.1 Research Publications

Attaining teaching and research excellence, core values underlying both MIT and Humboltian university models, has remained an important objective for IITs since their beginnings.[16] From a national perspective, the publication record of IITs at the international level (as seen through the count of publications in the SCI-extended database) shows them to figure among the top ten universities in India. To obtain an international perspective of the publication output from IITs, we take into account those

*Table 10.10 International publication profile of IIT Madras and IIT Bombay 1999–2007*

| IITs | 1999 | 2000 | 2001 | 2002 | 2003 | 2004 | 2005 | 2006 | 2007 | % increase (1999–2007) |
|---|---|---|---|---|---|---|---|---|---|---|
| IIT Madras | 414 | 395 | 351 | 382 | 467 | 510 | 639 | 750 | 834 | 101 |
| IIT Bombay | 342 | 424 | 413 | 476 | 503 | 604 | 635 | 743 | 798 | 133 |
| Total five IITs | 1650 | 1708 | 1693 | 1864 | 2261 | 2639 | 2928 | 3498 | 3652 | 121 |

*Source:* Science Citation Index–Expanded (accessed 28 February 2008 and 24 June 2008).

*Table 10.11 International publication profile of the most prolific academic research institutes in India 2001–02*

| Institutes | Total publications | Average impact factor | Average citations per paper | Total collaborative papers | Total collaborative papers at international level |
|---|---|---|---|---|---|
| IIT Bombay | 699 | 1.22 | 2.575 | 388 | 193 |
| IIT Madras | 575 | 0.96 | 1.550 | 300 | 136 |

*Source:* Adapted from Gupta and Dhawan (2006, pp. 115, 116, 165).

publications that are listed in the *Web of Science* – the Science Citation Index Expanded Version (SCI–Expanded) and the Social Science Citation Index (Table 10.10).

### 10.6.2 Quality of Publications and Impact Factor

From the scientometric perspective, the quality of research publications is generally examined from citation counts, impact factors and so on. Notwithstanding the direction (positive or negative), citation counts are generally recognized in the scientometric literature as giving some indication of quality. To give an indication about the nature and quality of publications, the publication profile of two IITs is given in Table 10.11, in terms of total number of publications, average impact factor, average citations per paper, total collaborative papers and total collaborative papers at an international level.

*Table 10.12 Intellectual property filed at IIT Madras and IIT Bombay*

| Institute | | 2000–01 | 2001–02 | 2002–03 | 2003–04 | 2004–05 | 2005–06 | 2006–07 | 2007–08 |
|---|---|---|---|---|---|---|---|---|---|
| IIT | Filed | 1 | 2 | 3 | 11 | 12 | 12 | 18 | 23 |
| Madras* | Issued | 1 | – | – | 3 | 2 | – | 15 | 14 |
| IIT | Filed | 10 | 13 | 13 | 22 | 18 | 15 | 13 | 13 |
| Bombay | Issued | 1 | 4 | 2 | – | – | 3 | 18 | 14 |

*Note:* Intellectual Property patent rights (IPR) filing includes Indian and foreign patent applications, copyrights, designs and trademarks.
* IPR filed in the calendar year.

### 10.6.3 Disciplinary Trends in Publications

Evaluating the subject profile of IIT Madras, it was found that chemistry dominated publications (17 per cent), followed by mechanical sciences (12 per cent), civil engineering (10 per cent) and metallurgy and material science (9 per cent). Subjects such as physics, electrical engineering and chemical engineering also accounted for the most prolific fields. In the case of IITB, chemical and physical sciences seem to be the main focus of research, accounting for 33 per cent of publications, followed by mechanical and material sciences (10 per cent), electrical engineering (7 per cent) and maths (6 per cent).

### 10.6.4 Patenting Activities

The technology transfer offices at IITs are involved in protecting intellectual property (IP) mainly in the form of patents, copyrights and designs. The filing of IP applications from academic institutions is gradually increasing (Table 10.12). The availability of data for the number of invention disclosures is unclear but the trend shows that it is increasing. As per talks with TTO personnel, the annual number of invention disclosures received by a typical IIT is 60.

The usual time lag between filing and granting of a patent varies from two years to as long as eight years. The average time lag works out to be 61 months or a little over five years. In IIT Madras, up to March 2006, 86 patents had been granted out of the 190 that had been filed.[17]

One presumes that researchers at IITs must be patenting in areas in which they have good publication track records. We have listed the five most prolific departments at the IITs in terms of publication counts as well as the number of patent applications filed (Table 10.13).

*Table10.13 Publications versus patents at the two IITs*

| IIT Bombay | | IIT Madras | |
|---|---|---|---|
| Publications (1999–2005) | Patents filed | Publications (1999–2005) | Patents filed |
| Chemistry (139) | Mechanical sciences (24) | Chemistry (81) | Mechanical sciences (37) |
| Physics (53) | Chemical sciences (21) | Mechanical sciences (47) | Chemistry (26) |
| Chemical sciences (51) | Chemistry (15) | Civil engineering (39) | Metallurgy (23) |
| Civil engineering (44) | Electrical engineering (7) | Physics (36) | Electrical engineering (21) |
| Electrical engineering (43) | Metallurgy (5) | Maths (36) | Chemical sciences (11) |

*Note:* Figures within parentheses are the publication/patent application count.

## 10.7 KNOWLEDGE TRANSFER

### 10.7.1 Institutional Arrangements for Knowledge Transfer

The institutional framework, consisting of policies, practices and appropriately trained human resources, is essential for meaningful knowledge transfer to occur between ARIs and industry. While ARIs reap direct economic benefits stemming from their involvement in sponsored research projects and consultancy assignments, and from their IP (protected or not), there are also high spill-over advantages germinating from the public-private collaborations in the form of increased economic activity, such as start-up firms and job creation. Often the intellectual assets developed by researchers in academic institutions – their inventions, technologies and know-how – are not transferred to industry and are rarely put to any practical or commercial use such that they can be employed in activities to stimulate economic growth. However, the institutional framework necessary for the transfer of technologies/know-how from academia to the private sector is not well developed in many developing countries, including India. IIT Madras has developed such institutional arrangements – more so in the last decade – for facilitating knowledge transfer.

### 10.7.2 Industrial Consultancy to Technology Transfer

#### 10.7.2.1 IIT Madras

In the early 1970s in some IITs (Kharagpur and Madras), efforts were made to formalize industry interaction through research projects and consultancies. As a result, industry liaison agencies were established. In the other IITs such functions were carried out by other departments because the quantum of research collaboration with industry was not large. In the last two decades, as Indian industry witnessed considerable growth and technological sophistication, the demand for knowledge and know-how from leading institutions such as IITs increased. This has led IITs to build institutional processes and mechanisms to promote knowledge transfer. One of the organizational innovations that became very common was the establishment of technology transfer offices (TTOs) or technology licensing offices (TLOs) or industry liaison offices in IITs and universities in India. This has generally followed the US model of creating TTOs.

In the past the R&D department of IITs normally undertook the task of commercializing IP generated within the institute, but now new systems have evolved. At IIT Madras, industry interaction and other functions of a TTO are undertaken by the Centre for Industrial Consultancy and Sponsored Research (IC&SR), established in the early 1970s. This centre is headed by the dean, while the key personnel involved are the chief techno-economic officer, who is supported by senior technical officers and research staff. The entrepreneurial infrastructure at IIT Madras has been developed since the 1990s. The centre has facilitated the MS programme in Entrepreneurship and successfully coordinated technology development and management projects. There has also been the creation of the Cell for Technology Innovation, Development and Entrepreneurship Support (C-TIDES) at IIT Madras, a student-focused establishment primarily to develop the spirit of entrepreneurship and provide mentorship from industry to young entrepreneurs.

Even though IIT Madras established a formal IC&SR unit, the Telecommunication and Computer Networking (TeNeT) group, which has been constituted from different science and engineering faculties, has been the focal point for incubation and enterprise creation through spin-offs in the last decade. It may be said that IC&SR and TeNeT operate at IIT Madras in a complementary mode. The TeNeT group which promotes incubation and enterprise creation at IIT Madras is described as an unconventional model[18] or a non-formal model, as its members, faculty and students are drawn from the engineering, electrical and computing departments, and are housed in close proximity to the Department of Electrical Engineering.[19]

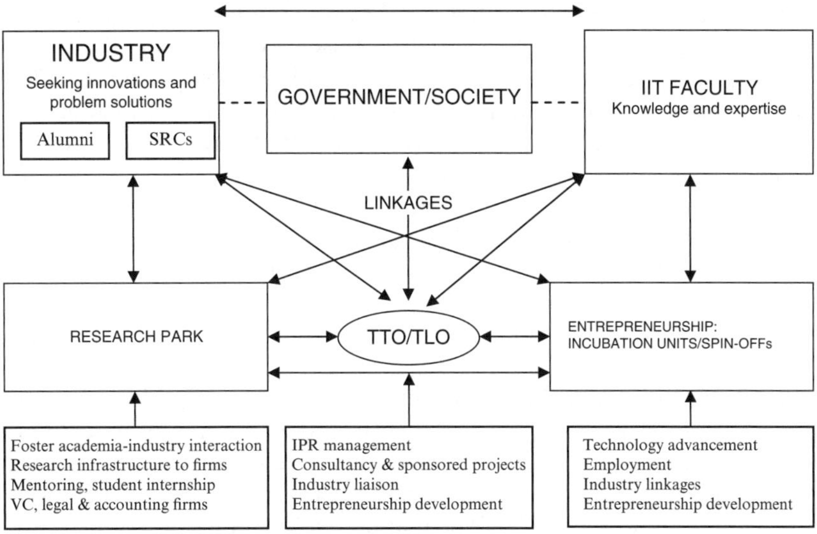


*Note:* IPR: intellectual property rights; VC: venture capital; SRCs: Strategic Research Coalitions.

*Figure 10.1 Schematic diagram of knowledge transfer infrastructure at IITs*

### 10.7.2.2 Research Park at IIT Madras

IIT Madras Research Park, a recent initiative, has been promoted by IIT Madras and alumni with the mission of creating a collaborative environment between industry and academia to enable, encourage and develop cutting-edge technology and innovation that exceeds global standards. The Research Park intends to leverage IIT Madras' technological capabilities to innovate and promote entrepreneurship by navigating research into ideas, developing the ideas into products/processes, incubating products/processes into ventures and nurturing ventures into enterprises.[20] This is expected inter alia to enable companies from IIT Madras to realize the commercial value of their IP.

In short, the knowledge transfer from IITs can be understood through a schematic diagram, as shown in Figure 10.1. There will, however, be some variations among different IITs.

### 10.7.2.3 IIT Bombay

The Industrial Research and Consultancy Centre (IRCC) was set up in IITB in the 1970s to foster the overall R&D growth of the institute, and to promote and manage academia-industry interaction and all externally funded R&D projects. It undertakes financial management and recruitment of researchers to work on funded projects as well as handling activities related to IP protection/management and technology transfer. The IRCC has an advisory committee with several faculty as members of the committee. The activities undertaken at IRCC are spread across its technical, administrative and accounts sections. The technical section has a chief technical officer and two technical officers along with other supporting staff. The accounts section has an administrative officer, while a deputy registrar heads the administrative section. The institute also has an information management cell, the main objective of which is to assist performance evaluation and internal decision making through collection of data and trend analysis.

IITB set up an incubator in 1999 at the Kanwal Rekhi School of Information Technology. This incubator, more than any other achievement in its initial years, was an important source for creating an environment for entrepreneurship.[21] This has further catalyzed the institute's effort to set up a fully fledged incubator covering other areas of science, engineering and technology. This effort received support from the government via the Department of Science and Technology which had created a 'window' for incubation support schemes in institutions. In an effort to further boost the transfer of technologies and commercialize R&D, IITB set up an independent Society for Innovation and Entrepreneurship (SINE) in 2004. On behalf of IITB, SINE manages equity in incubated firms or spin-offs.

### 10.7.3 The Policy of Revenue Sharing

One of the key policies in knowledge transfer is the policy that specifies distribution or sharing of revenue earnings from IP of the academic institute. Interestingly the sharing patterns for the five IITs are different. At IIT Madras, the revenue-sharing policy states: '50 per cent of the revenue is credited to IIT Madras while remaining revenue is divided equally among inventors as per the royalty sharing agreement. Out of IIT Madras's share, 5 per cent is transferred to the concerned department's development fund and 2 per cent to IC&SR overhead and balance to the institute corpus fund.' In the case of IITB, as Table 10.14 indicates, inventors receive a share of 70 per cent and the institute receives the remaining 30 per cent.

### 10.7.4 Consultancy and Sponsored Projects

Institutions of higher education engaged in teaching and research are now seen to occupy a place in the knowledge production system that distinctly

*Table 10.14 Revenue sharing policy of IITs from institute-owned intellectual property*

| Institution | Revenue sharing |
|---|---|
| IIT Bombay | Inventor(s) receive a share of 70% while IIT Bombay receives 30%. This holds as long as the net earnings do not cross a threshold amount for any inventor. |
| IIT Madras | 50% of the revenue is credited to IIT Madras while the remaining revenue is divided equally among inventors as per the royalty-sharing agreement. Out of IIT Madras' share 5% is transferred to the concerned department's development fund, 2% to IC&SR overhead and the balance to the institute corpus fund. |

reflects a shift towards playing other significant roles, for example, capitalizing on their knowledge assets, contributing to economic progress, aiding in regional development and so on. One of the most important channels that drive knowledge generation is the sponsored research and industrial consultancy (SRIC) projects. The trend in IITs is to embrace such activities and programmes that in effect give them a clear competitive advantage over other knowledge-producing institutions in creating and fostering innovations. Consultancy services and sponsored research projects have an exchange value that earns revenue for IITs. The recognition of an exchange value for such projects and assignments is generally referred to as capitalization of those assets.

Combined SRIC earnings have increased from as low as 12 per cent in 1999–2000 to 60 per cent in 2004–05 of the total government grant-in-aid in the case of IIT Madras, and from 29 per cent to 37 per cent in the case of IITB for the same period (Table 10.15). If we add the income from other sources (tuition fees, endowments) this share will decrease slightly, but the important point to note is that the earnings through SRIC, technology transfers, licensing and spin-offs has witnessed growth in a short span of five years and this trend is likely to continue (Table 10.16). However one should also note that the majority of earnings are from government-sponsored research projects funded through national agencies.

### 10.7.5 Sponsored Research Projects

In addition to the primary objective of teaching and research, the faculty members and research personnel of IITs undertake several sponsored research projects. These projects are generally funded by government agencies, national research councils and both public and private industry

*Table 10.15 Combined earnings of Sponsored Research and Industrial Consultancy (SRIC) projects as a percentage of the total grant given to IITs*

| IITs | Earnings through SRIC (1999–2000) | Government grants to IITs in 1999–2000 | Earnings through SRIC as a percentage of total grants (1999–2000) | Earnings through SRIC (2004–05) | Government grants to IITs in 2004–05 | Earnings through SRIC as a percentage of total grants (2004–05) |
|---|---|---|---|---|---|---|
| IIT Bombay | 197.6 | 671.5 | 29 | 380.0 | 1024.0 | 37 |
| IIT Madras | 112.8 | 943.7 | 12 | 435.5 | 1100.0 | 40 |
| All five IITs | 701.3 | 4066.2 | 17.2 | 2291.2 | 5154.0 | 44.4 |

*Note:* All values of earnings and grants in Rs million. IIT Kharagpur, grant-in-aid in 2000–01.

*Source:* Computed from the *Annual Reports* of respective IITs and websites of five IITs (IIT Bombay, IIT Delhi, IIT Kanpur, IIT Kharagpur and IIT Madras).

*Table 10.16 Increase in sponsored research projects*

| | 1999–2000 | 2004–05 | Percentage increase (%) |
|---|---|---|---|
| IIT Madras | 45.0 | 351.6 | 681 |
| IIT Bombay | 145.6 | 280.0 | 92 |
| All five IITs | 576.9 | 1669.5 | 189 |

*Note:* All values in Rs million.

*Source:* Compiled from *Annual Reports* of respective IITs (1999–2005).

(national and international). At IIT Madras the number of projects increased from 42 to 112 during 1999 and 2007 with a corresponding value of Rs45 million to 450 million for the same period. At IITB the number of projects increased from 130 to 200 during 1999 and 2004 with a corresponding value of Rs145.6 million to 280 million for the same period. These projects bring in new resources to the institute and also permit technical staff to be employed for specific durations to carry out the research.

The maximum Earnings per Project (EPP) at IIT Madras was achieved by electrical engineering at Rs 3.3 million while for chemical engineering the EPP was Rs 1.5 million. In the five-year period from 1999–2000 to 2004–05, out of the total of 388 faculty/researchers at IIT Madras, an average of 214 were involved in sponsored research, which is nearly 55 per cent of the total faculty.

### 10.7.6 Industrial Consultancy

Undertaking consultancy jobs has been an effective way of making available the expertise of IIT personnel for the benefit of industry, government and others. Its value to IITs in stimulating further interaction and research collaboration has been well recognized, in addition to the professional and financial benefits obtained by the academics themselves. Consultancy jobs also show a significant rise at IIT Madras from Rs67.8 million to 83.9 million, and at IITB from Rs52 million to 100 million during 1999 and 2005. This is an increase of 24 per cent and 92 per cent for IIT Madras and IITB, respectively, over this period.

### 10.7.7 Incubation and Enterprise Creation or Spin-offs

Incubation and enterprise creation, or what is known as spin-offs (we define spin-offs as companies that evolve from academic institutions through commercialization of IP and transfer of technology developed within academic institutions), has come into prominence and sharp focus in the literature on the triple helix. It is regarded as one of the main indicators for entrepreneurial universities. In our study, while IITs at Kanpur, Delhi and Bombay have adopted the conventional approach of creating formal incubation units, the spin-offs at IIT Kharagpur and IIT Madras have been created without a formal incubation set-up. This phenomenon of enterprise creation without the benefit of formal structures may be regarded as an unconventional mode of spin-off creation (Basant and Chandra 2007). However in October 2006 a formal incubation unit was set up at IIT Madras – the Rural Technology Business Incubation (RTBI) – that promotes ventures having a rural development focus. Up to March 2009, 12 companies have been residents of RTBI and three are about to complete their residency tenure and graduate successfully. IIT Madras has been quite active in incubation and enterprise creation, on par with other IITs. The TeNeT group at IIT Madras comprises faculty members from electrical and computer faculties who came together in 1994. Even though the main objective of the TeNeT group is to fulfil the socio-economic agenda of innovation in ICT for development, the group has incubated over 15 enterprises from the knowledge generated at the institute (Table 10.17). At IIT Madras telecommunications account for 36 per cent, computer science for 32 per cent, mechanical sciences for 7 per cent and 25 per cent for the remainder.

*Table 10.17 Incubation and entrepreneurial Infrastructure at IIT Madras and IIT Bombay*

| Institution | Incubation unit and year | No. of incubatee/ spin-offs (from 1994 until 2008) | Prominent areas of expertise of incubatee units | Other entre-preneurial infrastructure |
|---|---|---|---|---|
| IIT Bombay | Incubator at Kanwal Rekhi School of ICT (1999) and SINE (2004)* | 33 | IT, computer science, electronics, design, earth sciences, energy & environment, electrical, chemical, aerospace | Entrepre-neurship cell |
| IIT Madras | Rural Technology Business Incubation (2006) and non-formal dynamic group like TeNeT group (1999) | 12<br>16<br>Total: 28 | Telecommuni-cations; information technology; computer science; physics | C-TIDES; Research Park |

*Note:* An IT business incubator was set up at Kanwal Rekhi School of Information and Technology, IIT Bombay in 1999 prior to the existence of SINE.

*Source:* Compiled from official documents and newsletters at SRIC Centre, IIT Madras.

Most of the firms incubated from IIT Madras have been formed by the students and partners from outside the institute. Among the earliest firms incubated is Midas Communications which designed integrated circuits (ICs) and in collaboration with US-based Analog Devices manufactures and supplies to consumers all over the world. Midas has offices and outlets in some 25 countries around the world. Other incubated firms from IIT Madras have collaborated and partnered with global ICT firms such as Polaris Software and Intel Corporation and national firms such as the Electronics Corporation of India.

In the case of IITB, among the most successful stories include Geosyndicate Power Private Limited. The technology transferred to the firm currently involves the use of non-conventional energy mechanisms such as geothermal to deliver high-efficiency and low-cost electricity to its markets in the Indian power sector. Other firms fall into the software solutions category such as creative ideas portal technology to M/s Voyager2 Infotech, wireless gateways and connectivity bridges technology to M/s Myzus Technologies, information rights and firewall solutions technology to M/s Seclore Technology Pvt Ltd, and broadband-based technologies to M/s Eisodus Networks Pvt Ltd. Up to March 2009 a total of 33 SINE companies have been formed, of which 14 have graduated, 15 are residents at the IITB campus while four incubation units did not succeed. At IITB 50 per cent of the spin-offs are from computer science, 18 per cent are from telecommunications, 14 per cent from energy, 7 per cent from mechanical sciences and 11 per cent from others.

The concept of a research park near IIT Madras, however, also has additional features. The RTBI unit would also be relocating with its 12 companies to the Research Park. Apart from the pattern of incubation, maturation and relocation which is seen to be a key element in the strategy for the expansion of the Research Park, there is a 'real estate' component of attracting established businesses and laboratories to the park. The businesses housed in the Research Park also propose to offer internships to students.[22]

The establishment of spin-off firms is seen as an important commercialization mechanism to hold and develop IP where a high return is foreseen from future sales. A comprehensive analysis of the firms which have begun life within IITs provides an interesting picture.

In the case of start-up firms, it is observed that serious consideration is given to the nature of the technology in terms of its applicability to several markets and the availability of firms that are capable of bringing the technology to the market place. At IITs there is an increasing eagerness among the academics with substantial research performance to see the potential outcomes of their research being realized. For some incubators in the campus it was possible to realize revenues directly, while for others the economic return was indirect, but for both categories it was reportedly found that the social return was considerable.

## 10.8 CONCLUDING REMARKS

IITs in India have evolved as the country's leading engineering institutions over the last six decades. From the very beginning, the structure of teaching courses and research in IITs have drawn inspiration from the MIT model. Over the last six decades Indian IITs have acquired international recognition and a 'brand name' for excellence in teaching and producing quality engineering and science graduates. However it is only in the last decade that IITs have come to be recognized for their R&D potential with the increasing graduate and post-graduate and doctoral programmes. Through their human capital and research contribution, IITs have come to occupy a significant position in the Indian NIS, particularly during the last decade. The two institutions explored in this study, namely, IIT Madras and IIT Bombay, are clear examples of the IIT system as a whole.

Our exploration of knowledge production and knowledge transfer in these two institutions was mainly concerned with the ways in which the institutional structures have progressed over approximately the last decade to foster and promote university-industry linkages and enterprise creation. Second, our concern has been to explore the extent to which the Indian IITs have embraced or even reflect the emergence of an entrepreneurial university 'culture' or style. Third, we sought to explore how the two institutions differ in the processes of knowledge transfer and enterprise creation.

From the perspective of IITs as a whole, specifically the two cases considered here, it is apparent that the two IITs assign greater importance to teaching and research (open science, peer-reviewed publications and advancement of science and engineering knowledge) compared with the objective of transferring knowledge to industry and engaging in entrepreneurial activities. Given the long-standing tradition of emulating MIT as the main reference model, it appears that IITs have somehow focused more on attaining excellence in teaching and research compared to knowledge transfer. In a way, this also shows the pre-eminence of upholding Humboldtian values. However this does not suggest that the latter dimension is not important. Our interviews with several IIT faculty reveal that the supply of highly skilled and excellent engineering graduates to industry both in India and abroad – mainly the USA – is indeed seen as an indirect contribution to industry in that it is non-targeted through institutionalized mechanisms.

The concepts and innovation policies which promoted and popularized the triple helix, university-industry relations and knowledge-based economies did not emerge until the 1990s. India began to liberalize its economy in 1991 and the impact of globalization and the emerging discourse of knowledge-based economies, coupled with the importance of new technologies (biotechnology, ICT-based technologies and more recently nano and material sciences), drew attention to science-based innovation in universities. IITs and some other leading Indian universities such as the Indian Institute of Science, Jawaharlal Nehru University, Delhi University and Madurai Kamraj University among others, began to respond in the 1990s to this new policy discourse. Even though the two IITs began to institutionalize some form of TTOs from as early as the 1970s to promote consultancy and sponsored research, it was only from the late 1990s that the signs of a shift towards embracing an entrepreneurial university model became apparent with dynamic institutional and organizational factors. Clear indicators of this development are the establishment of organizations to promote enterprise creation (spin-offs). These include the formation of an incubator at the Kanwal Rekhi School of Information Technology in 1999 and the subsequent creation of SINE in 2004 at IIT Bombay, along with the establishment of the TeNeT group at IIT Madras and also the RTBI. It is observed that the two IITs are no longer content merely to provide human resources and new knowledge as a traditional ARI, but have added a more visible and proactive role in knowledge commercialization through increased sponsored research and industrial consultancy assignments, the number of spin-offs generated and a moderate rise in IPR activities.

With new initiatives given to faculty and research staff for technology commercialization, the institutionalization of certain IPR policies and the establishment and strengthening of organizational mechanisms, there has been a significant increase in knowledge transfer activities from IIT Madras and IIT Bombay in the 1990s compared with earlier decades. However these institutions are still facing challenges in this new role, particularly in faculty attitudes towards engaging in entrepreneurial activities in an academic environment.

Knowledge transfers at IIT Madras and IIT Bombay have shown considerable increase through two modes. The first is through the traditional forms of knowledge transfer, comprising sponsored research, industrial consultancy, training of industry/government personnel and such activities. The second mode is building an entrepreneurial culture via incubation and entrepreneurship programmes and training as noted above, which have fostered and promoted enterprise creation and spin-offs. The first mode has been by far the most successful mode as it involves many faculty members and researchers at the two IITs. More than 70 per cent of faculty in both the IITs are involved in industrial consultancy and sponsored research in departments such as ocean engineering, ICT, the composite technology centre, civil, mechanical and electrical engineering and applied mechanics.[23] Our empirical research and quantitative data reveal that the traditional forms or mode of knowledge transfer via sponsored research and industrial consultancy are considered more important relative to patenting and licensing. The other important finding that is clearly revealed from the two case studies of IITs is that IPR and patenting are not critical for enterprise creation and spin-offs. It may be the case that the two institutions in our study are focused more on engineering sciences and ICT-based disciplines rather than biotechnology and biomedical where IPRs assume greater significance.

Our comparison of IIT Bombay and IIT Madras has also revealed an important finding for the role of enterprise creation and spin-offs. While IIT Bombay institutionalized a formal incubator and TTO and was responsible for a cumulative total of 33 SINE companies, the TeNeT group and RTBI at IIT Madras reflected a mix of non-formal and conventional sources of incubation and enterprise creation. It is now responsible for 28 spin-offs. In a large measure, the two IITs reflect two different patterns or models of enterprise creation. Closely associated with these different patterns is the development of indirect spin-offs from IITs. In other words, there are several leading Indian software firms such as TCS and Infosys which are either managed or owned by engineers trained in IITs and who have spent several years abroad in Silicon Valley before returning to India to establish such firms. However despite such indicators it is too early to suggest that IIT Madras and IIT Bombay are emerging as entrepreneurial universities. IITs are rather a special case of the Indian university system, having institutionalized knowledge transfer modes as compared with other universities. We can thus say that the triple helix-based entrepreneurial mode is emerging and is seen as a special case in the Indian academic system in IITs.

From an overall policy perspective which fosters entrepreneurial culture and enterprise creation in ARIs, we may say that IITs and other universities in India lack the appropriate 'innovation ecosystem' and a well-developed venture capital support structure as is seen in the case of MIT, Stanford and Silicon Valley in the USA. The government has evolved several schemes to promote university-industry relations but again, it will take a few more years before they could have any visible impact. Further, India is still in the process of framing a national regulation and law equivalent to the Bayh-Dole Act of the USA which governs IPR in universities.

## NOTES

* This chapter was written when V.V. Krishna was associated as Visiting Senior Research Fellow, Asia Research Institute, National University of Singapore and Visiting Fellow, Center for Industry and Innovation Studies, University of Western Sydney, Australia.

1. IITs are recognized worldwide for the outstanding quality of engineers, scientists and managers they produce. This is evident in the *Times Higher Education* (2006) ranking of the world's top 100 technology institutions. IITs were ranked third after MIT and the University of California Berkley, http://www.timeshighereducation.co.uk/hybrid.asp?typeCode=163, accessed 1 July 2008.
2. The other important theoretical concept is the NSI framework. To a certain extent, it also explains why certain actors are important to the overall dynamism in the system of innovation.
3. For details, see Etzkowitz and Zhou (2006).
4. This committee was set up with N.R. Sarkar as Chairman and S.R. Sengupta as Secretary and had 22 members. In its introduction, the representatives of the committee expressed

that the existing facilities for higher technical education in India were inadequate – in both quantity and quality – to satisfy India's post-war needs for high-grade technologists.

5. L.S. Chandrakant and Biman Sen in the Education Ministry also played a significant role in producing a blueprint for a truly autonomous academic institution.
6. In the states of Andhra Pradesh, Bihar and Rajasthan.
7. See note 1.
8. Knowledge production at IITs is organized through the number of courses offered at the undergraduate (UG) and post-graduate (PG) level. There are in total 3537 UG and 3797 PG courses at the five IITs, offering 22 degrees in 409 disciplines (IIT, Review Committee 2004, p. 94).
9. Based on computing data on student enrolment from *Annual Reports* of the respective IITs in 2004–05. The total number of B.Tech students in the five IITs was 2699 while the number of M.Tech/MSc/PhD students was 4627.
10. For instance, in the seven IITs, the total number of vacant positions is 866.
11. Complaints were made during the first five years of the assistance programme stating that German experts had limited roles in the academic development of the SHTI. It was reported that efforts were being made without considering developments taking place in comparable German technical institutes. In order to resolve this issue, an informal committee was formed known as the 'Madras Committee', in which a German expert at SHTI would maintain a closer liaison with Indian counterparts (Sebaly 1972, p. 77).
12. This viewpoint has been pointed out by several scholars, for example, Professor S. Sampath in his address 'Aims and tasks of IIT Madras' delivered at the DAAD seminar on 20 January 1968, p. 4; Professor B. Sen Gupto, the first director of IIT Madras, in his report on a visit to the Federal Republic of Germany, 22 April to 16 May 1967; and Professor Dr Ing H.A. Havemann in an interview to K.P. Sebaly on 24 September 1969 (see Sebaly 1972, p. 79).
13. The IITB campus at Powai extends over 200 hectares and is situated in picturesque surroundings with green hills and Vihar and Powai Lakes on either side. In 1961 the institute was declared an institution of national importance by an Act of Parliament and was accorded the status of a university with the authority to award its own degrees and diplomas.
14. Krishna Menon, then Defence Minister under Prime Minister Nehru, recommended this appointment.
15. See Wikipedia and the website of IITB.
16. As early as 1947, at the second meeting of AICTE proceedings, it was highlighted that the establishment of high-grade technological institutes in India was envisioned to be along the lines of MIT – an institution known all over the world where technology is studied, and where the advantage of training in technology is appreciated (*AICTE 2nd Proceedings* 1947, p. 20 – address by Sir A. Ramaswami Mudaliar, Dewan of Mysore).
17. This includes patents/copyrights that were granted but were not renewed after expiry of their term.
18. See Basant and Chandra (2007).
19. It is headed by Professor Ashok Jhunjhunwala of the Department of Electrical Engineering, who has given leadership and directed its research and incubation for the last 12 years. The other leading actor at TeNeT is Professor B. Ramamurthy. There are over 200 professionals, engineers, technicians and other staff including 14 core IIT faculties who form the TeNeT group at IIT Madras. These professionals are working in diverse areas and fields of research from eight to ten dedicated laboratories of the institute engaged in wireless communication, computer networking, fibre optics, digital systems architecture, integrated voice, video and data communications, computing and translations for rural development in India.
20. IITM Research Park will have a built-up space of 1.5 million square feet, one Innovation cum Incubation Centre (IIC) that will be the fountainhead of R&D and entrepreneurship development, three R&D towers housing about 100 companies and organizations pursuing serious R&D activities which would be large, medium and small

industries and enterprises from India and overseas. The park also plans to have over 10000 engineers, scientists, researchers and innovators in diverse technologies.

21. See Basant and Chandra (2007).
22. We can draw a parallel with the University of Cambridge which has indirectly played a key role in the development of the area through the Science Park, since it has been at the origin of virtually all the new companies in one way or another. Some 17 per cent of the firms were formed by people coming straight from the university, while others were indirect spin-offs of research conducted at the university. Other start-ups owe their existence to the presence of the university nearby. Most of the companies are very small, with an average of 11 employees. The success of the Cambridge Science Park is widely recognized and is part of what has come to be known as the 'Cambridge phenomenon', which is regarded as a symbol of the innovative milieu.
23. As per our study, taking into account the number of faculty involved in consultancy projects across departments from 1999 to 2005. The case of sponsored research is similar, where the composite technology centre, department of physics and metallurgy have over 75 per cent faculty involved in such projects.

## REFERENCES


Abrol, D.K. (1983), 'Scientific research in Indian universities: problems in building centres of excellence and relevance in management of Indian science for development of self reliance', *Proceedings of the SYS Symposium*, 6–9 February, New Delhi: Allied Publishers, pp. 155–82.

Basant, R. and P. Chandra (2007), 'Role of educational and R&D institutions in city clusters: an exploratory study of Bangalore and Pune regions', *World Development*, **35** (6), 1037–55.

Chandra, N. (2007), 'Academia-industry linkages: technology commercialisation at Indian Institutes of Technology', Paper presented at the 6th Triple Helix Conference, Singapore, 16-18 May.

Chandra, N. and V.V. Krishna (2006), 'Knowledge production and transfer at Indian Institutes of Technology', Paper presented at the Fourth GLOBELICS International Conference, Trivandrum, India, 4-7 October.

Chidambaram, R. (1999), 'Patterns and priorities in Indian research and development', *Current Science*, **77** (7), 859–66.

Edquist, C. (1997), *Systems of Innovation: Technologies, Institutions, and Organizations*, London: Routledge.

Etzkowitz, H. and L. Leydesdorff (eds) (1997), *Universities and the Global Knowledge Economy: A Triple Helix of University-Industry-Government Relations*, London: Pinter.

Etzkowitz, H. and L. Leydesdorff (2000), 'The dynamics of innovation: from national systems and 'Mode 2' to a 'Triple Helix' of university-industry-government relations', *Research Policy*, **29**, 109–23.

Etzkowitz, H. and C. Zhou (2006), Triple Helix twins: Innovation and sustainability, *Science and Public Policy*, **33** (1), 77–83.

Etzkowitz, H., A. Webster, C. Gebhardt and B.R.C. Terra (2000), 'The future of the university and the university of the future: evolution of ivory tower to entrepreneurial paradigm', *Research Policy*, **29**, 313–30.

Freeman, C. (1987), *Technology Policy and Economic Performance: Lessons from Japan*, London and New York: Frances Pinter Publishers.

Freeman, C. (1995), 'The 'National System of Innovation' in historical perspective', *Cambridge Journal of Economics*, **19**, 5–24.
Fujisue, K. (1998), 'Promotion of academia-industry co-operation in Japan', *Technovation*, **18** (6–7), 371–81.
Gibbons, M., C. Limoges, H. Nowotny, S. Schwartzman, P. Scott and M. Trow (1994), *The New Production of Knowledge: The Dynamics of Science and Research in Contemporary Societies*, London: Sage.
Gu, S. (1999), 'Implications of national innovation systems for developing countries – managing change and complexity in economic development', *UNU/INTECH Discussion paper 9903*, Maastricht.
Gupta, B.M. and S.M. Dhawan (2006), *Measures of Progress of Science in India: An Analysis of the Publication Output in Science and Technology*, Report by National Institute of Science, Technology and Development Studies for the Office of the Principal Scientific Advisor to the Government of India, New Delhi.
Gupta, A. and P.K. Dutta (2005), 'Indian innovation system: perspectives and challenges', in *National Innovation Systems (NIS) in the Asia-Pacific Region*, Paper presented in the Asia-Pacific Forum on National Innovation Systems (NIS) for High-Level Policy Makers, New Delhi, 28–29 April.
Hicks, D. (1993), 'University-industry research links in Japan', *Policy Sciences*, **26** (4), 361–95.
IITDAA (2004), *Vision 2020: The IIT in the New Millennium*, [New Delhi.] IIT Delhi Alumni Association.
Indian Institute of Technology, Bombay website, available at http:// www.iitb.ac.in (accessed several times from September 2002 to August 2008).
Indian Institute of Technology, Delhi (various years), *Annual Reports*, IIT Delhi.
Indian Institute of Technology, Delhi website, available at http://www.iitd.ac.in (accessed 20 March 2009).
Indian Institute of Technology, Kanpur (1961–62), *Annual Report*, Kanpur.
Indian Institute of Technology, Kanpur (various years), *Annual Reports*, IIT Bombay.
Indian Institute of Technology, Kanpur website, available at http://www.iitk.ac.in (accessed 20 March 2009).
Indian Institute of Technology, Kharagpur (various years), *Annual Reports*, IIT Kharagpur.
Indian Institute of Technology, Kharagpur website, available at http://www.iitkgp.ac.in (accessed 20 March 2009).
Indian Institute of Technology, Madras (various years), *Annual Reports*, IIT Madras.
Indian Institute of Technology, Madras website, available at http://www.iitm.ac.in (accessed 20 March 2009).
Indian Institutes of Technology, Review Committee (2004), *Report of the Review Committee*, Chairman, P. Rama Rao, available at http://www.iitk.ac.in/infocell/iitk/newhtml/reviewcom.htm (accessed several times from 17 October 2007 to 18 August 2008).
Indiresan, P.V. (2000), 'W(h)ither IITs', *Seminar*, **493**, 26–30.
Indiresan, P.V. and N.C. Nigam (1993), 'The Indian Institutes of Technology: an experience in excellence', in S. Chitnis and P. Altbach (eds), *Higher Education Reform in India: Experience and Perspectives*, New Delhi: Sage, pp. 344–64.
Interim Report: Sarkar Committee (1946), *Higher Technical Institutions for the Post-war Industrial Development*, National Archives, New Delhi.

Jayaraman, K.S. (2005), 'Biotech boom', *Nature Biotechnology*, **23** (9), 1183–4.
Kim, L. (1993), 'National system of industrial innovation: dynamics of capability building in Korea', in R.R. Nelson (ed.), *National Innovation Systems: A Comparative Analysis*, New York: Oxford University Press, pp. 357–83.
Kneller, R. (1999), 'Intellectual property rights and university-industry technology transfer in Japan', *Science and Public Policy*, **26** (2), 113–24.
Krishna, V.V. (2001), 'Reflections on the changing status of academic science in India', *International Social Science Journal*, **53** (168), 231–46.
Krishna, V.V. (2007), 'Dynamics at the sectoral system of innovation: Indian experience in software, biotechnology and pharmaceuticals', in T. Turpin and V.V. Krishna (eds), *Science, Technology Policy and the Diffusion of Knowledge*, Cheltenham, UK and Northampton, MA, USA: pp. 199–233.
Lundvall, B.-Å. (ed.) (1992), *National Systems of Innovation: Towards a Theory of Innovation and Interactive Learning*, London: Pinter.
Menon, P.K.B. (2002), 'Technology incubation systems in India', Department of Science and Technology, available at http://www.technology4sme.com/ecoach/countries/india (accessed 12 February 2006).
Nath, N.C.B. (1992), 'A reformulation of issues based on Indian experience', Paper presented at an international workshop on Industry-University Interaction on the theme of technology, commercialization and R&D, Bangalore, 24–26 June.
Nelson, R.R. (ed.) (1993), *National Innovation Systems: A Comparative Study*, Oxford: Oxford University Press.
Raju, V.S. (1995), 'Building bridges between IITs and Indian industry', Paper presented at the Indo-British Seminar on Industry-Institute Interaction, British Council Division, 6–7 March.
R&D Spectrum (2005), IRCC (Industrial Research and Consultancy Centre), IIT Bombay.
Sebaly, K.P. (1972), 'The assistance of four nations in the establishment of the Indian Institutes of Technology, 1945–1970', PhD Thesis, School of Education, University of Michigan, Ann Arbor, Michigan.
Sengupta, A.K. (1999), 'Marketing of technological competence of academic institutes in technology transfer and in-house R&D', in B.K. Pattnaik (ed.), *Indian Industry*, New Delhi: Allied Publishers, pp. 359–66.
Sharif, N. and E. Baark (2005), 'The tamest of Tigers? Understanding Hong Kong's innovation system and innovation policies', *International Journal of Technology and Globalization*, **1** (3–4), 462–79.
Swaminadhan, D. (1995), 'Making engineering education relevant to industry', Paper presented at the Indo-British Seminar on Industry-Institute Interaction, British Council Division, New Delhi, 6–7 March.
Times Higher Education (2006), *World University Ranking 2006: The World's top 100 Technology Universities*, http://www.timeshighereducation.co.uk/hybrid.asp?typeCode=163 (accessed 1 July 2008).
Wong, P.K. (1996), *National Systems of Innovation: The Case of Singapore*, Seoul: Science and Technology Policy Institute.
Wong, P.K., Y.P. Ho and A. Singh (2007), 'Towards an 'entrepreneurial university' model to support knowledge-based economic development: the case of the National University of Singapore', *World Development*, **35** (6), 941–58.

# 11. University technology transfer and commercialization: the case of Multimedia University, Malaysia

**Ming-Yu Cheng***

## 11.1 INTRODUCTION

Traditionally, teaching and learning are the two primary activities taking place in universities. As teaching and learning interaction continued, the demand for new knowledge intensified. This led to the pursuit of research to discover and acquire new knowledge. While upholding the basic roles of teaching and training, universities have mounted their emphasis on the importance of allocating time and resources to conduct research, especially basic research, in order to add to the knowledge stock of the universities. The nexus of teaching and research (T*R) matured and has become an established model for almost all universities ever since Wilhelm von Humboldt, founder of the University of Berlin, proposed the idea of unifying teaching and research as the base of the university in the early nineteenth century (Yusuf 2007).

Following the development of the T*R nexus, there was also a growing concern that universities should do more than just conduct research; universities should also engage in practical activities that could benefit the economy and society more directly and explicitly. When government's support for university research (especially in the form of funding) grew in size and became institutionalized into the university's system, the linkage between university and socio-economic development was expected to intensify. Thus, policy makers have put an increasing emphasis on formulating policies to transfer the wealth of knowledge generated within the university to industry in order to benefit society and businesses more directly. Public expectation has caused the burgeoning of terms such as "university technology transfer" (UTT) and "university technology commercialization" (UTC). From the T*R model, universities are taking on additional roles and transforming to the T*R*D*C nexus (Teaching-Research-Development-Commercialization) (Farhan 2004).[1]

Research and development (R&D) activities are not exclusively taking place in universities. Corporations, especially large firms, also spend a substantial amount of resources in pursuing basic research and technology development, as basic research is the foundation for technology development. Moreover there is an increasing need for businesses to engage in innovation in order to remain competitive. However as the cost of technology continues to escalate and complexities increase, firms find it hard to catch up with the dynamism of technology and to engage in research independently. Hence firms are seeking opportunities to form partnerships and collaborative research efforts with other firms in the corporate sector and/or with the "knowledge factory" at research institutes and universities. When the roles of technology in development expanded, the link between universities and the business sector intensified (Yusuf 2007).

University-industry linkage (UIL) is seen as an instrumental way for both universities and industry to catch up with the fast-changing knowledge-based economy. On the one hand, industries rely on universities to provide ideas, expertise and support for exploring potential business ventures. On the other hand, universities have increased their dependence on firms to commercialize their research output. Ultimately, UILs enable firms and researchers to collaborate in the development of new technologies (Yusuf 2007).

This chapter provides a review of the development of university technology transfer and commercialization (UTTC) in Malaysia. Research development as well as technology commercialization are costly activities to be conducted at universities. This chapter examines the model of a private university in Malaysia in delivering its mission to be an entrepreneurial university.

## 11.2 SCIENCE AND TECHNOLOGY (S&T) DEVELOPMENT IN MALAYSIA

Science and technology (S&T) is a critical component of socio-economic development in Malaysia. During the early years of post-independence in the 1960s, advances in scientific and technological knowledge contributed significantly to the development of an agriculture-based economy in Malaysia. Discovery of new high-yielding clones, diversification of crop varieties, higher income and improved productivity were made feasible through agricultural research and technology development. As the nation moved to an industrial economy in the 1980s, greater emphasis was placed on research and S&T to fortify industrial development. In 1975 the government set up the National Council for Scientific Research and

Development (NCSRD) to promote S&T for economic development; however the results were not apparent as the pursuit of S&T was rather fragmented at that time.

Starting from 1986 a more concerted effort was initiated by the government. For the first time, a chapter was devoted in the Fifth Malaysia Plan (1986–90) for S&T to provide a more guided framework for S&T and national development. NCSRD was revamped to play a more active role. Also in 1986 a national S&T policy was launched to outline the development plan for S&T. The S&T policy was reviewed in 2000 and the Second National S&T Policy (STP II) was announced in 2003. The main goal of STP II is to speed the pace of progress in S&T in both research and technological capability and capacity, in order to improve the long-term performance of the economy as well as to ensure national competitiveness. Also emphasized in the policy was the strengthening of the partnerships between government, industry, universities and research institutions for better utilization of S&T and technology transfer as well as commercialization (Malaysia Plan, various issues).

The establishment of institutions such as the Malaysian Institute of Microelectronic Systems (MIMOS) for microelectronics-related research and the Nuclear Energy Unit (UTN) for nuclear science and technology during the Fifth Plan period has shown the government's determination to expand industry capacity and to enhance the technological capability of the nation, as well as to balance research efforts in categories other than agriculture. Other institutions such as the National Information Technology Council (NITC), the Academy of Sciences, the Malaysian Science and Technology Information Center (MASTIC) and the Space Science Studies Center have been established since the Sixth Plan to strengthen institutional support for S&T development and to promote research related to advanced technologies, such as information technology and communications, microelectronics, biotechnology and life sciences, advanced manufacturing technology, advanced materials and environmental as well as energy-related research. Currently the Ministry of Science, Technology and Innovation (MOSTI) is the ministry leading the development of S&T in the nation.

The intensification of science policy was accompanied by a boosted budget allocated to S&T. Prior to the Fifth Malaysia Plan (5MP), resources allocated for S&T were rather limited and the main contributor was the government; for instance, in 1982 total R&D expenditure accounted for 0.5 percent of the gross national product (GNP) with the public sector as the largest contributor (85 percent) while the private sector contributed only about 10 percent and universities 5 percent (Fifth Malaysia Plan 1986). In 1989 total R&D expenditure went up to 0.8

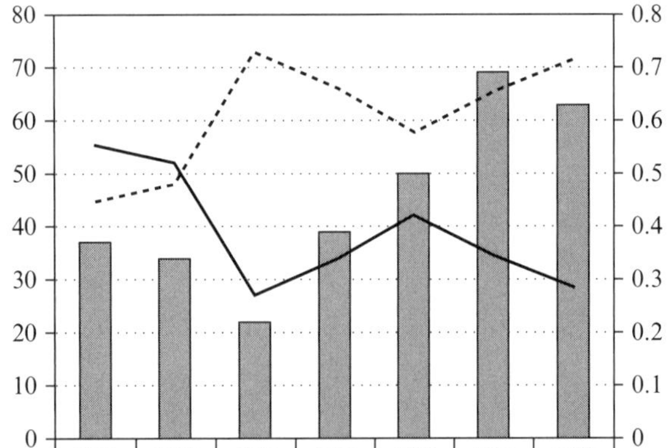


| | 1992 | 1994 | 1996 | 1998 | 2000 | 2002 | 2004 |
|---|---|---|---|---|---|---|---|
| GERD/GDP | 0.37 | 0.34 | 0.22 | 0.39 | 0.5 | 0.69 | 0.63 |
| ---- Private | 44.7 | 48.1 | 72.9 | 66.2 | 57.9 | 65.3 | 71.5 |
| —— Public | 55.3 | 51.9 | 27.1 | 33.8 | 42.1 | 34.7 | 28.5 |

*Source:* National Survey of Research and Development, 2006 Report.

*Figure 11.1 GERD/GDP ratio and R&D expenditure by public and private sector*

percent of GNP, with more than 80 percent of the contribution coming from the public sector (Sixth Malaysia Plan 1991). In 1992, although the intensity of research in terms of R&D expenditure as a proportion to GNP shrank to 0.4 percent (most probably due to an increase in GNP), the private sector's contribution increased to 45 percent of R&D expenditure (Seventh Malaysia Plan 1996). In 2002 the ratio of gross expenditure on R&D to gross domestic product (GDP) (GERD/GDP) was 0.7 percent, of which more than 70 percent of the funding came from the private and business sector (Ninth Malaysia Plan 2006). In fact the contribution from the private sector to R&D expenditure has shown a steady increase since the Fifth Plan and the private sector has become the major contributor in the 2000s. Figure 11.1 shows the GERD/GDP ratio as well as public and private sector expenditure on R&D from 1992 to 2004.

A special budget for R&D has been allocated since the Fifth Plan. Table 11.1 shows the federal government development allocation for S&T for the period 1986–2010.

In allocating resources for R&D activities, the Fifth Plan explicitly spelt out the roles played by universities and government as well as private research institutes in conducting different types of research. The primary area for universities to develop is basic research while public sector

*Table 11.1 Development allocation for science and technology (direct R&D)*

| Malaysia Plan (MP) | Approved (RM million) | Expenditure (RM million) |
|---|---|---|
| 5MP (1986–90) | 413.8 | 285.4 |
| 6MP (1991–95) | 629.0 | 567.1 |
| 7MP (1996–2000) | 755.0 | 718.1 |
| 8MP (2001–05) | 1000.0 | 926.6 |
| 9MP (2006–10) | 1581.6 | |

*Source:* Malaysia Plan, various issues.

*Table 11.2 Distribution of research allocation to types of research (percentage) in 1986*

| Research institution | Type of research (1986) | | |
|---|---|---|---|
| | Basic | Applied | Developmental |
| Universities | 40 | 50 | 10 |
| Government research institutes | 10 | 35 | 55 |
| Private research bodies | 5 | 20 | 75 |
| Total allocation | 18 | 35 | 47 |

*Source:* Fifth Malaysia Plan, 1986.

research institutes would focus on applied and development research. The allocation for basic, applied and developmental research in 1986 was based on the ratio of 18:35:47 (Fifth Malaysia Plan 1986) as shown in Table 11.2, while Table 11.3 shows the distribution in 2004. The emphasis has changed to focus more on applied research rather than basic and experimental research.

The funding arrangements under the Intensification of Research in Priority Areas (IRPA) program which was initiated in 1987 has become the main source of funding for R&D, especially for higher learning institutions and public research institutes in Malaysia. A total funding of RM836.9 million was approved under the IRPA program during the Eighth Plan, indicating an increase of more than 100 percent as compared with the allocation given during the Fifth Plan. The focus of the IRPA funding switched from agro-based research (about half of the total allocation in the Fifth Plan was for agriculture, 34 percent for industry and 17

*Table 11.3 Distribution of research expenditure to types of research (percentage) in 2004*

| Research institution | Type of research (2004) | | |
|---|---|---|---|
| | Basic | Applied | Experimental |
| Universities | 31.7 | 46.2 | 22.1 |
| Government research institutes | 17.9 | 64.1 | 18.0 |
| Private research bodies | 12.1 | 56.2 | 31.7 |
| Total expenditure | 16.1 | 55.4 | 28.5 |

*Source:* National Survey of Research and Development, 2006.

percent for other sectors) to applied research in sciences and engineering during the Eighth Plan (Ninth Malaysia Plan; MOSTI 2006b). More specifically, universities' focal point was on research related to engineering and natural sciences, while government research institutes focused on agriculture sciences and the private sector invested mostly in applied research for commercial purposes and spent less on basic research. Besides IRPA, other grants were also available for different and specific purposes, for example, the Industrial Research and Development Grant Scheme (IGS) was introduced in 1997 to encourage the private sector to participate in R&D; the MSC Research and Development Grant Scheme (MGS) was established in 1998 to promote innovative multimedia products, processes and services within the Multimedia Super Corridor (MSC); and the Demonstrator Applications Grant Scheme (DAGs) was launched in 1998 to promote usage of information and communications (ICT) among Malaysians (Eighth Malaysia Plan 2001).

The momentous increase in research funding has yielded a fairly positive outcome in terms of number of patents applied for and granted to residents and non-residents from 1986 to 2005, as shown in Figure 11.2. Although internationally Malaysia ranked rather low in the number of patents granted (far behind Singapore, but ahead of Thailand, the Philippines and Indonesia), the growth rate from year to year is encouraging. Based on the data collected by MOSTI in '*Facts and Figures 2006*' with the original source of the data being the Intellectual Property Corporation Malaysia, most of the patents granted were chemistry or metallurgy related (26.7 percent), followed by electricity (17.8 percent), performing operations (16.6 percent), human necessities (16.5 percent) and others. Another important R&D output as measured in terms of the number of scientific publications is shown in Figure 11.3.

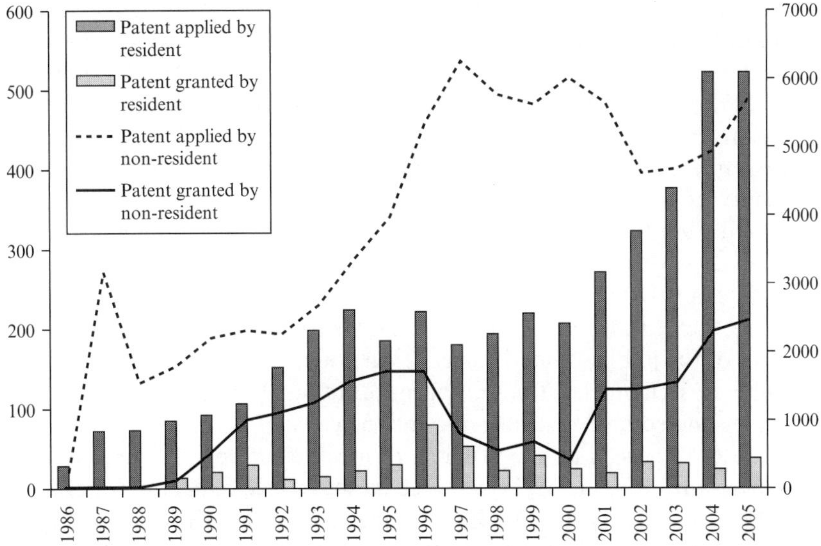


*Source:* Intellectual Property Corporation of Malaysia, quoted from *MOSTI Facts and Figures 2006*.

*Figure 11.2 Patent applications and grants (1986–2005)*

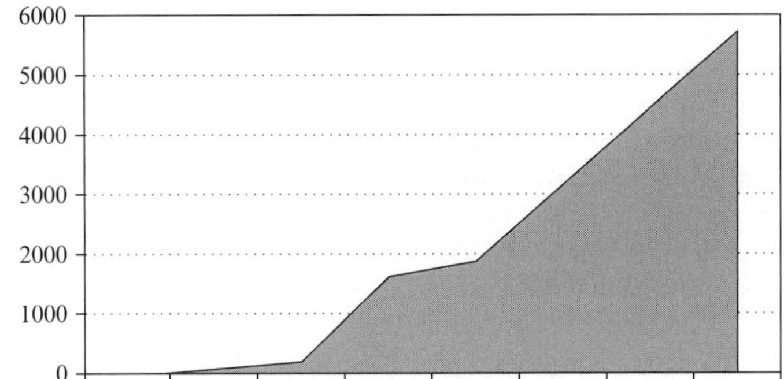


| | 1MP (1966–70) | 2MP (1971–75) | 3MP (1976–80) | 4MP (1981–85) | 5MP (1986–90) | 6MP (1991–95) | 7MP (1996–00) | 8MP (2001–05) |
|---|---|---|---|---|---|---|---|---|
| Publications | 14 | 59 | 204 | 1634 | 1888 | 3183 | 4466 | 5688 |

*Source:* MOSTI 'Science and technology knowledge productivity in Malaysia: Bibliometric Study 2003', quoted from *MOSTI Facts and Figures 2006*.

*Figure 11.3 Scientific international publications by Malaysians (First Malaysia Plan to Eighth Malaysia Plan)*

Despite the spiraling research activities, commercialization of R&D output remained rather low across all sectors; only about 3.4 percent of projects were commercialized during the Eighth Plan (Ninth Malaysia Plan 2006). There are a few reasons for this phenomenon. Firstly, most university research is far-off from industrial practice and is not relevant nor commercially useful; secondly, there is a lack of financial support to fund the various stages of commercialization; and thirdly, there is a lack of ability to identify and market intellectual property with commercial potential (Seventh Malaysia Plan). Malaysia still depends heavily on technology transfer from abroad to enhance its competitive edge in export markets. Thus in the Ninth Plan there was a target to commercialize at least 10 percent of the public-funded projects. An allocation of 67.9 percent of the funding was channeled for this purpose, to support downstream R&D activities and commercialization (Ninth Malaysia Plan).

In order to accelerate the progress of local technology R&D as well as commercialization, the Malaysian Technology Development Corporation (MTDC) was established in 1992. Starting from the Seventh Malaysia Plan, special-purpose grants were allocated to provide financial support to spin-off and start-up companies to commercialize local technologies. To start with, a total of RM161 million was made available for two specific funds, that is, the Commercialization of Research & Development Fund (CRDF) for commercialization of R&D work and the Technology Acquisition Fund (TAF) for acquisition of new technology to enhance the technology level and production processes. Other key functions of MTDC include providing consultancy services in licensing and intellectual property protection, technology marketing and promotion as well as technology transfer. To strengthen the UILs, several incubations parks were established in close proximity to universities to assist the start-ups venturing in specific S&T development. Three incubation parks which have been set up so far are the UPM-MTDC Technology Incubation Center (focuses on ICT, multimedia and agri-bio), UTM-MTDC Technology Incubation Center (focuses on advanced engineering and life sciences) and UKM-MTDC (for biotechnology).

An analysis of S&T development from the Fifth to Ninth Plans reveals the following trends: (a) the government has made serious efforts to drive S&T as the engine of growth in the knowledge-based economy; (b) active participation of the private sector in R&D has contributed to the expansion and diversification of research activities, especially in applied and experimental research; (c) universities are committed to continue their obligation in teaching and conducting basic research; and (d) the level of university technology transfer and commercialization is still scanty.

## 11.3 THE DEVELOPMENT OF UNIVERSITY-INDUSTRY LINKAGES (UILs) IN MALAYSIA

Generally, universities in Malaysia have gained credit for successfully performing their duty in nurturing talents and training highly skilled professionals for domestic as well as multinational corporations. However their role as a key driver of innovation and economic growth as well as an agent of technology diffusion to transfer and commercialize university science is still not well recognized as so far only marginal linkage is observed between universities and industry in Malaysia.

In general, the number of patents filed and granted is used as a measure of the extent of technology transfer from research activities. However patenting alone is a necessary but not sufficient indicator. Even in countries like Japan, output of technologies from universities as measured by patents is relatively small (Yusuf 2007). Universities contribute more significantly in publications in refereed scientific journals. Based on Sampat's (2006) observation, patenting and licensing are among the least important channels for technology transfer. Basically, knowledge transfer from universities takes the form of publication of research papers, conference presentations or lectures via short-term courses. Collaboration with industry in an informal and short-term manner such as co-authoring papers, joint research projects, participation of industry speakers in certain universities' courses and consulting arrangements are more prevalent than formal contractual relations with industry.

Despite the limited UILs in Malaysia, there are signs that the partnership between university and industry is flourishing, especially after the 1990s. A few factors contributed to the development. Externally, an expanding borderless economy demands that nations fortify the utilization of domestic technological and innovative capability in order to remain competitive in a more integrated global market. At the same time, universities in developed countries, especially the USA, Japan and Europe have successfully demonstrated the benefits of greater linkages between university and industry, thus providing an incentive and encouragement for Malaysian universities to follow suit. Internally, the formulation of government vision and policies has called for the enrichment and enlargement of the S&T base in Malaysia. More specifically, (i) the Vision 2020 announced in 1991 that aimed at building a scientific and technologically advanced industrial society by the year 2020 has required the expansion of the industrial base; (ii) the establishment of the MSC in 1996 to spearhead information technology (IT) and multimedia development has subsequently led to the creation of firms, particularly small and medium-sized enterprises (SMEs) seeking complicated scientific and technological knowledge; (iii) the release of the Knowledge-based Economy Master Plan in 2002 to foster the transformation of a production-based economy to a knowledge-based economy emphasizes the importance of knowledge utilization in addition to knowledge production in sustaining economic growth; (iv) institutionalization of the Patents Act of 1983 in Malaysia and protection of other intellectual property rights similar to those under the Bayh-Dole Act encourages the commercialization of university research; and (v) creation of commercial arms and technology transfer offices or research management centers in universities (Universiti Malaya, Universiti Malaysia Sarawak, Universiti Malaysia Sabah, Universiti Kebangsaan Malaysia, Universiti Teknologi Malaysia, Universiti Putra Malaysia, Universiti Sains Malaysia and Multimedia University) since the early 2000s to support and generate interactions with industry had facilitated the diffusion and commercializing of university technology.

The creation of proper physical infrastructure and institutional, legal and administrative frameworks had worked to promote the university and industry partnership. At the same time, attempts such as the creation of technology parks to incubate start-ups; setting up of special funds such as venture funds and direct funds; and establishment of government organizations such as Standards and Industrial Research Institute of Malaysia (SIRIM) to provide patent information services and formulation of intellectual property policy and guidelines for university and research institutions are critical in creating a favorable climate to nurture innovation and the transition from laboratory to commercialization.

One major development in the 1990s which propelled a closer linkage between the university sector and industry was the education reform which took place in 1996. The corporatization of public higher education institutions following the amendments of the Universities and University Colleges Acts (1971) in 1996, and the privatization of higher education institutions as a result of the Private Higher Educational Institutions Act (1996), depict the government's effort to develop a more business-oriented education system in Malaysia in order to tap the greatest benefits from this knowledge business. In 1998 five public universities, including the oldest university in Malaysia (Universiti Malaya) were corporatized and eventually all other public universities in the country are to follow suit. With this corporatization, public universities are expected to have greater freedom in their management and system of governance while reducing their reliance on government funding to finance their operations, including R&D expenditure. Public universities are expected to run as corporations to generate a greater proportion of their revenue from the fruit of their research via partnerships with private businesses, entering into business enterprises, and providing consultancy services for industry or by greater student enrollments. The commercial arms established in public universities are among the measures taken to foster entrepreneurship in order to actively hunt for potential revenue from the market.

The setting up of private universities and branches of foreign universities in Malaysia was first approved by the government in 1996. Prior to the Private Higher Educational Institutions Act 1996, the 1969 Essential (Higher Education Institution) Regulation was in effect to prohibit private institutions from offering degree-granting programs and foreign universities from setting up branch campuses in Malaysia. Thus business for private institutions was limited to pre-university courses or twinning and franchise programs. The establishment of private universities created a "market" for higher education in Malaysia and has intensified competition among public and private universities to bid for students, faculty and other resources. One of the key bases for the competition is the reputation of institutions in terms of teaching and research as well as entrepreneurship and innovation. Private universities have more flexibility and autonomy in adjusting to rapidly changing market needs in terms of scope of courses offered and are more entrepreneurial in linking up with industry to search for additional revenue from the market. In order to close the gap, public universities have no choice but to be more managerial and entrepreneurial in their approach and operations in order to generate sufficient funding to support their enterprises. In 1992 the number of private colleges in Malaysia was 156 (Lee 2002); however by 2009 the number of private higher education institutions had increased to 454, which included five branch campus of foreign universities, 21 private universities, 20 private university colleges and 404 private colleges (Ministry of Higher Education).[2] The mushrooming number and variety of higher education institutions (both public and private) led to a growing movement towards greater linkages with industry.

## 11.4 THE CASE OF MULTIMEDIA UNIVERSITY

When the government upgraded the Institute for Telecommunications and Information Technology (ITTM), the training institute under Telekom Malaysia Berhad (TM) in the historical city of Melaka, to University Telekom (UNITELE) in 1996, the first private university in the country was born. With the establishment of the MSC in the same year, the government decided to set up a scholarship and research hub at the heart of the MSC in Cyberjaya to support and facilitate the development of information and multimedia technology within the MSC. Once again, Telekom Malaysia Berhad was given the mandate to build the university in Cyberjaya. Located about 50 km south of Kuala Lumpur, the capital of Malaysia, Cyberjaya is the country's first "intelligent city", housing cutting-edge high-tech companies and spearheading the development of the MSC.

In March 1997 the university in Cyberjaya was constructed, and in July 1999 it was officially launched by the former Prime Minister of Malaysia, Tun Dr Mahathir Mohamad. The university was renamed Multimedia University (MMU). MMU with its twin campuses in Melaka and Cyberjaya (MMU-Melaka and MMU-Cyberjaya) was the first private university approved by the Private Higher Educational Institutions Act 1996 in Malaysia. Under the leadership of Tun Dr Siti Hasmah Hj Mohd Ali as the Chancellor since 1997, the university has grown to become one of the leading universities in Malaysia producing much-needed knowledge workers excelling in the fields of engineering, IT, multimedia and management.

Starting with a total enrollment of 1300 students at the Melaka campus in 1996, the total number of students has grown significantly to about 20000 students 13 years later, of which about 4000 students are international students from more than 70 countries.[3] Besides this MMU has its franchised programs conducted in more than ten countries, and has formed smart-partnerships with international institutions spanning from Africa to America, Europe, the Middle East and Asia Pacific. Currently, courses in the core areas of engineering, IT, management, law, creative multimedia, economics accounting, business studies and finance are offered at different levels ranging from Diploma, Foundation, Bachelor, and Master to PhD degrees. The total number of graduates from 2000–09 was 22840.[4]

Geared up with the vision to be a world-class university providing quality education in multimedia and IT and nurturing talents for the knowledge society and economy, excellence in teaching and research are the core businesses. To create an innovative, entrepreneurial and intellectual as well as lifelong learning environment in MMU, strategic partnerships and networks with industry and government agencies are greatly emphasized. Market-driven programs are uniquely designed and updated frequently to suit the needs of industry and to prepare graduates to meet the challenges in multimedia and IT. English was made the medium of instruction at MMU and remains one of the key competitive advantages of the university.

### 11.4.1 Research Initiatives in MMU

Being Malaysia's first private university, and also being a relatively young institution, MMU has to strive through many challenges to balance the need to provide quality education and maintain a sound business model at the same time. Despite all limitations, the pursuit of research excellence has always been the central concern.

Since its inception, MMU has allocated a provision of RM10 million in the form of internal research funding to serve as a seed fund for young researchers to initiate their research activities in MMU. In addition, the university has sourced external research grants from government agencies such as IRPA, MGS and IGS, and from industry especially Intel, JICA, Motorola and MACRES to support its R&D. The total amount of external grants received from 1997 to 2009 was close to RM60 million. Achieving excellence in cutting-edge research is expensive, and the amount of funding available for R&D at MMU is indeed limited and at an extremely low level by international standards.

Despite its budget constraints, the university has struggled to utilize its restricted resources to build its research capability in key research areas in ICT and multimedia. Its efforts in R&D were recognized by the Ministry of Higher Education of Malaysia which awarded MMU the 'Excellence in Research and Development, Private Higher Education Institution for University Level' award in 2005. In addition MMU staff and students have won many international as well as national awards for their excellent achievements in education and research. Among the awards are: HSBC Young IT Entrepreneur Awards in 2003; Gold Award at the ComGraph Asia Pacific Digital Art Animation Competition and Asian Television Technical and Creative, ComGraph, Singapore in 2002; 2002 and 2005 URSI Young Scientist Award in the Netherlands and India; 1999 and 2002 Malaysia Toray Science and Technology Award; Gold Medal at the 32nd International Exhibition of Inventions, Geneva in 2004; SCIN Best Practice Award in 2004 in the Netherlands; the 2006 World Congress Achievement Award in Computer Science, Computer Engineering, and Applied Computing, USA, and CISCO Networks Innovation Awards 2007.

Prior to 2008, the Center for Research and Postgraduate Programs (CRPP) served as the one-stop center to coordinate research activities and postgraduate programs in MMU. Research centers, known as Centers of Excellence (COEs) (listed in Table 11.4) were created. The COEs were aimed to group researchers with common interests to work together in order to broaden their scope of research and to diversify research activities in different core competence areas.

In 2008 the CRPP was dissolved and its functions taken over by the R&D Division to re-strategize the R&D activities in MMU. The Research Management Center (RMC) was formed under the R&D Division to manage all activities related to research and grants in the university, and

*Table 11.4 Research centers at MMU*

| Campus | Faculty | Research centers |
|---|---|---|
| Cyberjaya | Faculty of Engineering | • Center for Electric Energy and Automation<br>• Center for Applied Electromagnetics<br>• Center for Cryptography and Information Security<br>• Center for Multimedia Communication<br>• Center for Microelectronics, Materials and Systems |
| | Faculty of Information Technology | • Center for Artificial Intelligence and Intelligent Computing<br>• Center for Multimedia Computing |
| | Faculty of Creative Multimedia | • Center for Innovative Education<br>• Center for Interpretation and Expression |
| | Faculty of Management | • Center for Borderless Markets and Economies<br>• Center for Multimedia Banking, Investment and Accounting<br>• Center for Multimedia Technology Management<br>• Center for Ergonomics |
| Melaka | Faculty of Engineering and Technology | • Center for Image Processing and Telemedicine<br>• Center for Robotics and Automation<br>• Center for Thermo-Fluids and Energy System<br>• Center for CAD and K-Manufacturing |
| | Faculty of Information Science and Technology | • Center for Network Computing |
| | Faculty of Business and Law | • Center for Cyberpreneurship Development<br>• Center for Cyberlaw |
| Center for Research and Postgraduate Programs (CRPP) | | • Center for Mathematical Sciences<br>• Center for Virtual Reality and Immersive Technology<br>• Center for Biometrics and BioInformatics<br>• Center for E-Learning<br>• Institute of Photonics Research and Application |

*Source:* Multimedia University *Annual R&D Report*, July 2005–June 2006.

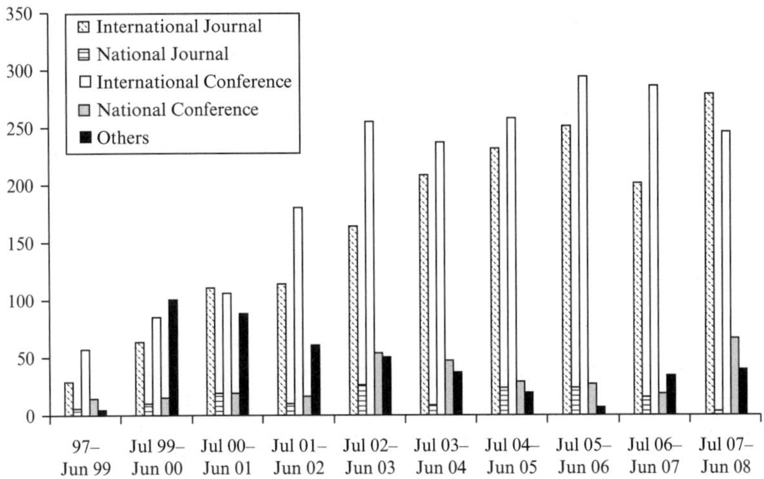


*Source:* Multimedia University, *Annual R&D Report*, various years.

*Figure 11.4 MMU's research output (1997–2008)*

to lead research activities in MMU. The MMU R&D Roadmap was launched on 22 May 2008 and serves as a blueprint to focus on research in high-impact and niche areas, especially in multimedia communication technology, business and management. Research clusters in the fields of engineering, IT and management were formed and led by the respective deans of research.

One common yardstick used to evaluate a university's research intensity is the number of publications produced in the university. In an effort to gain an international reputation in the shortest possible time, MMU has stressed the importance of staff publishing in high-quality refereed international journals. Thus, as shown in Figure 11.4, international journal publications and international conference papers far surpass national publications and other forms of publications. In terms of international journal publications, MMU secured a rating of 0.29 ISI-rated international journal publications per staff in 2007, an achievement which is on par with other local universities and many international universities.

### 11.4.2 Technology Transfer and Commercialization in MMU

Besides being an 'idea factory' to create new knowledge for research and teaching, MMU aims to build an environment that will encourage researchers to translate ideas into practical implementation and products.

University technology transfer and commercialization is not uncommon in Malaysia, however it is critical for MMU to quickly transfer and commercialize its information and multimedia technology as these products have a relatively short life span. To meet the challenges, ever since its establishment, much effort and resources have been committed to actively promote and facilitate a culture and environment of commercialization of research and innovation among the MMU community. Being a latecomer in technology transfer and commercialization, MMU enjoys the advantage of learning from other institutions in Malaysia and abroad in order to model its "entrepreneurial university".

The Center for Commercialization and Technopreneur Development (CCTD) was set up in 1999 to specifically cater for the needs of technopreneur development and commercialization in MMU. In 2008 a new organization, MMU Cynergy, was established to replace the functions of CCTD.[5] In this section, the key functions and activities of CCTD will be discussed as it was the initial model adopted in MMU for commercialization and technopreneur development.

The key functions of CCTD include:

- Providing a one-stop center for patent filing, technology licensing and commercialization.
- Initiating start-ups from ideas and innovations generated by faculty and students.
- Facilitating industry linkages and partnerships for start-ups to identify and locate appropriate markets for their products.
- Initiating funding for entrepreneurship and innovation.
- Promoting technopreneur development by providing amenities and equipment for the commercialization of research ideas/products.
- Providing training and mentoring services to young entrepreneurs and pre-start-ups.

The scope of services provided by CCTD is illustrated in Figure 11.5.

#### 11.4.2.1 Technopreneur development

The fundamental function of the technopreneur development program at CCTD is to incubate start-ups for the creation of new enterprises among the MMU community. Table 11.5 shows the number of start-ups in MMU at the various stages of development from 2001–06. As of 21 January 2008 there were 24 active start-ups, four start-ups that have graduated and 14 being incubated. A majority of these start-ups originated from students' projects and some are the result of team work between faculty and students. Only a small portion of the start-ups are initiated purely

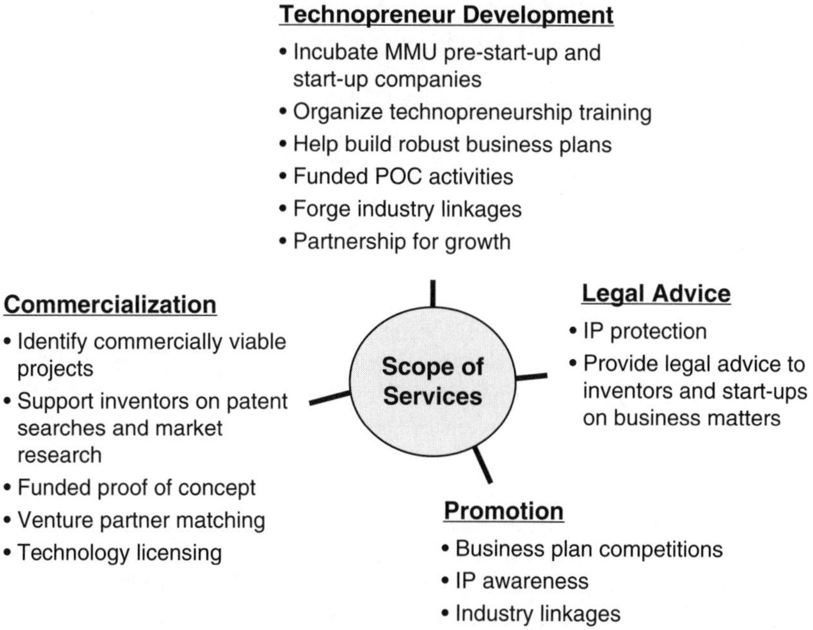


*Source:* CCTD at MMU.

*Figure 11.5 Key services provided by CCTD*

*Table 11.5 MMU start-ups*

| Year | 2001 | 2002 | 2003 | 2004 | 2005 | 2006 |
|---|---|---|---|---|---|---|
| No. of start-ups | 6 | 12 | 4 | 15 | 9 | 4 |
| No. of discontinued start-ups | 3 | 7 | 2 | 7 | 2 | 1 |
| No. of graduated start-ups | – | – | 1 | – | 2 | 2 |
| No. of active start-ups at end of period | 3 | 8 | 9 | 17 | 22 | 23 |

*Source:* CCTD at MMU.

by faculty. This indicates the importance of nurturing entrepreneurial ventures among MMU students. In this regard, CCTD introduced the "Student Entrepreneurial Ventures Grant Scheme", a program to support pre-start-ups. A funding of up to RM2500 was provided to each prospective venture. Successful pre-start-ups that have demonstrated core competencies and the ability to meet the targets are then considered for graduation to start-up status.

The average failure rate of about 50 percent is considered high but still within the acceptable range for start-ups. On the other hand, there are a number of success stories, including: (i) Meishio Interactive, which was initiated by a student from the Faculty of Creative Multimedia, graduated from the start-up program, merged with an external company (Unrealmind Interactive) and was listed on Mesdaq (a listing of mainly technological companies in the Malaysian Exchange) in June 2004. Meishio Interactive creates local content through 3D visualization; (ii) E-transact, which was created by three students from the Faculty of Management, provides an online computerized ticketing system for express bus operators and is currently capturing about 50 percent of the market nationwide. The business will soon be expanded to neighboring countries in Thailand, Singapore and Indonesia; (iii) Xirien, which was founded by two students from the Faculty of Engineering and one from the Faculty of Management, develops wireless solutions for home automation; (iv) Inventsys, which was a start-up created by MMU faculty, provides computerized medical and dental clinic management solutions for small and medium-sized healthcare centers. The start-up has been incorporated and is now known as Exsoft Enterprise.

#### 11.4.2.2 Legal advice

The legal service program manages the intellectual property of MMU. Intellectual property (IP) is protected through IP filing and registration. For the period from 2001 to 2004, the number of invention and innovation disclosures averaged about six per year while the number of patent applications averaged about four per year. As of 2008, only one patent had been granted the certificate, while 12 other applications were pending for patent registration in Malaysia.[6] It is expected that the number of disclosures and applications will increase over time as more R&D projects are completed. CCTD is also taking proactive measures to search for patents through constant discussions and meetings with the inventors. However, due to limited budget available for patenting, CCTD must be selective in patent application and will only choose the most promising inventions for IP protection and commercialization. This is to ensure that the funds spent for IP protection are cost-effective, as patent application and processing is an expensive process. However for MMU to realize benefits from commercialization and technology licensing in the long term, it must allocate more resources not only for innovation and inventions – to increase the level of activities and to groom more high-quality research, development and design output – but also for expenditure on IP protection.

#### 11.4.2.3 Commercialization

Commercialization of university technology takes many different forms depending on the types of IP involved (such as patents, designs, trademarks, copyright, know-how) and the area of the technology or business method or product involved. As MMU has few patents granted so far, commercialization in the form of technology licensing is limited in MMU. Consultancy is a common way for universities to transfer their know-how directly to society. MMU considers consultancy a vital activity for the university, not merely for income generation but also to broaden the skills and expertise of MMU staff and to promote linkages between the university and industry. MMU has formulated the University Consultancy Policy and made it compulsory for all academic staff to be involved in consultancy. Administrative staff are encouraged to conduct consultancy in their areas of specialization. For academic staff, consultancy is one of the criteria for performance evaluation. Through its commercialization program, CCTD is actively assisting faculty in seeking consultancy opportunities.

### 11.4.3 Entrepreneurship Education

One unique feature of the MMU entrepreneurial attempt is its strong emphasis on cultivating entrepreneurial talents among students through structured and built-in entrepreneurship education. Courses on entrepreneurship are made compulsory for every student at different levels of study, for instance, 'Introduction to Cyberpreneurship' at the undergraduate level and 'Cyberpreneurship and Technology Business Strategy' at the postgraduate level. At the undergraduate level, all students at MMU have a chance to be a 'real entrepreneur' when they undertake the university course 'Introduction to Cyberpreneurship'. This one-credit course is a hands-on course in which students are not required to sit for any examination, including a final exam. They are however expected to run a real venture within and outside the university campus during a specified time period (about eight weeks). The types of businesses range from selling food and consumer products to high-tech products like computer hardware and software. Students are required to form groups with other students from different faculties and having different backgrounds and expertise. As the name of the course implies, students are encouraged to utilize online tools, such as email or developing a website to promote their products and services online. At the end of the course, students are evaluated based on a number of criteria including team work, creativity, marketing and financial aspects, presentation of a business plan and ideas, as well as on the profits made during the business period. This course enables students to understand the real meaning of "entrepreneurship" and have their own experience as an entrepreneur in starting, developing, implementing and managing a business.

## 11.5 CONCLUSIONS

As the first private university in the nation, MMU has encountered many challenges which are not pertinent to other local government-funded universities. As a private university, MMU has the flexibility to respond speedily to industry needs and changes. This gives MMU the opportunity to experiment with its model to build a world-class education system in Malaysia. MMU celebrated its tenth year anniversary in 2007, having made remarkable achievements in teaching, learning and research. However financial constraints remain a challenge for the university in its effort to continue support for revolutionary research and technology development, as well as for technology transfer and commercialization. Unlike other renowned private universities in the world such as Harvard or Stanford, being a young university, MMU has no endowments to rely on. Thus closer academia-industry collaboration is essential for providing all kinds of support including sponsorship of lab facilities, consultancy, R&D grants and internships.

## NOTES

* The author would like to thank SPF for sponsoring this project and is grateful for the cooperation rendered by CCTD at MMU, especially Simon and Mazwan for collecting and analysing the information related to technopreneur development and commercialization in MMU prior to 2008. The research work was conducted in 2007–09 when the author was attached to MMU.

1. http://www.symbiosisonline.com/oct04_challenge.htm (accessed 1 January 2008).
2. Private university, http://www.portal.mohe.gov.my (accessed 13 December 2009).
3. *Facts and Figures*, http://www.mmu.edu.my (accessed 12 December 2009).
4. MMU website, http://www.mmu.edu.my (accessed 13 December 2009).
5. MMU Cynergy, http://www.mmu-cnergy.com/index.html.
6. CCTD's website, http://www.mmu.edu.my/~cctd/intelectual_property.html (accessed 30 January 2008).

## REFERENCES

Farhan, A. (2004), "Facing the challenge: universities & technology development in Malaysia", *Symbiosis*, 4 October–5 January available at http://www.symbio sisonline.com/oct04_challenge.htm.

Lee, M.N.N. (2002), "Malaysian universities: towards equality, accessibility, and quality", in *Restructuring Higher Education in Malaysia,* M.N.N. Lee (ed.), School of Educational Studies, University Science Malaysia, Monograph Series, 4, 37–42.
Malaysia, Ministry of Higher Education, "Private university", available at http://www.portal.mohe.gov.my/portal/page/portal/ExtPortal/IPT/IPTS.
Malaysia, The Economic Planning Unit (1986), *Fifth Malaysia Plan*, Kuala Lumpur: Government Printers.
Malaysia, The Economic Planning Unit (1991), *Sixth Malaysia Plan*, Kuala Lumpur: Government Printers.
Malaysia, The Economic Planning Unit (1996), *Seventh Malaysia Plan*, Kuala Lumpur: Government Printers.
Malaysia, The Economic Planning Unit (2001), *Eighth Malaysia Plan*, Putrajaya: Government Printers.
Malaysia, The Economic Planning Unit (2006), *Ninth Malaysia Plan*, Putrajaya: Government Printers.
Ministry of Science, Technology and Innovation, Malaysia (MOSTI) (2006a), *MOSTI Facts and Figures 2006*, Putrajaya, Malaysia: MASTIC.
Ministry of Science, Technology and Innovation, Malaysia (MOSTI) (2006b), *National Survey of Research and Development, 2006*, Putrajaya, Malaysia: MASTIC.
Multimedia University (various years), *Annual R&D Report*, various issues.
Multimedia University, Center for Commercialization and Technopreneur Development (CCTD), available at http://www.mmu.edu.my/~cctd/.
Multimedia University, university website, available at http://www.mmu.edu.my/).
Sampat, B.N. (2006), "Patenting and U.S. academic research in the 20th century: the world before and after Bayh-Dole", *Research Policy*, **35** (6): 772–89.
Yusuf, S. (2007), "University-industry links: policy dimensions", in S. Yusuf and K. Nabeshima (eds), *How Universities Promote Economic Growth*, Washington, DC: World Bank, pp. 1–25.

# 12. University technology transfer and commercialization: the case of Mahidol University, Thailand

**Thanaphol Virasa***

## 12.1 INTRODUCTION

It is becoming increasingly apparent that in many countries, innovation has become a central theme of national development. The emergence of the innovation-driven economy has brought changes in public policy such as public sector reform, education reform and privatization. It has also stimulated increased relationships and interactions among knowledge producers, transfer agents and knowledge users (Jacob et al. 2003). The various institutions in a national system (and particularly the relationships and interactions between these institutions) play an important role in determining the rate and direction of innovation activities. These institutions greatly determine how people relate to each other and how they learn and use their knowledge (Johnson 1992).

In this policy context, universities – as knowledge producers and transfer agents – play a larger role in industrial innovation as providers of human capital and as the seed from which new firms grow (Jacob et al. 2003). This is especially true for developed countries. In the USA, universities are gradually extending their activities deeper into the technology transfer process (Etzkowitz et al. 2000; Siegel and Phan 2005). They do this by identifying gaps in the technology push process and filling these gaps by establishing incubators to assist the formation of firms from campus research. Concurrently they look to venture capital firms to fill gaps in the provision of seed funding. This study also points out that in the UK, the central government has reduced university research funding and demanded that universities seek alternative sources of funding. This has forced universities to undertake activities that attract industrial funding or find innovative ways of generating income. At the same time, the universities have become involved in exchange activities such as licensing patents and establishing innovation centers. In Continental Europe and Latin

America, where many universities are state-controlled, there has been a gradual shift towards an increased autonomy of the university from the state and closer engagement with industry. Funding programs are geared to provide resources for creating intermediary mechanisms such as liaison offices and industrial research.

This transition illustrates the new institutional set-up and the development of mechanisms that support entrepreneurial processes. This requires establishing new institutional forms (such as industrial liaison offices and incubator firms) that facilitate the entrepreneurial process and create new forms of knowledge diffusion. Academic entrepreneurs are encouraged to secure formal rights to their intellectual property (Jacob et al. 2003). The growth of university spin-off firms also denotes their relationship to their parent university (Etzkowitz et al. 2000). The emergence of the innovation-driven economy will encourage universities to reconfigure themselves with structures which maximize their capacity for innovation.

Similarly universities in Thailand are shifting towards an increased autonomy from the government and closer engagement with industry (Brimble and Donor 2006). At the same time, Thai universities are increasingly being viewed by policy makers as engines of economic growth via the commercialization of intellectual property through technology transfer. They have been pressured by the government to seek alternative sources of funding and commercialize research and development (R&D) output. However management of an intellectual property portfolio is still a relatively new phenomenon for many universities in Thailand. This has led to the question of how Thai universities will manage their technology transfer and commercialization process to play the critical role of knowledge diffusion in the national innovation system (NIS).

To answer this question I present Mahidol University, a renowned research university of Thailand, as a representative case of the development of university technology transfer and commercialization in Thailand. This empirical chapter aims to examine agents involved in the university-industry technology transfer process and institutions that have emerged to facilitate commercialization, such as technology licensing offices (TLOs), industry-university cooperative research centers, incubators and related institutions. Key issues and challenges in managing university-industry technology transfer and commercialization are identified and compared with other universities in Asia to provide policy recommendations to the university.

The remainder of this chapter is organized as follows: The second section provides a review of the changing framework conditions of the Thai NIS and university-industry linkages in Thailand. This section uses previous studies to provide a background of the current status and major changes that influence the development of the NIS in Thailand. The third section uses a literature review to provide an understanding of university-industry linkages in Thailand. The fourth section provides the background of Mahidol University and an overview of university technology transfer and commercialization. The fifth section describes the university-industry technology transfer process, institutional set-up, policy and organizational practice. The sixth section highlights three case studies of STANG Holding Co., Ltd (STANG), the Thailand Center of Excellence in Life Science (TCELS), and the Nanotechnology Center of the Faculty of Science. These latter two sections use primary data from interviews with key personnel. They include the Associate Director and Board members of Mahidol Applied Technological and Service Center, the Managing Director of STANG Holding, and faculty and students at the Nanotechnology Center, Faculty of Science. The seventh section identifies issues and challenges in managing the university-industry technology transfer process. The last section gives conclusions and recommendations.

## 12.2 CHANGING FRAMEWORK CONDITIONS IN THE THAI NATIONAL INNOVATION SYSTEM

Since the 1990s innovation has been accepted by many countries as an engine of socioeconomic development. An innovation-driven economy places great importance on the creation, diffusion and use of knowledge. National economic success is ever more determined by a country's effectiveness in mobilizing and utilizing knowledge. In the context of a knowledge-based economy for developing countries like Thailand, knowledge diffusion and exploitation can be viewed as more important than knowledge generation.

In the past, developing countries have to a very large extent relied on knowledge created elsewhere. The copycat approach to 'creativity' of certain Asian countries is a cliché based on reality. This strategy, while adequate for initial stages of development, is not sufficient for sustained development and growth over the longer term. Developing countries are beginning to realize that copying existing knowledge (for example, as captured in product innovations and accessed by reverse engineering) can only help take an economy so far down the road towards development. Sooner or later (preferably sooner) developing countries need to understand that a key success factor is the ability to create new knowledge through high value-added products and services. Sustaining this knowledge edge over time is critical if developing countries are to participate in global competition.

*Table 12.1 R&D expenditure in Thailand, 2000–06*

| | 2000 | 2001 | 2002 | 2003 | 2004 | 2005 | 2006 |
|---|---|---|---|---|---|---|---|
| Gross Expenditure on R&D (GERD) | 12406 | 13486 | 13302 | 15499 | 16571 | 16667 | 19548 |
| Public sector | 8087 | 8202 | 8138 | 8694 | 10548 | 9988 | 11550 |
| Private sector | 4319 | 5284 | 5164 | 6805 | 6023 | 6679 | 7999 |
| GERD/GDP (%) | 0.25 | 0.26 | 0.24 | 0.26 | 0.25 | 0.24 | 0.25 |
| Public: Private ratio | 65:35 | 61:39 | 61:39 | 56:44 | 64:36 | 55:45 | 60:40 |

*Note:* Unit: Million baht.

*Source:* NSTDA (2009).

Over the last three decades, Thailand has experienced major structural shifts. Beginning in the 1970s, exports shifted from agricultural products towards manufactured goods. The 1990s saw exports shift from labor-intensive products (such as textiles, manufactured apparel and food processing) to technology-intensive products (such as computer components and electronic goods). A further development has been a rapid growth in exports of computer equipment and electronic spare parts. However the provenance of these products is better described as 'Assembled in Thailand' than 'Made in Thailand'.

Moreover even though the quantity of electronic goods exported is increasing, the local technology content in the products themselves has increased little. For a developing economy, the existence of limited technological capabilities is a major impediment to becoming a major producer of high value-added products (Arnold et al 2000). From Table 12.1, it can be seen that during the period 2000–04, the total R&D expenditure of Thailand was around 0.25 percent of gross domestic product (GDP). It is likely that to date this ratio remains unchanged. The key R&D performers are in the public sector, especially public research and technology organizations (RTOs).

A wealth of research indicates that most Thai manufacturing firms have grown without deepening their technological capabilities (TDRI 1989, 1998; Arnold et al. 2000). This is confirmed by the findings of the Thailand *R&D/Innovation Survey* in 2002 (NSTDA 2003). For such firms, technological learning has mainly (almost exclusively) relied on available imported technologies as shown in Table 12.2. These have mostly been in the form of machinery and turn-key installations from abroad or joint ventures with foreign partners. Nevertheless R&D activities in business firms, especially in large Thai companies such as Siam Cement Group,

Chareon Pokhaphan Group, PTT Group, Betagro Group and Mitr Phol, are increasing substantially.

*Table 12.2 Value of Thai imports, 1998–2003*

| | 1998 | 1999 | 2000 | 2001 | 2002 | 2003 |
|---|---|---|---|---|---|---|
| All manufacturing product import | 25.56 | 29.31 | 33.35 | 33.98 | 36.14 | 38.29 |
| High-technology manufacturing product | 14.49 | 17.24 | 22.91 | 22.69 | 25.96 | 31.72 |
| High-tech/All manufacturing product (%) | 57 | 58 | 69 | 67 | 72 | 83 |

*Note:* Unit: Billion US dollars.

*Source:* NSTDA (2006).

### 12.2.1 Government Policies

Over the past three decades, Thailand's national macroeconomic policies have shifted from import substitution towards export-led growth. In the manufacturing sector, investment promotion was the driving mechanism used to attract foreign direct investment. This reflects a change in the national economic structure dominated by exports of industrial products such as computer equipment and components, integrated circuits, and automobiles and auto-parts. In 1992 the Thai government, realizing the emerging trend of globalization and the global impact of trade liberalization, formulated macroeconomic policies towards an open economy and liberalization in an effort to create an increased competitive environment in Thailand's domestic market.

The government attempted to extend the privatization agenda in state-owned enterprises. In the wake of this policy, many state-owned enterprises were privatized and publicly listed in the stock market; examples of such enterprises are the Petroleum Authority of Thailand, Thai Airways International, the Telecommunication Authority of Thailand and the Communication Authority of Thailand. The Mass Communication Organization of Thailand is the most recent state enterprise to follow the privatization route.

The public sector reform begun in 2002 has transformed the structure and practices of the Thai government system. Many government departments have been restructured into new functions. The roles and duties of staff and their lines of control and reporting have also been radically changed. A social contract has been introduced, in which public organizations and public universities are required to state their commitment towards achieving desirable goals. Performance-based budgeting has been introduced and is now used in budget allocation. Across all departments, government officers have been offered early retirement or have been retrained and moved to new assignments. This reform process has increased the pressure on all employees to achieve greater effectiveness and make efficiency improvements.

The recent Thai government formulated a new strategic direction for Thailand and set policies towards an innovation-driven society. The government has also allocated resources through a new budgeting system for new public investment. Preference for allocation of resources is given to either national strategic projects or integrated projects undertaken by various government agencies.

After the economic crisis in 1997, the government established several agencies such as the Institute of SME Development, the SME bank, the SME Promotion Office and the National Innovation Agency. These agencies were established as autonomous bodies. They undertake performance-based budget allocation with the aim of encouraging entrepreneurship and giving support for entrepreneurial activity. These new agencies are committed to perform certain missions within their terms of contract and are required to deliver their promises. After completion of the contracted term, these agencies are required to seek private resources and compete in acquiring public resources. A key feature of their role is to act entrepreneurially; that is, to maximize value creation by minimizing the resources committed and coordinating non-controlled resources.

### 12.2.2 Public Research and Technology Organizations (RTOs)

The major public RTOs in Thailand are the Department of Agriculture, the National Science and Technology Development Agency of Thailand (NSTDA), the four National Research Centers (respectively, biotechnology, electronics and computer technology, material technology and nanotechnology) and the Thailand Research Fund (TRF). These semi-government agencies have greater autonomy and flexibility in designing their own management systems as compared with traditional government agencies.

However these research institutions place emphasis on traditional R&D rather than technology diffusion and commercialization. The formation of new firms has been placed in incubation centers such as at the Thailand Science Park and the Thailand Software Park. Despite the recent establishment of the Technology Management Center of the NSTDA, the allocated resources, supporting infrastructure and incentive mechanisms for encouraging entrepreneurial activities are limited. In addition the linear model of innovation is still a dominant school of thought in Thailand.

### 12.2.3 Education System

Education reform in Thailand has long been a high-profile political issue. The new Education Act (introduced as statutory law in 2001) reformed the role of various public institutions in the education system. Public universities are required to move away from bureaucratic systems in order to have autonomy in management and increase their capacity for academic development.

In the past decade a new type of public university has been established, such as Suranaree University of Technology, Naresuan University and the King Mongut University of Technology Thonburi (KMUTT). These universities partially rely on government funding. University staff, who are employed under limited-year contracts, enjoy a higher salary than in the government system. In contrast, the supporting infrastructure and incentive mechanisms for encouraging entrepreneurial activities in these universities have not fully developed. To fill this gap, the Office of Higher Education Commission implemented new schemes such as the University Business Incubation (UBI) Project in 2004 and Technology Licensing Offices (TLO) Project in 2006.

## 12.3 UNIVERSITY-INDUSTRY LINKAGES (UILs) AND ECONOMIC DEVELOPMENT IN THAILAND

UILs have become more important in light of the innovation-driven economy. This is especially true for Thailand in sustaining its economic growth while heavily depending on natural-based resources, skilled labor, macroeconomic policies and provision of physical infrastructure. Intense competition from emerging economies through lower wages and/or higher technological skills has been intensified by trade liberalization and economic integration. Increasingly Thai companies have to improve their operations in order to keep up with higher product and production standards of multinational companies. The structure and incentive systems of Thai universities are not favorable for meeting these increasing needs of the business community.

In Thailand the principal mechanism by which educational resources are transferred to the production sector is personal contacts (Brimble and

Donor 2006). Most interactions between universities and industry are through informal channels. The study of Brimble and Donor (2006) identified three important issues of UILs in Thailand. Firstly, the linkages tend to involve fairly low levels of technology. Secondly, they tend to be weakly institutionalized, as reflected by the fact that many are of a short duration, operate through individuals rather than through more organized 'linkage mechanisms' and lack significant linkages to local or regional institutional assets. Thirdly, there are only a few instances of linkages with significant benefits to both sides, such as improvements in curricula, creation of new academic units or upgrading of firms.

The survey results of Schiller (2006) show that modes of UIL in Thailand are mainly limited to services without deeper research involvement and to linear modes of knowledge transfer. Half of the UIL projects in Thailand consist of consulting services. Technical services such as sample testing and informal contacts are the second most important mode of cooperation. Such service-oriented UIL is followed by linear modes of UIL such as licensing, sales and contract research. The third most important group of linkages is teaching through internship and training. Both research-based and joint research projects are less than 10 percent of total activities. It is interesting to note that UIL in Thailand is driven by pecuniary benefits for earning additional income rather than academic benefits (Schiller and Hiefner 2007).

It has become increasingly apparent that many firms, which are increasingly dependent on innovation to sustain competitiveness, are coming to appreciate the merits of nurturing and tapping the capabilities of universities, national research centers and public RTOs (Hershberg et al. 2006). This is also true for Thailand. The government encourages firms by offering incentives such as tax deductions and investment promotion to expand the scope and depth of their research. Universities in Thailand, which saw their main objective as human resource development, are being drawn into the game by international demonstration effects, the push exerted by the government and interest on the part of firms.

## 12.4 BACKGROUND OF MAHIDOL UNIVERSITY AND UNIVERSITY TECHNOLOGY TRANSFER AND COMMERCIALIZATION

Mahidol University (MU) was the first institution of higher education in Thailand. MU was founded by His Majesty King Chulalongkorn the Great (Rama V) in 1890 as a medical school attached to Siriraj Hospital. The medical school was successively renamed the Phatayakorn School in

1893 and the Royal Medical School in 1900. In 1916 King Rama VI reorganized the higher education system in Thailand, and the Siriraj Hospital Medical School became the Faculty of Medicine.

The establishment of the University of Medical Science in 1943 was another major reorganization of higher education in Thailand. The Faculty of Medicine and Siriraj Hospital were transferred to the University of Medical Science along with the Faculties of Dentistry, Pharmacy and Veterinary Medicine from Chulalongkorn University (CU). During 1948–68 the University of Medical Sciences founded ten new faculties, namely: the Faculty of Public Health, the Faculty of Medicine at Chulalongkorn Hospital, the Faculty of Medical Technology, the Faculty of Medicine at Chiang Mai Hospital, the Faculty of Tropical Medicine, the Faculty of Medical Science, the Faculty of Graduate Studies, the Faculty of Medicine at Ramathibodi Hospital, a second Faculty of Dentistry at Phayathai Road and a second Faculty of Pharmacy.

Further educational restructuring transferred various faculties of the University of Medical Sciences to other public universities. On 1 March 1969 the *Royal Gazette of Thailand* announced the renaming of the University of Medical Sciences to Mahidol University (MU). MU at that time comprised nine faculties: Faculty of Science, Faculty of Pharmacy, Faculty of Dentistry, Faculty of Medical Technology, Faculty of Public Health, Faculty of Graduate Studies, Faculty of Tropical Medicine and two Faculties of Medicine at Siriraj Hospital and Ramathibodi Hospital.

Mahidol University takes its name from HRH Prince Mahidol of Songkla, the Royal Father of His Majesty King Bhumibol Adulyadej (Rama IX). In the early 1900s HRH Prince Mahidol promoted medical and scientific training and improved the state of medicine in Thailand. He established international scholarships and founded training schools in various medical fields. In recognition of his outstanding efforts in medicine, he is known as 'The Father of Modern Medicine and Public Health in Thailand'.

At present MU comprises 16 faculties, six colleges, eight research institutes, six centers, and over 70 sub-faculty centers and laboratories for directed research. MU offers 79 bachelor degree programs, 18 certificate programs, 66 graduate/higher graduate diplomas, 166 masters degree programs and 66 doctoral degree programs. Courses are offered in medicine, public health, nursing, pharmacy, dentistry, engineering, natural sciences, computer science, health sciences, social sciences, applied sciences, applied arts, humanities and arts. Three medical schools are producing around 1000 medical doctors annually and providing medical services to more than four million out-patients and 140 000 in-patients per year.

One goal of MU from 2002 to 2007 was to actively pursue dynamic

*Table 12.3 MU profile in 2007*

| Indicators | FY 2007/8 |
|---|---|
| Total number of academic and administrative staff (as of 31 October 2007) | 25 182 |
| Faculty members and researchers (27.5 %) | |
| • Professors | 151 |
| • Associate Professors | 768 |
| • Assistant Professors | 855 |
| • Lecturers | 1301 |
| • Professional and research staff | 3840 |
| Total number of students (as of 31 October 2007) | 25 868 |
| • Undergraduate students enrolled | 15 816 |
| • Graduate students enrolled | 5950 |
| • PhD students enrolled | 1258 |
| • Medical residency | 1594 |
| • International postgraduate students (from 32 countries) | 144 |
| Total research funding (million baht) (2007) | 1464 |
| • Government budget | 202 |
| • University revenue | 33 |
| • National funding agencies | 1059 |
| • International funding agencies | 170 |
| Journal publications in SCI/SSCI (1999–2007) | 8812 |
| Patents filing (as of 2 February 2008) | 95 |
| Patents granted | 9 |

research goals. This included having every academic staff member publish at least one research paper annually, having approximately 30 percent of all papers published internationally and assuring that at least 30 percent of all research results prove useful and effective to the local community and society at large. Another goal was to increase the number of higher academic degrees held by MU faculty members. The aim was to have no less than 60 percent of all lecturers holding a PhD or an equivalent degree, and at least 6 percent of all lecturers having the academic rank of full Professor in each of MU's various academic and research fields (Table 12.3).

### 12.4.1 R&D Performance of MU

Research is a mission and primary activity of all MU faculties. The total research expenditure of MU accounts for more than 1 billion baht annually. Funds come from national funding agencies, the government budget, the university's revenue and international funding agencies; 72 percent of

*Table 12.4 Research funding of Mahidol University by source of funds, 2004–08*

| | 2004 | 2005 | 2006 | 2007 | 2008 |
|---|---|---|---|---|---|
| National funding agencies | 505.2 | 795.5 | 818.7 | 1059.1 | 925.9 |
| Government | 186.7 | 192.5 | 202.5 | 202.5 | 207.8 |
| International funding agencies | 21.9 | 103.7 | 163.1 | 169.9 | 144.9 |
| University revenue | 28.4 | 31.8 | 21.7 | 32.9 | 48.5 |
| Total | 742.2 | 1123.5 | 1206.0 | 1464.4 | 1327.1 |

*Note:* Unit: Million baht.

*Source:* Mahidol University (as of 22 January 2009).

*Table 12.5 Number of scientific and technological publications by top five universities, 2002–05*

| | 2002 | 2003 | 2004 | 2005 | 2006 |
|---|---|---|---|---|---|
| Mahidol University | 433 | 597 | 509 | 712 | 711 |
| Chulalongkorn University | 341 | 464 | 428 | 597 | 706 |
| Chiang Mai University | 175 | 212 | 213 | 294 | 341 |
| Prince of Songkla University | 143 | 130 | 117 | 197 | 249 |
| Kasetsart University | 86 | 124 | 97 | 194 | 184 |

*Note:* Unit: Article papers.

*Source:* NSTDA (2007).

research funding comes from national funding agencies; 16 percent comes from the university's income and the government budget and 12 percent comes from international funding agencies (Table 12.4). Table 12.5 shows that out of the top five universities in Thailand, MU generates the highest research output in terms of journal publications. The increasing trend of MU publications can be seen from Table 12.4. From Figure 12.1 it can be seen that for much of 1999 to 2007 the number of MU articles cited increased at about the same rate as publications.

However it can be seen from Table 12.6 that research output in terms of patent applications and patent grants is not significant. Despite the increasing number of in-process patents and petty patents during the period 2005–07 as depicted in Table 12.7, the number is still far behind top foreign universities. This indicates that a primary activity of MU is knowledge generation and diffusion to the public domain and local com-

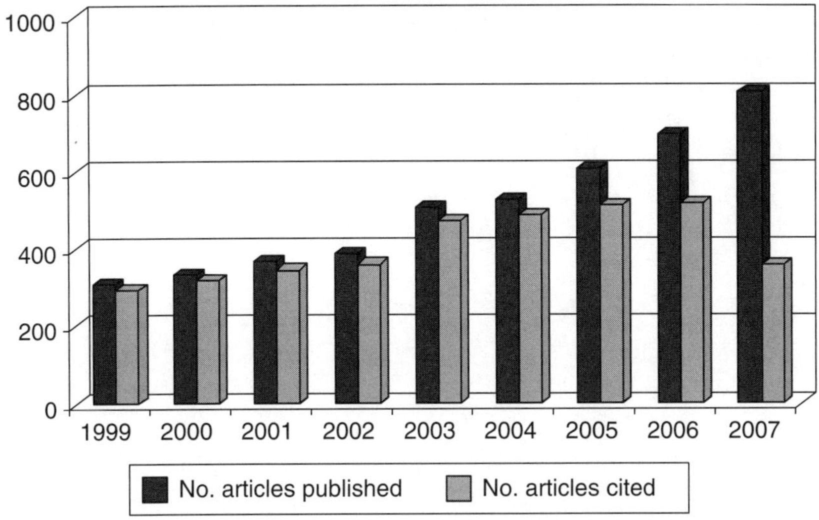


*Figure 12.1* *Number of MU articles published and cited, 1999–2007*

*Table 12.6* *Number of patent applications and grants by top five universities, 1979–2006*

| | 1979–2005 | | 2006 | |
|---|---|---|---|---|
| | Patents applied | Patents granted | Patents applied | Patents granted |
| Chulalongkorn University | 71 | 6 | 25 | 2 |
| Mahidol University | 28 | 4 | 13 | 2 |
| Kasetsart University | 66 | 3 | 7 | 3 |
| Prince of Songkla University | 14 | 2 | 6 | 1 |
| Chiang Mai University | 16 | 1 | 10 | 0 |

*Source:* Department of Intellectual Property.

munity rather than commercialization. This is not surprising given the fact that MU originated from a medical school and most of its research is conducted with the goal of improving the quality of life of Thai society.

To improve the quality of life of people in Thailand and other developing countries, MU research gives priority to medical science. For instance, the main research areas of the Faculty of Medicine at Siriraj Hospital include infectious diseases (AIDS, tuberculosis, melioidosis, diarrhea and so on) and non-infectious diseases (cancer, coronary heart disease and so on),

*Table 12.7 Number of MU patents and petty patents, 1998–2007*

| | Patents | | Petty Patents | |
|---|---|---|---|---|
| | In process | Granted | In process | Granted |
| 1998 | 4 | – | – | – |
| 1999 | 3 | – | – | – |
| 2000 | 1 | – | – | – |
| 2001 | 12 | 2 | – | 1 |
| 2002 | 4 | 1 | 1 | 2 |
| 2003 | 8 | 1 | 1 | 4 |
| 2004 | 13 | – | 5 | 3 |
| 2005 | 6 | 1 | 5 | 7 |
| 2006 | 24 | 2 | 6 | 11 |
| 2007 | 20 | 2 | 10 | 22 |
| Total | 95 | 9 | 28 | 50 |

*Source:* Applied and Technological Service Center (ATSC), Mahidol University (as of 2 February 2008).

caused by social conditions and poor environment (accidents, air pollution, poisoning and so on), as well as molecular genetics. Another location of medical research is the Faculty of Medicine at Ramathibodi Hospital. Its research focuses on traffic accidents, osteoporosis, stem cells, genomics, the epidemiology of coronary disease and HIV/AIDS in pregnancy.

In order to support medical research and strategic industry research, the Faculty of Science built up national research capacity by establishing research centers in strategic areas for Thailand. These include the Center of Excellence in shrimp molecular biology and biotechnology, the Center of Excellence in protein structure and function, the Center of Excellence for vectors and vector-borne diseases, the Center of Excellence for rubber technology, the Center of Nanoscience and the nanotechnology research unit, and the Center of bioinformatics and the applied genomics research unit. Research is also conducted in the chemistry of natural products; molecular pathology of tropical diseases such as malaria, liver fluke and hepatitis; insect carriers of diseases and biological control; economic plants, reproductive biology; and conservation biology.

### 12.4.2 UILs at MU

To bridge the wide gap between the absorptive capacity of Thai companies and knowledge generation of universities, MU categorizes its UILs into three models as depicted in Table 12.8, which include training- and

*Table 12.8 Models of Mahidol University-industry linkages*

| Training and Education-related activities | Research-related activities | Innovation-related activities |
|---|---|---|
| • Training of industry researchers in MU labs<br>• Industrial training for undergraduate students<br>• Cooperative student training program<br>• Entrepreneur training programs | • Joint research projects<br>• Government-funded research projects (initiated by industry) | • Intellectual property rights<br>• Technology licensing<br>• MU Holding company<br>• Spin-off companies |

education-related activities, research-related activities and innovation-related activities. Training programs involve industry researchers in MU laboratories, and MU students and researchers in industrial training programs. Joint research projects and government-funded research projects are UILs through research-related activities. Innovation-related activities such as licensing and spin-off formation are a recent mode of technology transfer being promoted by the university. Currently innovation-related activities at MU are in the early stage of development and are based on a linear model of innovation. The implementation of the university's structure and incentive systems will be discussed in the next section.

The results from interviews show that personal contacts are the principal mechanism by which education resources are transferred to the production sector. This is confirmed by the study of Brimble and Donor (2006) on UILs in Thailand. More formal UIL mechanisms of MU include specialized research centers, research institutes, incubators and university holding companies.

### 12.4.3 Challenges Ahead

In October 2007 the Mahidol University Act was promulgated and the university became autonomous from the Ministry of Education. Since 2008 the university has been in a transitional period, transforming its administration from that of a public university to an autonomous university. The new management team has been appointed, and is preparing for organizational transformation and planning for the future direction of the university.

In fact, over the past few years, the university has prepared itself by making necessary changes which make university administration more flexible and on a par with other world-class universities. The administration has recently embarked on an ambitious Quality Assurance Program and established the Department of Quality Development to ensure the university continues to improve in all areas. This includes the development of university technology transfer mechanisms such as a technology licensing office, venture fund, university business incubator and entrepreneurship development.

Moving 23 000 staff towards a new paradigm of autonomy, however, is a major challenge for university administrators and the new management team. The university has to improve its teaching and research capabilities in order to meet the future needs of economic and social development in Thailand. It is necessary for management to effectively communicate the university's policy direction and strategic planning to the MU community. With regards to the pressure from the government's innovation-driven economic policy, MU's technology transfer policy also needs to be revisited to improve the existing university support mechanisms.

## 12.5 THE UNIVERSITY TECHNOLOGY TRANSFER (UTT) AND COMMERCIALIZATION SYSTEM OF MU

### 12.5.1 University Policies on UTT

'True success is not in learning, but in its applications for the benefit of mankind.'

This statement of HRH Prince Mahidol asserts that the role of MU in the NIS with regard to knowledge diffusion has been recognized since the university's establishment in 1900. The previous MU management team followed this philosophy and translated the university's mission into implementation. In the past few years, various supporting mechanisms within the university such as a technology licensing office, business incubator and venture fund, as well as entrepreneurship training, technology ventures and spin-offs have been developed.

Figure 12.2 shows the technology transfer and commercialization system of MU. Like many universities MU disseminates its research output in the public domain through journal publications, conference presentations, teaching activities and patenting. Results of research are selected by the university for further development. Two channels of commercialization are through research collaboration with existing companies in industry and licensing technologies to start-up companies.

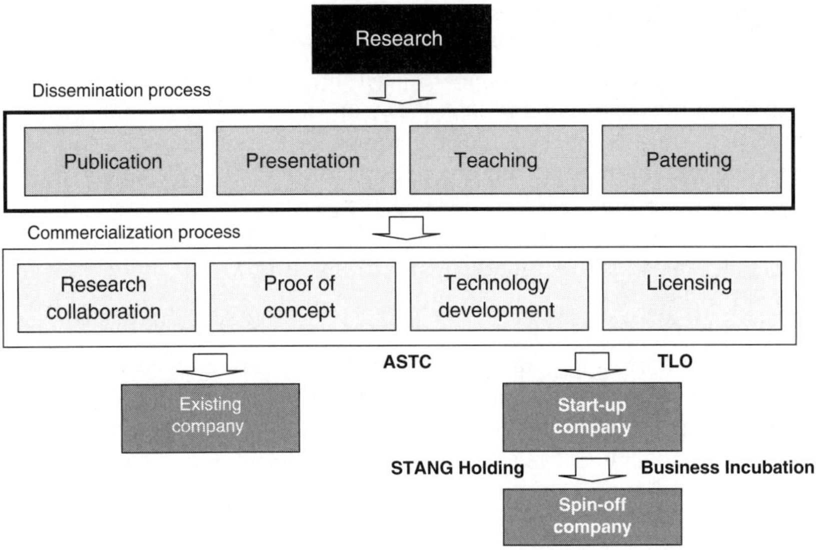


*Figure 12.2 The MU technology transfer and commercialization system*

A key player in the UTT system of MU is the Applied and Technological Service Center (ATSC). ATSC was founded in 1976 under the name Applied Academic Services for Development. The unit was set up to lay a foundation for the university to be self-reliant. The unit was renamed the MU ATSC in 1985. The center is responsible for promoting invention of equipment, products and processes. Research output or inventions that are considered to have commercial value are manufactured for customers. The returns are used to support further R&D activities of the university. The center also provides academic services to the government and the private sector with regard to selection of appropriate technologies, improvement of existing technologies and adaptation of technology to local conditions.

ATSC fulfills the role of promoting and implementing UTT and commercialization. It works with university inventors to promote inventions of products, processes and equipment for their commercial value. The center coordinates with private companies for development and production of medical equipment. It also provides services from a pathological diagnostic laboratory, as well as examination and analysis for the improvement of agricultural and industrial products, production of pharmaceuticals and nutritional services. Other services include computer services, educational aids and media services, and social services in music, art and culture.

To facilitate the implementation of supporting mechanisms for technology transfer and commercialization, the MU management team reorganized related institutes in the university and established new institutes to promote UTT and commercialization. In 2001 ATSC was refocused to facilitate the technology transfer and commercialization process. University business incubation was set up under ATSC in 2004 with a three-year grant from the Office of Higher Education Commission. This initiative was followed by funding to establish a TLO in the university.

At the same time, MU established the Science Technology and New Business Innovation Group Co., Ltd (STANG Holding Company) in December 2004 as a private company to invest in MU technology start-ups and other technology ventures related to the MU mission. STANG Holding has more flexibility than ASTC in terms of administration and venture investment. In addition a model of Centers of Excellence has been established, in areas such as shrimp molecular biology and biotechnology, protein structure and function, nanoscience and nanotechnology, bioinformatics and applied genomics. These Centers of Excellence function as a vehicle of research collaboration with other research units and with industry.

### 12.5.2 Intellectual Property Policies of MU

MU's intellectual property (IP) policy clearly states by law that MU claims ownership of all employees' and students' IP rights resulting from university research activities. This policy is applied in other Thai universities and other universities around the world.

Benefit sharing from IP revenues (after deduction of all expenses such as administration fees, application fees, R&D) is outlined in the university's regulations as shown in Table 12.9. MU and CU have the same benefit sharing structure. The inventor is the largest beneficiary, receiving 50 percent of the shared benefit.

*Table 12.9 MU and CU benefit sharing structure for IP revenues*

| | Mahidol University (%) | | Chulalongkorn University (%) |
|---|---|---|---|
| | Patent | Copyright | IP |
| Inventor/owner | 50 | 70 | 50 |
| Production manager | – | 20 | – |
| University | 30 | 4 | 25 |
| Faculty | 10 | 3 | |
| Department | 10 | 3 | 25 |

## 12.6 CASE STUDIES OF TECHNOLOGY TRANSFER AND COMMERCIALIZATION AT MU

### 12.6.1 STANG Holding Co., Ltd

STANG Holding Company is a private company owned by MU. Its current registered capital is 200 million baht. MU is the major shareholder (60 percent), with the other two major shareholders being the SMEs Development Bank (20 percent) and the National Innovation Agency (20 percent). The primary objective of the company is to manage IP and innovations from the university's research (both by research agencies and by individuals) in order to develop and upgrade research output for commercial purposes. The company's services also include venture capital investment, business plan development and training on technology business development. Table 12.10 shows a portfolio of STANG's business ventures during the period 2004–07.

Despite being initiated as a private company, the operation of STANG Holding is heavily dependent on the decisions of the company's Board of Directors whose backgrounds are in academia and the civil service. The dominant interest of the company's policy is technology commercialization of MU research output. The university requires that STANG

*Table 12.10 Business ventures of STANG Holding Co.*

| Companies | Est. year | % holding | Main businesses |
|---|---|---|---|
| Cosmetic Research Center | Exited | | |
| Xcellerex Inc. | 2004 | 2 | Biopharmaceutical manufacturing based in the USA |
| International Bio-Service Co., Ltd | 2005 | 10 | Services in the study, analysis and testing of drugs such as bio-efficacy tests, pharmaceutical equivalence tests |
| International Drug Development | 2006 | 20 | Services in clinical research |
| ACLIRES Holding Ltd | 2006 | 7 | Services in clinical research |
| MU Trading Company (2007) | 2007 | 49 | Selling MU patented products |
| Shrimp Biotechnology Business Unit | – | – | Currently in pilot scale for producing Ezee Gene |

Holding owns more than 50 percent of total equity shares. In the past four years only three technical service companies were established by STANG Holding's venture fund, and the Faculty of Science owns 80–90 percent of the total equity shares of these companies. Most of the companies' activities are still operated by university staff. STANG Holding's Managing Director asserts that there is a lack of marketing efforts being devoted to matching university resources with external demand. Currently the management team is proposing a long-term plan to reposition the company's business model.

### 12.6.2 Collaboration with Thailand Centre of Excellence for Life Sciences (TCELS)

The recent Thai government recognized an opportunity for Thailand to progressively build a global network of life sciences R&D in collaboration with educational institutions and the private sector. Since the USA has started to outsource more clinical research to other countries, Thailand should take this opportunity to boost her credibility to win these outsourced projects.

To become part of the world's clinical research network it is important for Thailand to establish the appropriate infrastructure, specifically, recruiting life sciences experts and promoting human resource training. Once having achieved this goal Thailand will have more proficient human resources, as the scientists and physicians involved in life sciences discovery can acquire new knowledge and experience from life sciences research. Legal protection for these innovative discoveries is also an important factor in promoting investment in local research, attracting foreign researchers and enabling Thai scientists to apply the new discoveries in real life.

TCELS was established by a Royal Decree endorsed on 18 June 2004. Its establishment followed meetings and the development of public-private partnerships between the US-ASEAN Business Council and Thai representatives. The Thai representatives comprised members of the Cabinet, and several concerned agencies which had foreseen the growth and investment potential of life sciences in Thailand. The establishment of TCELS is a crucial life sciences collaboration in the Asia-Pacific region. TCELS serves as an independent public organization under the supervision of the Royal Thai Government. It focuses on investment and development of life sciences business through partnership with foreign investors, as well as promoting life sciences R&D.

The goal of TCELS is to establish Thailand as the world's leader and center of life sciences research and investment by building team work among involved organizations, providing life sciences information to the public, and encouraging the government to support life sciences for health by building up the environment so that people, agencies and institutions are well equipped. Moreover there are incentives to invest in life sciences for health and infrastructure.

TCELS also aims to draw investment for the development of human resources as well as to promote national and international collaborations, ethical control and respect for IP rights. TCELS' support focuses on three main areas of life sciences: genetics such as pharmacogenomics; drug and medical supplies such as a national laboratory animal center; and stem cell development.

To build up the infrastructure for life science R&D in Thailand, TCELS supported the development of the National Laboratory Animal Center (NLAC) at MU-Salaya campus, and the development of disease management software at MU-Ramathibodi Hospital for decision support in the treatment of HIV/AIDS, cancer and other genetic diseases.

NLAC was officially established by the National Committee of Economic and Development Board of Thailand in 1987. Its mission is to breed and supply laboratory animals to meet any requests for gender, age, size and number of listed breeds while maintaining health, genetic and environmental quality control. It is important for NLAC to maintain the highest quality standards in order to support international businesses in their R&D activities and toxicity and safety tests. This requires human resource development, experts and an accredited quality system for its operation. TCELS and MU recognized the importance of NLAC as a national infrastructure for drug and medical supplies development and provided funding support of 120 million baht for increasing the center's capabilities development.

TCELS identified and contacted experts in the area of pharmacogenomics. An MU researcher whose expertise is in bio-markers was chosen and received support from TCELS for the development of disease management software at Ramathibodi Hospital, with the hospital being the information processing center. The software integrates the knowledge of molecular biology, pharmacogenomics and bioinformatics to provide an analysis of clinical information to monitor drug treatment of HIV patients. The program will support the decisions of medical doctors in assigning prescriptions. The project is currently under experiment. However there was no agreement on benefit sharing between MU and TCELS at the beginning. The interviewee mentioned that this resulted from the perception of MU administrators that the project is only at the laboratory stage and is too far from the commercialization stage.

### 12.6.3 Nanotechnology Center, Faculty of Science, MU

The Nanotechnology Center was established in 2001 as the first nanoscience and nanotechnology research unit in Thailand. Its purpose is to develop the necessary infrastructure for the new generation of nanoscientists. Current research topics are of both academic and commercial interest. The center's strategy is to transform skills and knowledge from traditional sciences into nanoscience and nanotechnology. Research projects performed at the center can be categorized into nine groups as follows: molecular electronics and photonics; nanoscale sensors; molecular machinery and robotics; carbon nanotubes; nano-pores; nano-machines and nano-storage; nano-transport; biomolecular surface science; biomolecular simulation; and nanoinformatics.

Dr Teerakiat Kertcharoen, a researcher at the center who is an expert in molecular sensors, is currently working on the 'Smart Vineyard' Project with a wine producer at Pakchong. He started this project by visiting the farm and having discussions with the farm's owner in order to understand the problems of the business. He then went back to the center to pursue the possibility of solving these problems in his research project. From the first day of his project, he aimed for technology commercialization. Instead of sitting in the lab and trying to develop a product, Dr Teerakiat reached out to industry and learned about their problems. He acknowledged that it would be very difficult for a researcher to understand a business. Thus from his point of view, researchers should work closely with businesses and learn from the experts. At the end of the project, researchers would be able to market their products by working with these business partners. Dr Teerakiat's viewpoint is different from many researchers who prefer to stay in their small laboratories all the time.

Dr Teerakiat used this project as a research training ground for his graduate students. Every year he received several requests from students abroad to send their interns to the center. Some students have also brought expertise together with industrial connections. In 2008 the center employed three full-time researchers, seven PhD students and 13 graduate students. Dr Teerakiat also plans to expand the 'Smart Vineyard' Project to other farms in different locations in Thailand and in Southeast Asia.

### 12.6.4 The Siriraj Dust Mite Center for Service and Research (SDMC), Faculty of Medicine, Siriraj Hospital

SDMC was founded in 2007 at the Faculty of Medicine, Siriraj Hospital, the first service and research center for studying house dust mites in Southeast Asia. The formation of the center came from the House Dust

Mites and Allergic Diseases Research Project conducted by Associate Professor Vanna Mahakittikun from 2001. The center's services include testing efficacy of anti-mite materials and agents, determining allergen concentrations of house dust mite samples, identifying mite species and producing lyophilized mites for research and vaccine production. The center's research focuses on the effectiveness of procedures for controlling dust mites and the development of guidelines for various encasing materials.

In 2001 Professor Vanna started her project by producing dust mites as raw materials for the research of a medical doctor in Singapore. At that time the project earned about one million baht per year. Professor Vanna saw an opportunity in this and wished to commercialize dust mite production and then export to international markets. The faculty's management team did not agree with her idea, because the administration focused on academic contribution rather than business-like projects as indicators of faculty performance. The faculty, because of its risk-averse nature, was reluctant to provide seed grants for development projects and did not attend to developing the financial support system. Professor Vanna then proposed the project to ATSC of MU, but was rejected because of the project's unclear business plan and operation.

Despite the lack in the university's financial support and commercialization supporting system, Professor Vanna continued developing her project. She conducted workshops in the World Allergy Congress and on World Allergy Day to present her studies. Her research used an experimental approach to answer relevant social and business questions on dust mites and allergic diseases and on the efficacy of anti-mite products in the market. She filed seven patents and received patent grants for inventions such as a mite separation equipment. In 2007, persuaded by Professor Vanna's efforts and proven track-record, the Faculty of Medicine, Siriraj Hospital agreed to invest 1.1 million baht in the project and established the SDMC for services and research. Recently the Faculty of Medicine, Siriraj Hospital signed a memorandum of understanding with a Thai pharmaceutical company to transfer technical know-how of mite vaccine production.

## 12.7 RESULTS AND DISCUSSION

The results of this study revealed a number of issues in MU's efforts to cope with increasing interaction between university and industry. There are two broad issues: external issues related to the development of the UTT system and internal issues within the university.

### 12.7.1 Pressures from the Innovation-driven Policy Context

The recent changes in national framework conditions have influenced interaction between universities and industry. In the innovation-driven policy context, Thai universities are being pressured by the government to generate income from research collaborations with industry and commercialization of their research output. In this regard MU has prepared itself to become autonomous and transform from a bureaucratic system to a performance-based system. The UTT system has been established for several years. Nevertheless this study shows that a linear model of innovation is a dominant interest in UTT policy. The focus of technology commercialization is on MU research and promotion of start-up companies. Business orientation is well taken by the management and administrative of UTT supporting units. The general feeling is that key stakeholders in the MU community such as academic faculties and research centers lack awareness and understanding of the UTT system.

### 12.7.2 Organization of the UTT System

Many of the structures for supporting UTT and commercialization at MU have been in existence for several years, yet the results from the interviews show that the majority of researchers at MU are not aware of their existence and/or function. Although all of the interviewees are involved with different parts of the infrastructure for promoting commercialization activities at MU, their knowledge of this infrastructure varied considerably. Some persons reported that they met different actors involved in the infrastructure and many of them had good working knowledge of its different components. Despite this overall knowledge, the interviewees pointed out that many researchers show some degree of ignorance as to how to use the different components of the system. This ignorance is reflected in the fact that no research-based company has been started within the MU infrastructure for commercialization of innovation.

The results from the interviews also show that there is a lack of concerted effort in promoting UTT and commercialization. Interviewees pointed out that information flows among key stakeholders in the UTT system are limited. Many researchers are not aware – or ignore the existence – of the university's commercialization infrastructure or availability of pertinent information. The case of the Nanotechnology Center is a good example of limited information flows within the UTT system.

### 12.7.3 Organization Capabilities

The management of the supporting infrastructure pointed out that in the process of creating new ventures, there is a financing gap at the proof-of-concept stage which the government and the university are not able to fill. This is because of the risky nature of this stage and the attitude towards risk held by administrators in public funding agencies.

The results show that the university's venture fund by STANG Holding is underutilized. The university's policy in venture investment (own more than 50 percent) impedes business formation and creates conflict between the university and inventors. In point of fact, it is very difficult to sell and commercialize MU products due to risk aversion towards business start-ups. The backgrounds of administrative staff also show that the supporting units lack marketing ability. It should be noted that the size of the current fund is too small for investing as venture capital.

### 12.7.4 University Culture

MU has been rooted in a science and research orientation for 100 years. The university culture is one of knowledge generation and diffusion through research for the wellbeing of mankind. It can be seen that MU administrators manage technology transfer and commercialization in a closed system rather than an open system; most of their efforts are related to the university's academic community and lack engagement from industry.

While the increase in knowledge exploitation and commercialization is a new phenomenon for the MU community, entrepreneurial culture is not a dominant interest of the MU community. Entrepreneurial training is partially conducted as an activity of the university's business incubator. The Entrepreneurship Program is a teaching program at the College of Management and has not been integrated into the UTT system. It is difficult for the TLO, having a lack of business and management skills, to run the university incubation center.

## 12.8 CONCLUSION

In summary, the MU case is representative of universities in many developing countries. The climate of the government's innovation-driven policy creates pressures for universities such as MU to put its efforts into increasing interaction between university and industry. In light of education reform, the university has developed its UTT system to cope with this challenge. This study provides some insight into how top-down policy affects processes at the bottom.

While the shift from a growth-oriented policy to an innovation-oriented policy has brought a legitimizing aspect, a number of issues arise from this ongoing effort of the government. These are not only issues for MU, but for other universities in emerging economies. They include a dominant interest in the linear model of innovation by the university's policy makers; lack of awareness and understanding of the university's technology transfer and commercialization system; lack of concerted effort in promoting UTT and commercialization; lack of marketing and business experience; and fragmentation of organization structure for supporting science-based entrepreneurship.

The above discussion points to a number of issues that universities such as MU encounter in their efforts to respond to increasing interaction between university and industry, while transforming from an existing bureaucratic system into an autonomous system. A number of implications for university and policy makers can be drawn from the MU case. The policy implications fall into two broad categories: communicating and implementing value changes; and designing an integrated structure for the supporting UTT system.

It is necessary for university administrators, researchers and academic professors to understand the concept of the innovation system and the role of university in technology transfer and commercialization. There is a need for researchers to be knowledgeable about what comprises a marketable invention. It is also important to understand the differences between university and industry culture. Research for publication is viewed as a main performance indicator and less time-consuming than research for development. The university needs a balance between the promotion of university-based start-ups for research and traditional relationships with large or existing companies.

## NOTE

* My special thanks go to the College of Management, Mahidol University, interviewees who participated in this research and Professor Poh-Kam Wong, Director of the NUS Entrepreneurship Center, and his colleagues who coordinated the project.

# REFERENCES


Arnold, E., M. Bell, J. Bessant and P. Brimble (2000), *Enhancing Policy and Institutional Support for Industrial Technology Development in Thailand*, Washington, DC: World Bank.

Brimble, P. and R.F. Donor (2006), 'University-industry linkages and economic development: the case of Thailand', *World Development*, **35** (6), 1021–36.

Etzkowitz, H., A. Webster, C. Gebhardt and B.R.C Terra (2000), 'The future of university and the university of the future: evolution of ivory tower to entrepreneurship paradigm', *Research Policy*, **29** (2), 313–30.

Hershberg, E., K. Nabeshima and S. Yusuf (2006), 'Opening the ivory tower to business: university-industry linkages and the development of knowledge-intensive clusters in Asian cities', *World Development*, **35** (6), 931–40.

Jacob, M., M. Lundqvist and H. Hellsmark (2003), 'Entrepreneurial transformations in the Swedish university system: the case of Chalmers University of Technology', *Research Policy*, **32** (9), 1555–68.

Johnson, B. (1992), 'Institutional learning', in B.-Å. Lundvall (ed.), *National Innovation Systems: Towards a Theory of Innovation and Interactive Learning*, London: Pinter Publishers, pp.23–45.

National Science and Technology Development Agency (NSTDA) (2003), *Report on R&D/Innovation Survey in Manufacturing and Service Sectors in Thailand in 2001*, Bangkok: NSTDA (in Thai).

National Science and Technology Development Agency (NSTDA) (2006), *Thailand Science and Technology Profile 2006*, Bangkok: Ministry of Science and Technology.

National Science and Technology Development Agency (NSTDA) (2007), *Thailand Science and Technology Profile 2006*, Bangkok: Ministry of Science and Technology.

National Science and Technology Development Agency (NSTDA) (2009), *Thailand Science and Technology Profile 2008*, Bangkok: Ministry of Science and Technology.

Schiller, D. (2006), 'The potential to upgrade the Thai innovation system by university-industry linkages', *Asian Journal of Technology Innovation*, **14** (2), 67–91.

Schiller, D. and I. Liefner (2007), 'Higher education funding reform and university-industry links in developing countries: the case of Thailand', *Higher Education*, **54**, 543–56.

Siegel, D.S. and P.H. Phan (2005), 'Analyzing the effectiveness of university technology transfer: implications for entrepreneurship education', in E. Liebcap (ed.), *Advances in the study of Entrepreneurship, Innovation and Economic Growth*, Amsterdam: Elsevier Science and JAI Press, vol. 16, pp.1–38.

Thailand Development Research Institute (TDRI) (1989), *The Development of Thailand's Technology Capability in Industry*, vol. 2–5, Bangkok: TDRI.

Thailand Development Research Institute (TDRI) (1998), *Effective Mechanisms for Supporting Private Sector Technology Development and Needs for Establishing*

# Index